An introduction to the design of small-scale embedded systems

An introduction to the design of small-scale embedded systems

Tim Wilmshurst

palgrave

First published 2001 by
PALGRAVE
Houndmills, Basingstoke, Hampshire RG21 6XS and
175 Fifth Avenue, New York, N. Y. 10010
Companies and representatives throughout the world

PALGRAVE is the new global academic imprint of
St. Martin's Press LLC Scholarly and Reference Division and
Palgrave Publishers Ltd (formerly Macmillan Press Ltd).

ISBN 0–333–92994–2

This book is printed on paper suitable for recycling and made from fully managed and sustained forest sources.

A catalogue record for this book is available from the British Library.

Typeset by Ian Kingston Editorial Services, Nottingham, UK

10   9   8   7   6   5   4   3   2   1
10   09   08   07   06   05   04   03   02   01

Printed and bound in Great Britain by
Antony Rowe Ltd, Chippenham, Wiltshire

This book is dedicated to my former colleagues, drawn from diverse countries and backgrounds, in the Department of Electrical Engineering, Polytechnic of Botswana.

Contents

Appendices

Abbreviations

AC	Alternating Current
ACIA	Asynchronous Communication Interface Adaptor
ADC	Analogue to Digital Converter
ALU	Arithmetic Logic Unit
AMD	Advanced Micro Devices
ASCII	American Standard Code for Information Interchange
BCD	Binary Coded Decimal
CISC	Complex Instruction Set Computer
CCR	Condition Code Register
CLCC	Ceramic Leaded Chip Carrier
CMOS	Complementary Metal Oxide Semiconductor
CPI	Cycles per Instruction
CPU	Central Processing Unit
C/T	Counter/Timer
BDM	Background Debugging Mode
DAC	Digital to Analogue Converter
DC	Direct Current
DTE	Data Circuit-Terminating Equiment
DIL	Dual Inline
DTE	Data Terminal Equipment
DVM	Digital Voltmeter
EEPROM	Electrically Eraseable Programmable Read-Only Memory
EIA	Electronic Industries Association
EM	Electromagnetic
EMC	Electromagnetic Compatibility
EMF	Electromotive Force
EMI	Electromagnetic Interference
EPROM	Eraseable Programmable Read-Only Memory
FNT	Fowler–Nordheim Tunnelling
FPGA	Field Programmable Gate Array
HLL	High-Level Language
IDE	Integrated Development Environment

IEEE bus	Parallel bus protocol approved by the Institution of Electrical and Electronic Engineers
I/O	Input/Output
IC	Integrated Circuit
ICE	In-Circuit Emulator
I^2C	Inter-Integrated Circuit
IP	Intellectual Property
ISO	International Organization for Standardization
ISR	Interrupt Service Routine
JSD	Jackson's Structural Design
K-	(as in Kbytes), 2^{10}, i.e. 1024_{10}
LED	light emitting diode
LCD	Liquid Crystal Display
lsb	least significant bit
MISRA	Motor Industry Software Reliability Association
msb	most significant bit
op-amp	operational amplifier
op code	operation code
OSI	Open System Interconnect
OTP	One Time Programmable
PCB	printed circuit board
PLCC	Plastic Leaded Chip Carrier
ppm	parts per million
PSW	Program Status Word
PWM	Pulse Width Modulation
RISC	Reduced Instruction Set Computer
ROM	Read-Only Memory
RTOS	Real-Time Operating System
SFR	Special Function Register
STN	Supertwisted Nematic (in context of LCD displays)
TN	Twisted Nematic (in context of LCD displays)
TSOP	Thin Small Outline Package
TTL	Transistor–Transistor Logic
UART	Universal Asynchronous Receiver Transmitter
UPS	Uninterruptable Power Supply
UV	Ultraviolet
VCO	Voltage-Controlled Oscillator
WDT	Watchdog Timer
WWW	World Wide Web

Near equivalent terminologies

Instruction cycle	Machine cycle	
Status register	Condition code register	Program status word
V_{CC}, V_{DD}, V_s	(Circuit power supply)	

Number bases

The base (radix) of a number is indicated by the subscript which follows, for example

128_{10} (Decimal) = 80_{16} (Hexadecimal) = 10000000_2 (Binary)

Introduction

Microprocessors, and microcontrollers in particular, have revolutionised the designer's ability to add 'intelligence' to a small-scale engineering product. You don't have to look far, in the modern household or workplace, to find a microcontroller. Alternatively, as you go through your daily schedule, you probably won't have to wait long before you use a microcontroller. A car journey, a phone call, a grocery purchase, a credit card sale, a doctor's visit, or use of a washing machine or cooker may all invoke microcontrollers. Their presence is all-pervading. Because the microcontroller is hidden inside the product it controls, we call the product overall an embedded system.

To electronic or systems designers the microcontroller represents another system element, like an op-amp or DC motor but on a much grander scale, that they must know about and be able to apply where necessary. To software or computer specialists they represent a way of applying their expertise to control things that really happen, instead of just numbers or images on a screen.

This book is an introduction to the design of the embedded system. Because there are many, many types of embedded system, whose scale and complexity vary enormously, the qualifier *small scale* is added to the title. While even this is open to some interpretation, it allows us to concentrate in a focused way on a particular type of embedded system. In general, the title is meant to imply design of embedded systems which are of low to moderate hardware and software complexity, *possibly* physically small (in which case portable and battery-powered), with single or few processors in the system. Hence the smaller, 8-bit, microcontrollers are described, and assembler remains an important programming tool. Moreover, because the overall project is small, developers require all-round capability. This means that they work on all aspects of the project development, and are not just specialists in software or hardware, or one of their aspects.

One of the pleasures of designing embedded systems is indeed the multi-disciplinary nature of the work. Analogue and digital electronics, sensors and actuators, software design and computer architecture all come into play, and the skill of the designer often lies in orchestrating the capabilities of each. The book attempts to do justice to each of these fields, as far as they impinge upon the central subject, presenting them at what is considered to be the most suitable level of complexity and formalism. It is important to remember that each is developed much further as its own discipline elsewhere, in applications which are not dependent on the constraints of the embedded system. I hope that in bringing these different disciplines together here it can be said of this book that 'the whole is greater than the sum of the parts'.

To make sense of the material in this book, it is necessary for the reader to have a reasonable knowledge of both analogue and digital electronics. In analogue electronics this should include the principal electronic devices (for example diodes and transistors) and knowledge of standard op-amp applications. In digital electronics it should include a good grounding in combinational and sequential logic, and some understanding of logic technologies (in particular CMOS), as well as digital sub-systems such as counters, latches, adders, and shift registers. A knowledge of binary arithmetic is also required, but this is reviewed in Appendix A. In particular, readers should have some prior general knowledge of microprocessors. This includes some knowledge of a typical 8- or 16-bit device, such as the Motorola 6800 or 68000, together with the principles of data and address buses and memory mapping and addressing. However, because it is expected that readers will come to this subject from a wide variety of backgrounds, reviews of important aspects of electronics are sometimes made.

It is near impossible to discuss microcontrollers in the abstract, with no reference to actual devices. On the other hand, a book introducing the broad field is of limited use if it takes its examples from only one processor. Every controller family is different, and each one has its own unique features. Examples in this book are therefore taken broadly from *three* microcontroller families; the Microchip 16XXX series, with special emphasis on the 16F84; the Intel/Philips 80C51, with special reference to the 80C552; and the Motorola 68HC05, its cousin the 68HC11, and its successor the 68HC08. The book is not a manual on any of these devices, nor is it a particular endorsement of any of them. The coverage of each is moreover not equal. The early chapters have a strong emphasis on the 16F84, which is used as the vehicle to introduce some important hardware concepts. As the complexity of ideas increases, the source of examples is more widely drawn.

The book is aimed primarily at undergraduate students in the later stages of Electronic Engineering or Mechatronics programmes. It should also be of use to undergraduates in related disciplines, Higher National Diploma students, and the practising engineer. The book's usage should not just be restricted to lecture courses. One of my tasks in life over many years has been to supervise final year students in their projects, many of which are microcontroller-based. It is their 'frequently asked questions' which formed one impetus for the writing of this book.

The structure of the book is such that the first three chapters form an almost self-contained introductory unit, covering the simpler aspects of hardware and programming, with the intention that the reader should develop the competence to design a minimum system, programmed in Assembler. The subsequent chapters develop most of these early topics in much greater detail, and add a number of entirely new issues, such as low-power design.

My own association with embedded systems spans many of the developments that microprocessors and microcontrollers have been through. My first designs used the RCA 1802, a venerable CMOS device which is still sold. I claim, on the slenderest of evidence, that this is the first microprocessor ever to enter the British Houses of Parliament, when the company for which I worked undertook studies (in 1980) to automate the recording of parliamentary debates. This was in the early days of microprocessing, when the subject was swathed in mythology and mystique. There were very few textbooks on the subject, and knowledge was gained laboriously by trial and error. Those days seem long ago now, but it is surprising in a way how little the fundamentals have changed, and at the microcontroller level the fundamentals matter very much, as they are not masked by the clever interfaces and operating systems of their more sophisticated and powerful counterparts in the computing world.

Later, in between two major spells working in the Cambridge University Engineering Department, I spent six years in Botswana, Southern Africa. Among the many things we did, one was to develop, with participants from several countries, a low-cost data logger, based on the Motorola 6805. With this logger we recorded the soaring temperatures of the Kalahari desert, and the soaking humidity of Zimbabwe's Eastern Highlands. I claim, with only slightly more evidence than before, that the 6805 was the first microprocessor ever to enter the Kalahari desert.

Wherever your designing may take you, I hope that everyone reading this book will be able to share with me the fun and excitement of designing embedded systems.

Acknowledgements

If there is any merit in this book, it is due to the influence of many people. I would like to thank Roland Thorp and Peter Spreadbury of Cambridge University, from whom I learnt much electronics. Mike Robson of the University of Zimbabwe is the best teacher (of microprocessors and many other things) I have ever listened to; he enabled many microprocessor-based things to happen in Southern Africa in the late 1980s. Thanks also to Richard Taylor and Gary Bailey, who run with energy and expertise the Electronics Development Group at Cambridge University, and Paul Pedersen, who built much of the 8051 stuff we did in the mid-1990s. Ken Wallace, also of Cambridge University, is inspirational in his design teaching; much of Chapter 12 depends on that inspiration. Design teaching at Cambridge is epitomised in the celebrated Integrated Design Project, whose smooth running is ensured by Doug Isgrove. Participating in that project was both a privilege and a learning experience. Thanks finally and especially to Naomi, Jeremy, Imogen and Beate, my family, for supporting me during the writing of this book.

Trademarks

All trademarks are acknowledged within the text and are the property of their respective owners. Any omissions should be notified to the publisher (see p. iv) for inclusion in future reprints and editions of this book.

Figures supplied by Microchip Technology Incorporated are reprinted with permission of the copyright owner, Microchip Technology Incorporated © 2001. All rights reserved. No further translations, reprints or reproductions may be made without Microchip Technology Inc.'s prior written consent. Information contained in this publication regarding device applications and the like is intended as suggestion only and may be superseded by updates. No representation or warranty is given, and no liability is assumed by Microchip Technology Inc. with respect to the accuracy or use of such information, or infringement of patents arising from such use or otherwise.

Introducing embedded systems and the microcontroller

IN THIS CHAPTER

We are living in a second age of industrial revolution, when the availability and processing of information are causing untold changes in all our lives. While mankind has dreamed for many years of the possibility of building computing machines, the dream started to become a reality in the late 1940s. It was then that the first electronic computers, based on massive racks of thermionic valves, started their laborious calculations. In 1948 the transistor was invented, and the first integrated circuit was built in 1959. This set the stage for a spectacular process of electronic miniaturisation. Integrated circuits became more and more compact, enabling more and more circuitry to be placed on them.

In 1971 Intel produced the first microprocessor, the 4004, which handled data as 4-bit numbers, and contained 2250 transistors. It followed this soon with the 8008, and within a few years a number of companies were making their own microprocessor offerings. The age of the microprocessor had arrived! Very early in their development, and certainly by the end of the 1970s, two trends were emerging for these remarkable devices. One was to scale down, in size if not computing power, the general-purpose computer; this led quickly to the first desktop machines. The other, much more revolutionary, was to place the microprocessor in products which apparently had nothing to do with computing. They began to find their way into photocopiers, grocery scales, washing machines, and a host of other products, wherever there was a requirement to exercise some control function. While the first trend led to an inexorable demand for faster and bigger processors with increasingly sophisticated mathematical capability, the second placed lower demands on computational power and speed. It wanted physically small and cheap devices, with as much functionality of the system as possible squeezed onto one integrated circuit.

Such microprocessors became known as microcontrollers, and the systems they controlled, embedded systems.

Though humbler by far than their high-powered cousins, microcontrollers sell in far greater volume, and their impact has been enormous. To the electronic and system designer they offer huge opportunities.

This chapter starts our exploration of the fascinating and hidden world of embedded systems. We will meet first the embedded system itself, and discover something of its nature and characteristics. Then we will start our study of the intelligence inside the embedded system: the microprocessor or microcontroller.

Specifically, the chapter aims:

- to introduce the embedded system and describe its characteristics

- to review prerequisite microprocessor knowledge, thereby defining a starting point

- to consider certain fundamental choices in microprocessor design

- to introduce the features of a general-purpose microcontroller

- to introduce three microcontroller families, which will be used as examples in parts of the book

1.1   Embedded systems and their characteristics

1.1.1   The essence of the embedded system

The newspaper article of Fig. 1.1 describes a jet-propelled bicycle that appeared on TV and in the newspapers. Apart from its extraordinary novelty value, the design overcame some very difficult technical problems. One reason for the success of the project was the inbuilt microcontroller system which kept the engine under control. Yet in a substantial write-up, the reporter makes no mention of this at all. Why doesn't the headline shout 'Microcontroller Tames new Jet Engine', or similar? The answer is simple. The reporter almost certainly didn't even know the microcontroller was there. Her attention was drawn entirely by the novel combination of a jet engine and a bike. The microcontroller was not visible, there was no indication of the presence of a computer, and she had no reason to believe there might be one involved. Nevertheless, the engine could not have functioned for more than a few seconds without the continuous action of the control system, which not only enabled successful operation, but also provided the condition monitoring to eliminate situations of danger. With a miniature jet engine there are plenty of those!

Inventor fears noise and heat will dull appeal of jet cycle

Paul Ford on his jet powered bike PHOTOGRAPH: TONY JEDRE]

Intolerably noisy, something of a fire hazard and not fit for use on public highways – as mad inventions go, this one rates highly, *writes Amelia Gentleman.*

Cambridge engineer Paul Ford has fitted a home-designed jet engine to his bicycle and created a potentially record-breaking machine capable of travelling at 100 mph.

Aside from its speed, the vehicle does have a couple of advantages: there is no need to pedal and jet paraffin is affordably priced.

But even the inventor accepts that these attractions are outweighed by the problems the prototype bike poses.

Primarily there is the noise. It emits 102 decibels when stationary and when it gets going it sounds like an aeroplane on take-off.

Then there is the heat. "You have to be careful not to stand behind the bike because the exhaust emerges at about 480 degrees centigrade – hot enough to burn the hairs off your body," Mr Ford warned.

And this is not a bicycle to delight the environmentalists. "We weren't trying to be particularly green. The emissions are what you would expect from a jet engine."

Mr Ford, aged 37, co-owner of a model aircraft shop in Cambridge, invented the miniature gas turbine engine. Although it can turn out 22 lbs of thrust – about equivalent to a moped engine – this generates high speeds because the bicycle itself is so light.

During preliminary tests at a disused airfield, the vehicle reached 55 mph at half power and Mr Ford is confident that, with a bit of work, 100 mph will be reached easily.

"I've been too scared to go any faster. At the moment the steering is extremely sensitive, something else that needs refining. To begin with I was also concerned that it might actually take off but the design seems to have prevented that risk."

While happy to accept that his invention is not practical, Mr Ford remains uncomfortable with the mad professor status the creation has forced on him.

"I'm pretty certain that this is the first jet powered bike in Britain. People thought it couldn't be done and I wanted to prove them wrong."

He added: "A lot of people have told me that it's a crazy thing to try to do, but I don't think it's eccentric at all" he said.

Figure 1.1 A high-speed embedded system: the jet-propelled bicycle. (From *The Guardian*, 12 May 1998. Reproduced by permission of *The Guardian* newspaper and the *Cambridge Evening News*.)

We call this type of control *embedded control* – and the overall system an *embedded system*. A definition of an embedded system is as follows:

An embedded system is a system whose principal function is not computational, but which is controlled by a computer embedded within it.

The computer is likely to be a microprocessor or microcontroller. The word *embedded* implies that it lies inside the overall system, hidden from view, forming an integral part of a greater whole. One consequence of this is that the user may be unaware of the computer's existence. Another is that the computer is usually purpose designed, or at least customised, for the single function of controlling its system. If removed from the system it would be an odd assortment of printed circuit boards and/or integrated circuits, recognisable only to the specialist as something which might be called a computer.

Applying this definition tells us that a personal computer, even though it contains a microprocessor, is *not* an embedded system. Its end function is to compute. Even if the same computer was connected to a set of instruments, which it then controlled, that would not be an embedded system. If, however, the same computer was built permanently into an identifiable system, and customised so that its sole purpose was to control the one system (which may mean losing such apparently essential features as its case, keyboard, screen, or disk drives), then it would form part of an embedded system.

Embedded systems come in many forms and guises. They are extremely common in the home, the motor vehicle and the workplace. Most modern domestic appliances – washing machines, dishwashers, ovens, central heating and burglar alarms – are embedded systems. The motor car is full of them, in engine management, security (for example locking and anti-theft devices), air-conditioning, brakes, radio, and so on. They are found across industry and commerce, in machine control, factory automation, robotics, electronic commerce and office equipment. The list has almost no end, and it continues to grow.

Figure 1.2 re-expresses the embedded system definition as a simple block diagram. There is a set of inputs from the controlled system. Based on information supplied from these inputs, the controller computes certain outputs, which are connected to actuators within the system. There *may* be interaction with a user, e.g. via keypad and display, and there *may* be interaction with other sub-systems elsewhere, though neither of these is essential to the general concept.

In the jet-propelled bicycle the control system measures three variables from the engine (temperature, pressure and rotational speed), and also receives a control input from the driver. Its only output controls the fuel flow to the engine. From the inputs it first of all determines whether the engine is operating safely. If a danger condition is detected (for example the motor is running too hot or too fast), the controller takes emergency action. In the absence of a danger condition, it computes the appropriate drive signal for the fuel flow.

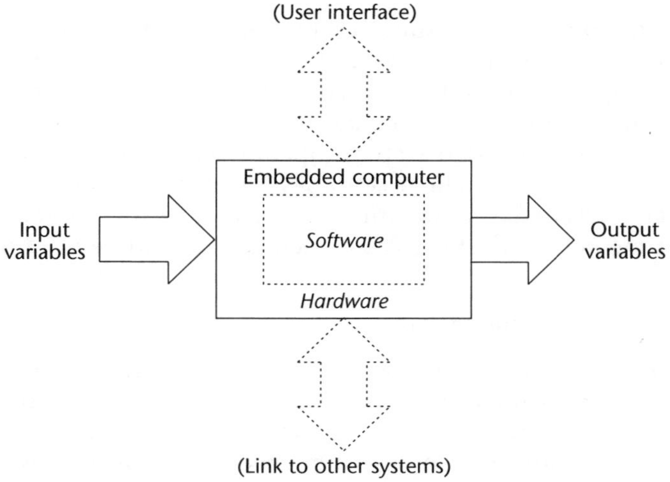

(User interface)

Embedded computer

Input
variables

Software

Output
variables

Hardware

(Link to other systems)

Figure 1.2 The essence of the embedded system.

1.1.2 Further features of the embedded system

1.1.2.1 *Constituents of the embedded computer: hardware and software*

As with all computer systems, the embedded computer is made up of hard-ware and software, as symbolised in Fig. 1.2. In the early days of micropro-cessors much of the design time was spent on the hardware, in defining address decoding, memory map, input/output and so on. When the hard-ware design was completed, a comparatively simple program was devel-oped, limited in size and complexity by restricted program memory size, and the development tools available. Since then there have been huge strides in hardware development. Much of the hardware system is now contained on a single chip, in the form of a microcontroller, and develop-ments in memory technology allow the use of much longer and more sophisticated programs. Hardware design of the computing core of the embedded system is now in many cases viewed as a comparatively straight-forward affair. The design attention has shifted to some extent towards soft-ware development, with advanced languages and tools available to develop sophisticated programs.

1.1.2.2 *Timeliness*

The example jet engine is able to change its speed extremely fast, and can easily self-destruct. The controller must be able to respond fast enough to keep its operation within a safe region. This is a characteristic of operating in 'real time'; the controller must be able to respond to inputs as they happen and make responses within the time-frame set by the controlled system. This style of operation is different from the mode of operation, for example, of a personal computer. While it may be annoying, you can tolerate waiting for your computer to refresh the graphics display or complete a computation. You cannot tolerate waiting while your car's anti-skid braking system decides whether or not to apply the brakes!

Some embedded systems operate within absolutely rigid time demands; for others the demands are less stringent. They all, however, exhibit the characteristics of timeliness: a need for the designer to understand fully the time demands of the controlled system and be responsive to them.

1.1.2.3   System interconnection

Figure 1.2 raises the possibility of interaction with other systems. While some embedded systems clearly need only one controller, others are likely to use several or many, each to control one sub-system. Necessary shared information is then passed between them by a simple network, devised to suit the needs of the overall system. A good example of this is the modern motor car. Though each of the 'embedded sub-systems' in it may be controlled by one microcontroller, they can all be linked together to form one overall interconnected system. This approach is made more attractive due to the extremely low cost of most commercial microcontrollers. A network of low-cost microcontrollers is often cheaper, and simpler to develop, than a single complex computer undertaking many tasks.

With the advent of the Internet, a generation of Internet-compatible embedded systems is emerging. The cooker, television and washing machine may soon be communicating together! It is anticipated that within a few years even the most simple of devices may be Internet-linked. The truly standalone device will then exist in a dwindling minority.

1.1.2.4   Reliability

Suppliers of software packages designed to run on Personal Computers release them on the market knowing that they are likely to contain software errors (bugs). It is vitally important to get them to market early, and fixes can always be distributed after the faults have been discovered. Suppliers of most embedded systems cannot afford this luxury. One significant software error in a car model could destroy the reputation of the manufacturer for ever. Therefore the embedded system designer must develop a good grasp of reliability issues, and how a reliable system can be achieved. This implies good design procedures in both hardware and software, coupled with systematic testing and commissioning.

1.1.2.5   The market-place

The market that the embedded system sells into is very competitive. As with other 'hi-tech' markets, the challenge is increased greatly by the very rapid advances of technology. New products may quickly be rendered obsolete by technological change, and thus potentially have very short life cycles. This lays the stress on excellent design and development strategy.

Adding all these features together, a second definition of the embedded system now follows, more descriptive and verbose. The technical features mentioned in this effectively lay down the agenda for this book.

An embedded system is a microcontroller-based, software-driven, reliable, real-time control system, autonomous, or human or network interactive, operating on diverse physical variables and in diverse environments, and sold into a competitive and cost-conscious market.

1.1.3 The skills of the embedded system designer

It is becoming clear that embedded systems have enormous variety, and call upon many technical disciplines. This is indeed one of the attractions of working with them. This multi-disciplinary nature is illustrated in Fig. 1.3.

A full understanding of the microcontrollers we will work with only comes with some knowledge of computer architecture and integrated circuit design and manufacture. The need for control, which inevitably implies measurement and actuation, leads us into further branches of electrical and electronic engineering. Associated with the measurement, we find a need for analogue as well as digital electronics. One could go on adding further disciplines, for example Digital Signal Processing or Electromagnetic Compatibility, to the diagram. These are also important to the embedded system, but will not claim much space in an introductory book like this.

1.2 Our starting point: the microprocessor

1.2.1 The microprocessor reviewed

Our approaching study of the microcontroller will rely on the reader having a reasonable knowledge of microprocessors. We will pause briefly to review this knowledge, to ensure a defined starting point. Figures 1.4–1.6

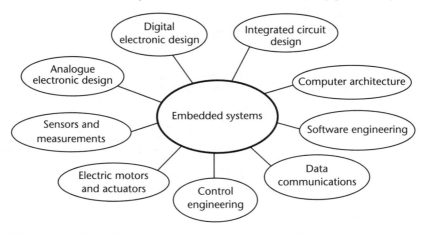

Figure 1.3 Embedded system design calls on many disciplines.

summarise what we need to know. The contents of each diagram will be briefly reviewed, but readers who need greater detail on these matters should consult an appropriate text, for example Refs. 1.1 and 1.2. Both of these have excellent introductory chapters on microprocessors. Appendix A contains a summary of binary arithmetic and counting schemes, which may also be worth reading at this stage.

A microprocessor is a simple computer, contained more or less in one integrated circuit (IC, also colloquially called a 'chip'). Like any computer it follows a sequence of instructions, known as a program. Each instruction causes a very simple action to take place, generally either a computation, a transfer of data or a decision. The microprocessor can perform each instruction extremely fast, so that by building on these very simple actions much more complex tasks can be undertaken.

A diagram of the hardware of a simple microprocessor-based system is shown in Fig. 1.4. The essential features are:

- the microprocessor
- a section of memory to store the program
- another section to store temporary data
- some contact with the outside world (through the input/output port)
- a means of interconnecting these elements (i.e. data and address bus, together with some control lines)

Program memory is usually stored in a form of memory called *ROM – Read-Only Memory*. Data memory is usually stored in a type of memory called *RAM – Random Access Memory*. ROM retains its contents when the system is powered down; RAM does not. Memories are defined according to size, generally in terms of numbers of bytes. For this the prefixes K- and M- (or Mega) have gained ubiquitous customary usage. These differ from the

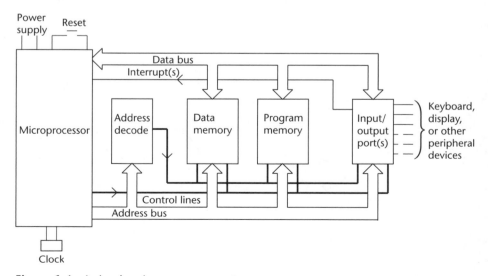

Figure 1.4 A simple microprocessor system.

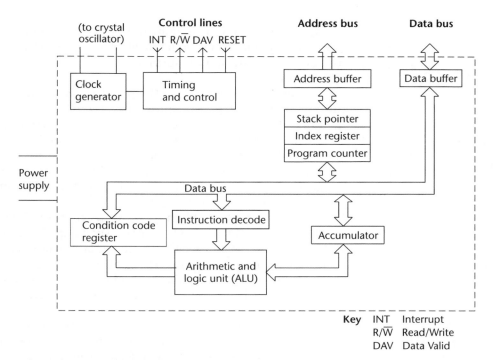

Figure 1.5 A typical microprocessor.

conventional decimal multipliers (e.g. the kilo- of kilometre or kilogram). K- indicates a multiplier of 2^{10}, i.e. 1024, while Mega is actually 1 048 576, i.e. 2^{20}. A memory of 4 Kbytes contains 4096 byte-sized locations.

A block diagram of a 'typical' imaginary microprocessor appears in Fig. 1.5. The computing function takes place in the *Arithmetic Logic Unit* (*ALU*), where arithmetic and logical operations take place. Part of the ALU is the *accumulator*. This is the register where the operand, the number on which the operation is being performed, is held. The size of the Accumulator, in number of bits, determines the size of number that the processor can operate on. It is reflected across the whole microcomputer system, for example in the size of the data bus and memory locations. The ALU, together with the control section around it, is known as the *Central Processing Unit* (*CPU*).

The action of the microprocessor is synchronised to the clock generator, often based on a quartz crystal oscillator. Any microprocessor can only operate within a certain range of clock frequencies, whose limits are set by the fabrication technology of the device and specified by the manufacturer. Each has a maximum (for microcontrollers usually in the range 4 MHz to around 30 MHz). Those based on *dynamic logic* (see Ref. 1.3 for further details) have a minimum as well. Those which can operate down to DC are known as 'fully static'.

The clock oscillator frequency is divided down within the microprocessor (generally by a factor between 4 and 12, depending on the microprocessor), giving a lower internal operating frequency. One period of this internal frequency is sometimes called a *machine cycle*, or an *instruction cycle*. All instruction execution is made up of integer numbers of machine or instruction cycles.

In normal system operation the processor works down the list of instructions which make up the program. It fetches each one from program memory, decodes it with its Instruction Decode circuit, and then executes it. The instruction is in many cases accompanied by further pieces of code, also stored in program memory, which are treated as operand data, or addresses where the operand data may be found.

The microprocessor 'keeps its place' in the program by means of the *Program Counter*, which always holds the address of the next instruction to be executed. In order to fetch the next instruction, the processor places the value held in the Program Counter on the address bus, and signals through the control lines that it wishes to read data. Memory corresponding to that address will, upon receiving the address and control signals, place the instruction word on the data bus, which the processor can then read. As each word is read from program memory, the Program Counter is incremented.

Figure 1.6 illustrates this sequence of activities, for the processor of Fig. 1.5 and for a certain instruction, as a timing diagram. It can be seen that there are four clock cycles in each machine cycle. The first cycle shown is an 'instruction fetch' cycle. The address of the instruction to be fetched is placed on the address bus, and the R/$\overline{W}$ line indicates that the data transfer is to be a 'read'. In response the addressed memory places data onto the bus. This is received by the microprocessor and decoded by the Instruction

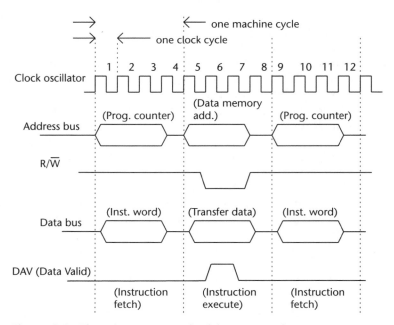

Figure 1.6 The microprocessor fetch/execute cycle.

Decode circuit. In the second machine cycle the instruction is executed; the example illustrates a data move from processor to memory. The processor sets values on the address and data buses, and signals a write by setting the R/W line low. The DAV line goes high to indicate that the bus data is valid. The falling edge of this signal is used to latch the data into memory. This particular instruction has taken two machine cycles to complete. It is then followed by the Instruction Fetch cycle of the next instruction. It follows that simple microprocessor operation can be seen as a relentless cycle of instruction fetch, decode and execute.

1.2.2 More on instructions, and the ALU

A typical 8-bit ALU is able to perform the operations shown in Table 1.1. Using combinations of these very simple operations, almost any other mathematical function can be implemented, albeit sometimes laboriously.

Each processor (or processor family) has its own instruction set, from which the program is written. Each instruction is a binary word, known individually as the *op code* (operation code), or collectively as *machine code*. The processor CPU can recognise and respond to these codes. The instruction set is the collection of all these op codes. It uses the basic ALU operations listed earlier, and adds to these certain data transfer and branch instructions. This gives an instruction set the following typical instruction categories:

Table 1.1 What an ALU can do.

Increment A	A = A plus 1
Decrement A	A = A − 1
Add A to M	A = A plus M
Subtract M from A	A = A − M
*AND A with M	A = A · M
*OR A with M	A = A + M
*Exclusive OR A with M	A = A ⊕ M
Shift A left	A = 2A
Shift A right	A = A/2
Rotate A left	
Rotate A right	
Complement A	NOT A
Clear A	A = 0

A represents the contents of the accumulator; M is a number held in memory.
The statement 'A = ' implies 'A becomes' (original value of the accumulator overwritten).
*The logical function is performed between corresponding bits of the two operands.

- *Data transfer*: instructions which move data from one register or memory location to another.
- *Arithmetic*: instructions which perform arithmetic operations between specified data words.
- *Logical*: instructions which perform logical functions between specified data bits or words, for example INVERT, AND, OR, Rotate.
- *Program branch*: instructions which cause a program to deviate from simple sequential execution of instructions held in program memory, for example as a subroutine call or return, or conditional branch.[1]

The result of an operation undertaken in an accumulator frequently exceeds the range of the number which can be held in the Accumulator. Therefore associated with the ALU is a 'Flag Register'; this contains a number of bits which give further information about the result of the previous instruction. It is known as the *Status Register* (Microchip Inc.), *Condition Code Register* (Motorola), or *Programme Status Word* (Intel and Philips). These bits may include:

- a **zero** bit, indicating whether the result was zero
- a **carry** bit, indicating whether there was a carry from the most significant bit (msb) of the accumulator, also used as a 'borrow' in subtraction
- a **sign**, or **negative** bit, indicating whether the result was negative (interpreting the result in *two's complement* arithmetic[2]) – hence this bit is simply set to the msb of the result
- a **half-carry** bit, indicating whether there was a carry between the lower and higher nibbles of the result – this is useful for *Binary Coded Decimal* (BCD) arithmetic
- an **overflow** bit, indicating whether the two's complement range has been exceeded. It is set if there has been a carry out of bit 7 but not bit 6, *or* a carry out of bit 6 but not bit 7
- a **parity** flag, indicating whether an odd or even number of 1 bits are in the accumulator

As there are not usually enough 'condition code' flags to fill an 8-bit register, many processors use the remaining few bits for other purposes, for example interrupt mask bits or register bank address bits. These will be considered later.

---

1 A conditional branch instruction tests a certain condition of the microprocessor system, for example a register bit. It transfers program operation to a different program section if the test condition is met, and continues the program in sequence if it is not.

2 'Two's complement' is a means of expressing negative numbers in binary. Read Appendix A if you are unfamiliar with it.

WORKED EXAMPLE   1.1

The hexadecimal numbers F8 (= 11111000 in binary) and 09 (= 00001001 in binary) are added in an ALU. Determine the result and what Condition Code Flags are set (assume the ALU has all the above flags except the Parity flag). Interpret the result in unsigned binary and in two's complement.

Solution: Figure 1.7 illustrates the addition, together with the flag settings. The 8-bit unsigned binary result is valid only if the state of the carry flag (C) is noted. The result is valid in two's complement (hence OV is zero), and is a non-zero (Z = 0) positive number, as indicated by the state of the N flag. A half-carry (HC) has also occurred.

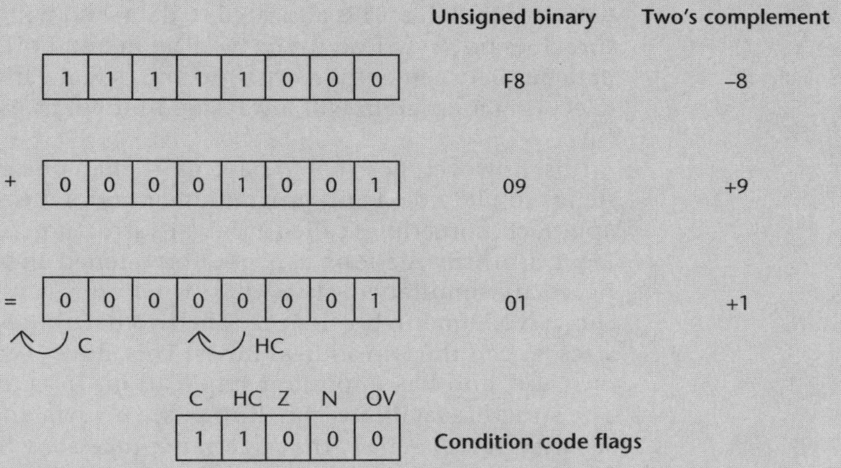

	Unsigned binary	Two's complement
1 1 1 1 1 0 0 0	F8	−8
+ 0 0 0 0 1 0 0 1	09	+9
= 0 0 0 0 0 0 0 1	01	+1

C   HC

C	HC	Z	N	OV
1	1	0	0	0

Condition code flags

Figure 1.7 Example addition of two binary numbers.

1.3 Some microprocessor design options

We go on to consider some aspects of microprocessor design which go beyond the basic structure assumed so far. These aspects are discussed at a level appropriate to the small-scale microprocessor or controller; their application in larger computers is far more sophisticated. Readers who wish to gain further background in these areas are referred to Refs. 1.4 and 1.5.

1.3.1 Von Neumann and Harvard

In the conventional von Neumann architecture, program and data memory share the same address and data buses, and are hence both within the same memory map. This is illustrated in very simple form in Fig. 1.8(a). This approach is simple, robust and practical, and has been widely and successfully applied. If data memory is being accessed, program memory lies idle,

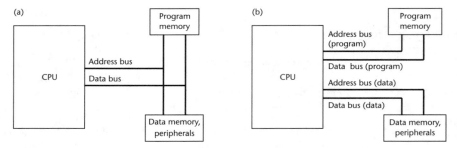

Figure 1.8 (a) The conventional von Neumann structure; (b) the Harvard structure.

and vice versa. Once the overall memory space is defined, it is up to the user to decide which area is allocated to data, and which to program. The structure does however lead to the 'von Neumann bottleneck'; time-sharing the data bus between both instruction and data means that maximum speed of executing a program will always be limited, as each has to use the bus in turn.

It is, however, possible to have more than one address and data bus, and hence to place data and program memory in different memory maps. This approach, sometimes called a Harvard structure, is shown in simple form in Fig 1.8(b). Instructions can now be fetched independently from, and if necessary simultaneously with instruction execution, thereby eliminating the von Neumann bottleneck. The two data buses can now be of different sizes, as can the two address buses. This allows each to be optimised for its own use, and has important implications in certain processor structures. The structure facilitates pipelining (see below), and also enhances program security. It is less likely that an errant processor will attempt to overwrite its own program, or jump into data memory and start interpreting data as instructions.

With its multiplicity of buses, this architecture does lead to a more complex hardware realisation than conventional von Neumann. Moreover, not every memory use is clearly divided into 'data' or 'program'. Look-up tables (i.e. tables of constant data, defined within the program), for example, may be embedded in program memory, but required for use as data.

1.3.2 Instruction sets – CISC and RISC

In simple terms, the operations that the designer of the microprocessor has at her initial disposal are those listed in Table 1.1. The microprocessor instruction set *could* be based on these, and thus they would be available to the programmer in their 'raw' form. Alternatively, it is possible to group them together in simple combinations, so that an instruction from the instruction set is actually interpreted by the CPU as a sequence of perhaps two or three of these primitive instructions. This practice is known as *microcoding*, and the task of interpreting each program instruction into instruction primitives is done by a ROM internal to the CPU.

Many early microprocessor designers adopted this practice, and tried to create instructions for every possible eventuality. This appeared to approach a sophisticated and 'ideal' machine. A processor of this type gained the name Complex Instruction Set Computer (CISC). One of its more obvious characteristics is that code for different instructions can be of quite different lengths, and have widely differing execution times. The CISC processor also occupies more space on the IC, due to the requirement for internal ROM.

Studies of CISC instruction usage revealed, however, that in a 'typical' program most of the instructions were not being used for most of the time (e.g. 80% of programs were made up of 20% of the instruction set). It was therefore reasoned that if the most-used instructions were optimised in terms of speed, and the others removed (but with their function still achievable by combinations of those that remained), then program execution time could be reduced and the CPU design simplified. In parallel with this, technological advances, especially in the area of high-density memory, meant that the pressure to minimise program code length was no longer so great.

The result was a 'back to basics' move, leading to the simpler but faster Reduced Instruction Set Computer (RISC). This has the following characteristics:

1. *The CPU does not make use of microcoding.*
2. *Memory is accessed via load and store instructions only*; the requirement to operate on memory contents is achieved by multiple instructions.
3. *All instructions are executed in one machine cycle*; this means that each instruction must be represented by one word only – hence all op codes must be equal to or within the instruction bus size, and must include the operand within them.

RISC machines have the advantages of simplicity and speed, but carry the apparent disadvantage that their program code is almost invariably longer and more complex. With memory becoming ever cheaper and of higher density, and with more efficient compilers for program code generation, this disadvantage is diminishing.

1.3.3 Instruction pipelining

As Fig. 1.6 showed, conventional microprocessor program execution is a relentless sequential cycle of instruction fetch, decode and execute. For a given processor the only way of speeding up this operation is by speeding up the clock.

Consider an alternative: that as one instruction is being executed, the next is already being fetched. If this is done, the instruction throughput can be dramatically increased without reducing the actual instruction execute time. This is the basis of pipelining – it's a simple idea which can make processors run much faster, but it does place certain strict requirements on the nature of the instructions.

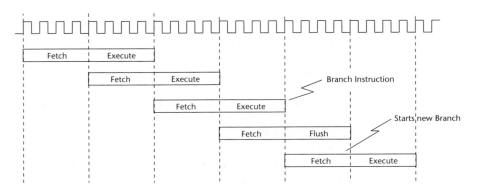

Figure 1.9 Pipelined instruction execution.

In order to work, *all* the instructions of the processor must have the same duration of execution, and it must be possible to split the fetch–decode–execute cycle for *all* individual instructions into a number of stages of equal duration. Then, as any one instruction enters its second stage, the following instruction enters its first. This is illustrated in Fig. 1.9, for instructions divided into just two stages (i.e. fetch and execute). As one instruction is executing, the next is already being fetched. It can be seen that the instruction throughput for the first three instructions is twice as fast as in Fig. 1.6. If the instructions had been broken into three stages, it would have been three times as fast, and so on.

Simple pipelining fails at conditional program branches. When the processor is executing a branch instruction, it is already fetching the next instruction in the program, but if the branch does take place that next instruction is no longer needed. So it must 'flush out' that instruction, and fetch the one where the branch starts. This is why branch instructions often take longer in a pipelined architecture. The example of Fig. 1.9 shows two instructions being successfully fetched and executed. The third is a branch, and the fourth instruction, though fetched, is never executed, and one machine cycle is lost. The fifth instruction shown is from the start of the program section to where the branch has taken place.

1.4 The microcontroller: its applications and environment

A microcontroller is a particular type of microprocessor, optimised to perform control functions for the lowest cost and at the smallest size possible. Generally microcontrollers are used in a recognisably 'embedded system' environment.

There is a huge range of microcontroller applications. Some are drawn from volume markets – the motor car, domestic appliances, mobile phones and toys. These applications are sold in such high volume that dedicated controllers are frequently developed for them. Others, like medical or scientific instruments, are sold in smaller numbers, and are more likely to make use of the wide variety of general-purpose controllers that are available. At one extreme of complexity, simple (and very cheap) controllers are used to replace 'glue logic' in a digital system. At the other extreme, advanced 32-bit controllers perform sophisticated signal processing activities.

1.4.1 Microcontroller characteristics

Arising from their 'embedded control' environment, microcontrollers usually have the following features:

- input/output intensive, i.e. they are capable of direct interface to a significant number of sensors and actuators
- a high level of integration, with many peripheral[3] devices included 'on-chip'
- physically small
- comparatively simple program and data storage requirements
- ability to operate in the real-time environment
- an instruction set optimised for the embedded environment, e.g. yielding compact code, limited arithmetic and addressing capability, strong in bit manipulation
- low cost

In many microcontroller applications either or both of the following features are also essential:

- an ability to operate in hostile environments, for example of high or low temperature, or high electromagnetic radiation;
- a low power capability, and features which ease the use of battery power.

In today's fast-moving world, both the manufacturers of microcontrollers and the people who design with them work under a complex set of sometimes conflicting forces. On the one hand, semiconductor technology is advancing inexorably. Every year it becomes possible to integrate more onto a single IC, and to do this more cheaply, with the chip operating at faster speed and lower power. Interacting with this technological change are powerful market forces, spurring on the development by demanding new capabilities from the microcontrollers. Against this, however, is set a certain conservative tendency. Companies using the microcontrollers have invested time and money in supporting work with a particular device, and don't want all this wasted when they move on to its more powerful successor.

---

3 A point of terminology: the term 'peripheral' in the computer world used to (and still does) refer to equipment which worked peripheral to the computer, such as printers and modems. In the microprocessor world it was adopted to describe such off-chip devices as I/O ports and serial links. Now these peripherals have become integrated onto the controller chip itself, but we still call them peripherals.

The outcome of all this is that the manufacturer usually develops a family of microcontrollers all based around one core, where the core contains the CPU and its surrounding control features (i.e. essentially the features of the early microprocessor, as in Fig. 1.5). The core defines the instruction set, and hence keeping the core design constant ensures software compatibility between different members of a processor family. To the core, and on the same IC, can be added the peripheral devices which seem best to meet a particular need. Even though the microprocessor world is one of such great change, many microcontrollers can trace their history very directly back for over 20 years! Once a company has committed itself to designing with a particular microcontroller family, it is reluctant to change, but looks to the manufacturer to supply it with the necessary technological advances, based around a familiar core. Infrequently, the manufacturer makes a step change by introducing a new core.

1.4.2 Features of a general-purpose microcontroller

Every microcontroller is different, and each has its own unique combination of core and peripherals. Figure 1.10 shows, in block diagram form and with no interconnections, the features which might be found in a simple general-purpose controller. The core is the element that remains constant for the whole family built around it. Ideally all memory is on-chip, and several different memory technologies may be applied to meet the differing needs of program and data storage. Interconnection to the outside world is through a number of parallel and serial ports. A counter/timer is available for event counting, or to measure or generate timing intervals.

1.4.3 Some example controllers

There are a huge number of different microcontrollers now on the market, and it is not easy to select a small number of representative devices to illustrate the hardware principles described in this text. Three example microcontroller families have been chosen, and in keeping with the introductory nature of the book they are all 8-bit devices. The families selected are well established in industry, and are chosen to illustrate the variety of approaches taken to solve common problems. Each one carries at least one

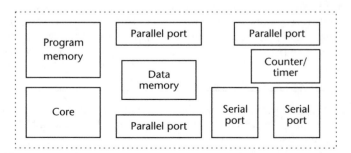

Figure 1.10 An example microcontroller block diagram.

feature not shared with the others. The example controllers are not, however, selected or treated as equals. First of all, the PIC 16F84 device is very small and low-cost, and used for some of the simplest possible embedded systems. The 80C552, on the other hand, is a comparatively complex (and more costly) device, rich in peripherals and with extensive memory addressing capability. The 68HC05 and 68HC08 lie in complexity between these two. Second, to encourage in-depth knowledge to be built up of one microcontroller, most early examples are of the PIC 16F84. Later examples are drawn more generally from the microcontrollers named, as well as one or two other close relatives.

It should be understood that the selection of these devices as examples is not intended to compare them in a competitive way (in the sense of seeking out 'the best'), nor does it necessarily represent an endorsement of any one of them.

It is recommended that the reader obtains at least the full data sheet of the 16F84 (Ref. 1.6). It will also be useful, but not essential, to have access to data on one or more of the others (Refs. 1.7 and 1.8). There are also many very useful Application Notes, published by the manufacturers, as well as books targeted towards individual devices, for example Refs. 1.9 and 1.10.

The remainder of this chapter introduces the three example families, looking particularly at their background, architecture and CPU. In the coming chapters we build on this introduction by looking at certain features of these microcontrollers in greater detail.

1.5 Microchip Inc. and the PIC™ microcontroller

1.5.1 Background, and meet the family

It was the General Instruments Corporation, back in the late 1970s, that first produced the PIC microcontroller (Ref. 1.11). In its early years it did not make a wide impact. The design was later taken over by Microchip Inc., and PICs are now one of the fastest moving families in the 8-bit arena, in more senses than one. First, they run very fast; second, the family is growing at a tremendous rate; and third, at the time of writing Microchip only operates with 8-bit controllers, and therefore has a special interest in making this controller size as attractive as possible. PICs cover a very wide range of 8-bit operation. At the lower end, they are simpler, cheaper and smaller than most devices that the competition can offer, and are thus used in situations where controllers would not be thought of as the right solution, even down to simple glue logic applications. At the high end, however, they are quite ready to take on the best of the 8-bit competition, with sophisticated devices equipped with excellent peripherals. PICs have made themselves particularly attractive to the student and low-budget developer. Development tools (both hardware and software) are cheap and readily available, and Microchip is very supportive of the novice designer.

Table 1.2 PIC microcontroller families.

Family	Program word size	Number of instructions	Minimum instruction execution time
12CXXX	12/14-bit	33/35	400 ns
16C5X	12-bit	33	200 ns
16C/FXXX	14-bit	35	200 ns
17CXXX	16-bit	58	120 ns
18CXXX	16-bit (enhanced)	77	100 ns

Microchip offers five closely related families of microcontroller, as shown in Table 1.2. PIC examples in this book are taken mostly from the 16F84, with a limited number from the more sophisticated 16C74.

All PIC controllers use a RISC-like structure, with Harvard architecture and pipelined instruction execution. This leads to one of the strengths of the PIC family: a very high instruction throughput.

1.5.2 A 16F84 microcontroller overview

The block diagram of this controller is shown in Fig. 1.11(a), and the IC pin connections in Fig. 1.11(b). The structure is radically different from other microcontroller families that we will see. The program memory area can be seen at the top left of the diagram. Its 13-bit address input is derived from the Program Counter, and its 14-bit data output forms the instruction word (transferred on the *Program Bus*). Note that although an address bus of this size can address 8K words, the actual program memory size is 1K. To the right of this is the RAM area, made up of 68 8-bit locations. Microchip calls these memory locations *file registers*. This memory has its own 7-bit address bus (again, the potential range of this bus is not fully exploited). Its data input/output is linked to the 8-bit microcontroller data bus. With the address and data buses to program memory and RAM being separate and independent (and of different sizes), we can conclude that this is a Harvard organisation. To add to the addressing complication, the EEPROM memory area has its own address source held in a dedicated RAM register (called EEADR). Data is transferred through another register EEDATA. The stack forms yet another distinct area of memory, with only eight 13-bit locations. It has no connection to the data bus, and cannot be used for temporary data storage.

As with all other 16XXX microcontrollers, the 16F84 makes no attempt to allow conventional memory expansion. Data and address buses are simply not 'bonded out' to external pins, and there is no chance to extend internal memory within the microcontroller memory map.

As peripherals, the 16F84 has:

- two parallel ports, one of 5 bits (Port A) and the other of 8 (Port B)
- one 8-bit Counter/Timer (TMR0)

(a)

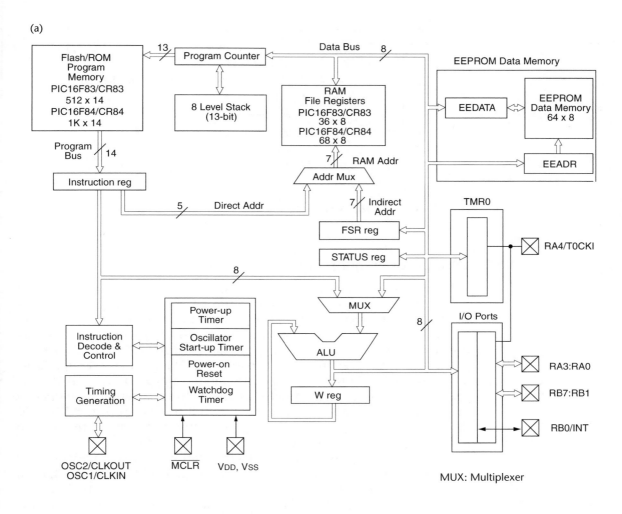

MUX: Multiplexer

(b)

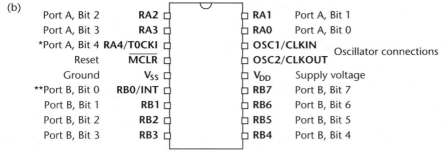

Port A, Bit 2	**RA2**		**RA1**	Port A, Bit 1
Port A, Bit 3	**RA3**		**RA0**	Port A, Bit 0
*Port A, Bit 4	**RA4/T0CKI**		**OSC1/CLKIN**	Oscillator connections
Reset	**MCLR**		**OSC2/CLKOUT**	
Ground	**VSS**		**VDD**	Supply voltage
**Port B, Bit 0	**RB0/INT**		**RB7**	Port B, Bit 7
Port B, Bit 1	**RB1**		**RB6**	Port B, Bit 6
Port B, Bit 2	**RB2**		**RB5**	Port B, Bit 5
Port B, Bit 3	**RB3**		**RB4**	Port B, Bit 4

*also Counter/Timer clock input
**also external interrupt input

Figure 1.11 The PIC 16F8X. (a) Block diagram; (b) pin connections: plastic dual-inline package (PDIP). Part (a) reprinted with permission of the copyright owner, Microchip Technology Incorporated © 2001. All rights reserved.

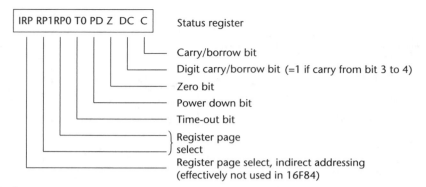

Figure 1.12 The 16F84 Status Register.

1.5.3 CPU, programming model and instruction set

Manufacturers of complex microprocessors usually supply a *programming model*. This is a simplified diagram of those internal microprocessor registers that are of direct interest to the programmer. Owing to its comparative simplicity, a programming model of the 16F84 is not normally necessary; the block diagram of Fig 1.11(a) is used in its entirety.

The ALU can be seen in the lower part of Fig 1.11(a). There is a single accumulator (the Working, or W, register), and most operations are performed between it and either the contents of a RAM file register, or with operand data embedded within the op code. The Status Register (Fig. 1.12) holds three bits (Z, DC and C) which give status information about the result of the most recent instruction executed.

The 16F84 has 35 instructions in its instruction set, a summary of which appears in Appendix B. These will be explored in detail in Chapter 3. It can operate at any clock frequency up to a maximum of 10 MHz. Each instruction cycle is made up of four oscillator cycles. The resulting fastest instruction execution time is therefore 400 ns.

1.6 The Philips 80C552 microcontroller

1.6.1 Background, and meet the family

As Intel was the first company to produce a microprocessor, it seems right that it was also the first to produce a microcontroller. It did this in 1976 with the MCS-48 (appearing in three versions, the 8035, 8048 and 8748; Ref. 1.11). In its time the MCS-48 was revolutionary. The 8748 had on-chip ultraviolet erasable programmable read-only memory (EPROM), 64 bytes of RAM, and three input/output ports. It attracted many adherents. In 1980 Intel launched its successor, the 8051. The 8051 took up where the '48 left off, and has also become firmly embedded, in more senses than one, in the microcontroller world. Though the 8051 itself is now an old device, many companies (for example Atmel, Dallas, Philips and Siemens) offer controllers based on the '51 core, and further developments are repeatedly being produced.

As one of the manufacturers who have adopted the 8051 design, Philips has developed many variants. The 80C51 is a CMOS version of the 8051, and Philips has extended this into a wide-ranging family. Its practice has been to use the *whole* of the 80C51 as core, and to add further peripherals to this.

1.6.2 Controller overview and architecture

The 80C552 is an advanced member of the Philips 80C51 family. It has a powerful collection of on-chip peripherals, and is one of three versions of the 8XC552, which differ only according to the program memory available on-chip. The main features are:

Within the 80C51 'core'
- Four 8-bit parallel ports
- Two 16-bit Counter/Timers
- One UART (Universal Asynchronous Receiver Transmitter)

Extra to 80C51 Core
- Two 8-bit parallel ports
- One 16-bit Counter/Timer with Capture and Compare enhancements
- Two PWM (Pulse Width Modulation) ports
- One I^2C (Inter-Integrated Circuit) serial port
- 256 bytes of static RAM (data memory)
- Eight-input 10-bit ADC (Analogue to Digital Converter)
- 8 Kbytes mask programmable ROM, 256 bytes RAM (83C552)
  or 8 Kbyte EPROM, 256 bytes RAM (87C552)
  or as 83C552, without ROM (80C552)

The 8XC552 block diagram is shown in Fig. 1.13. The 80C51 core can be seen to the left, and all the peripherals listed above can easily be distinguished. All sub-systems within the controller are served by a single data bus. By sacrificing two of the parallel ports, the data and address buses *can* be made available to the outside world. In this case, the lower 8 bits of the address bus are multiplexed with the data bus. This is essential for the 80C552, which has no internal program memory, but usually unnecessary for the 83C552 and the 87C552, which are self-contained.

The '552 is offered in three speed ranges, with clock frequencies of up to 16 MHz, 24 MHz and 30 MHz. In each case the minimum clock frequency is 1.2 MHz. The clock is divided internally by 12 to form one machine cycle.

1.6.3 CPU, programming model and instruction set

A programming model of all members of the 80C51 family is shown in Fig. 1.14. All the registers shown (excluding the Program Counter) are memory-

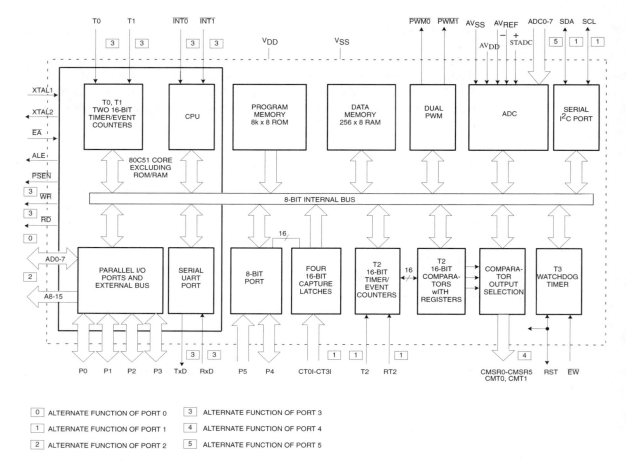

Figure 1.13 The 80C552 block diagram. (Reproduced by permission of Philips Semiconductors.)

mapped. There is a single 8-bit accumulator. There is also a B register, used only during 8-bit multiply and divide instructions. The Stack Pointer is 8-bit, and may be used to locate the stack theoretically anywhere in the 256 bytes of on-chip RAM. Available stack space is effectively restricted by other uses to which the programmer wishes to put the RAM. The Program Status Word (PSW, equivalent to the Status Register of the 16F84) has 4 bits (high-lighted in the figure) which indicate arithmetic status.

An interesting feature of the 80C51 CPU is that it includes a 1-bit Boolean processor, which uses the PSW Carry bit as the accumulator. This allows a complete set of instructions on single-bit operands, including Boolean operations, as well as move, set and clear. These operations can be used on a block of bits in the static RAM, as well as certain locations in the peripheral controls (including all parallel port bits).

All 80C51 variants use the same (CISC) instruction set, which contains 50 distinct assembler mnemonics. When adapted to the different addressing modes available, these amount to 111 instructions. These are listed in the

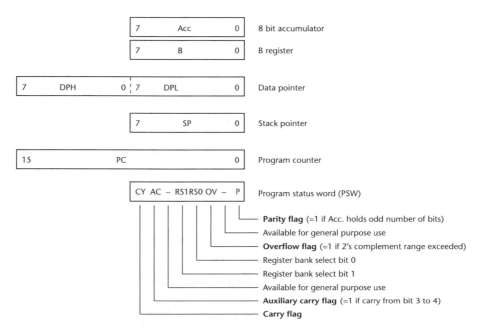

Figure 1.14 The 80C51 programming model.

categories: arithmetic (including 8-bit by 8-bit hardware multiply and divide), logical, data transfer, Boolean variable manipulation, and program branching. Instructions are encoded in one, two or three bytes, and execute in from one to four machine cycles (i.e. 12 to 48 oscillator cycles).

1.7 The Motorola 68HC05/08 microcontrollers

1.7.1 Background, and meet the family

Motorola was early in the microprocessor field, but was not the first. By the time it entered, with the 6800, it was able to offer a device which enjoyed remarkable longevity. From the 6800 it developed further 8-bit conventional microprocessors (e.g. the 6809), and also a number of single-chip controllers, starting with the 6801. These led to the 68HC11, a sophisticated and widely used microcontroller. The HC infix indicates the new high-speed CMOS (Complementary Metal Oxide Semiconductor) technology with which it is made. From the 68HC11 the 68HC12 and 68HC16, both 16-bit controllers, have been developed.

An indirect development of the 6800 family was the 6805 (M146805 in full), available initially in HMOS (High-Density N-Channel MOS) and CMOS versions. Here the CPU was simplified, for example by the removal of the second (B) accumulator of the 6800, reduction in addressing capability, and consequent reduction of certain register sizes. As one of the earlier CMOS controllers, the 6805 had a great impact on low-power applications. The 6805 was subsequently upgrade and reissued using 'HC' CMOS technology. This has enjoyed very widespread use as a simple and low-cost microcontroller. Motorola claims that over 2 billion (2×10^9) units of the 68HC05 have been sold. The number of variants are too many to list, but contain devices targeted specifically for automotive, computer, consumer, industrial, telecommunications, TV and video applications.

Since the late 1990s the 68HC05 has been in the process of being replaced by the 68HC08, which provides a direct upgrade. Both the '05 and the '08 use the '7' infix to indicate EPROM or OTP (One-Time Programmable) memory version (e.g. the 68HC705P) and the '9' infix to indicate Flash memory.

All of the 68HCXX microcontroller families have some similarity in architecture and instruction sets, so it is a comparatively easy task to move from one to another, selecting the device most appropriate for the job.

1.7.2 68HC05 controller overview

The block diagram of the general purpose 'B' version of the 68HC05 is shown in Fig. 1.15. The main features are:

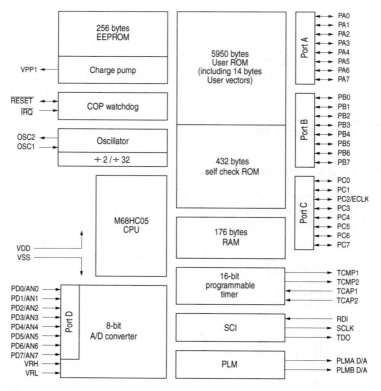

Figure 1.15 The 68HC05B6 block diagram. (Copyright of Motorola Inc. Used by permission.)

- three 8-bit parallel input/output ports and one 8-bit input-only port
- one 16-bit timer system
- one Serial Communications Interface (SCI)
- eight-channel 8-bit Analog to Digital Converter
- 'Computer Operating Properly' (COP) Watchdog System
- two Pulse Length Modulation (PLM) outputs, intended for digital to analogue conversion
- 256 bytes of EEPROM
- 176 bytes of RAM
- 5950 bytes of User ROM

The bus structure, though not shown, follows a simple von Neumann pattern. Like the 16F84 there is no external bus connection. There is a single memory map, within which *all* memory and addressable registers lie. Although it is not normal operation, programs *can* be executed from RAM or EEPROM, and there is a mechanism for loading either of these memory areas through the serial port.

The 68HC05 can operate with an internal clock frequency up to 2.1 MHz. This is divided down by 2 from the external clock. A SLOW mode of operation is also available, in which the user may insert a further divide-by-16 in the clock generator. This is useful for low-power applications. With a fully static design, it can operate down to DC.

1.7.3 CPU, programming model and instruction set

The 68HC05 programming model is shown in Fig. 1.16. There is a single Accumulator, an 8-bit Index Register and a 16-bit Program Counter. The Stack Pointer is initially set to $00FF_{16}$, and counts down as data is entered onto the stack. As the 10 most significant bits are fixed as shown, the actual range of the stack is from $00FF_{16}$ down to $00C0_{16}$, giving 64_{10} actual memory locations.

The CPU has 62 basic instructions. These are the same as the earlier M146805, with the addition of an unsigned hardware multiply, in which the contents of the Accumulator and Index Register are multiplied together. A wide range of addressing modes are supported, which when applied to the basic instructions gives a final total of 210. All instructions are encoded in one, two or three bytes, and execute in from two to eleven internal clock cycles.

With four bits in the Condition Code Register which indicate arithmetic status, program branching can take place on any of the conditions $<, \leq, =, \neq, \geq, >$, negative and positive. Bit set and clear, and bit test and branch instructions are supported for operands in the first page of memory (i.e. the first 256 bytes). Boolean operations between bits are not available.

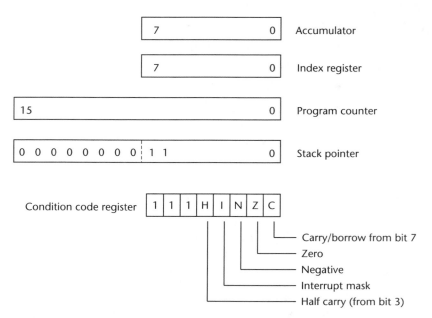

Figure 1.16   The 68HC05 programming model.

1.7.4   The 68HC08

The 68HC08 family of microcontrollers offers a direct upgrade to the 68HC05. The CPU has been expanded, some new instructions have been added (for example a 16-bit by 8-bit divide), and higher operating speeds are possible. The most significant advance that the '08 offers, however, is its inclusion of Flash memory, which allows greatly increased programming flexibility. Like the '05, the '08 is available in many versions, targeted towards specific areas of application. These are identified by an alphabetical suffix to the device number, which include **-GP** (general-purpose), **-AS** (automotive), and **-JL/JK** (low-cost general-purpose).

The programming model of the '08 is shown in Fig. 1.17. Index and Stack registers are now 16 bit, and the Condition Code Register is enhanced by the addition of a two's complement overflow flag.

SUMMARY

1.   An embedded system incorporates a computing element, typically a microprocessor or microcontroller, to perform a control function. Many embedded systems are small and low-cost, and are aimed towards the volume market. They apply recognised hardware and software principles to meet the particular requirements of the embedded environment.

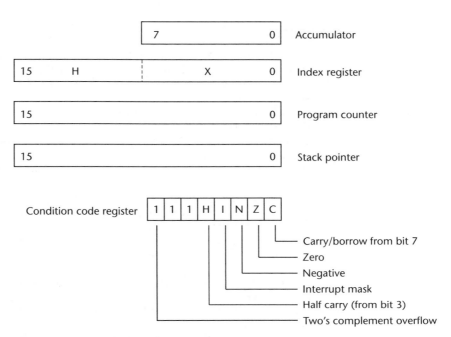

Figure 1.17 The 68HC08 programming model.

2. The microprocessor was one of the great technological revolutions of the 20th century. It is, however, based on principles that are now well established and stable. Owing to technological advances, faster and more powerful processors are continuously being introduced.

3. The microcontroller is a microprocessor intended for small-scale control applications. It integrates a conventional microprocessor core and a range of peripheral devices on a single IC, at the smallest size and lowest cost possible. A family of controllers is based around the same core, but with different peripherals and IC packaging, optimised for different applications.

4. While all microprocessors differ, there are some fundamentally different options in processor design, which have major significance for the final perfomance. These options include RISC vs. CISC, conventional von Neumann vs. Harvard, and the option of pipelining.

5. The PIC, 80C51 and MC68HC05/08 series of microcontrollers are all successful and well-established 8-bit controllers, each with their own unique attributes and advantages.

REFERENCES

1.1   Horowitz, P. and Hill, W. (1989) *The Art of Electronics*, 2nd edn. Cambridge: Cambridge University Press. Chapters 10 and 11.

1.2   Storey, N. (1989) *Electronics, a Systems Approach*, 2nd edn. Reading: Addison-Wesley. Chapter 12.

1.3   Sedra, A. and Smith, K (1998) *MicroElectronic Circuits*, 4th edn. Oxford: Oxford University Press.

1.4   Clements, A. (1991) *The Principles of Computer Hardware*. Oxford: Oxford Science Publications.

1.5   Hennessy, J. L. and Patterson, D. A. (1998) *Computer Organisation and Design*. San Mateo, CA: Morgan Kaufmann.

1.6   Data on Microchip PIC 16F84:
      `http://www.microchip.com/10/Lit/PICmicro/16f8x/index.htm`

1.7   Data on Philips 80C51 and family:
      `http://www.philips.semiconductors.com/mcu/products/`

1.8   Data on Motorola 68HC05/08:
      `http://www.mcu.motsps.com/hc08/index`

1.9   Peatman, J. B. (1997) *Design with PIC Microcontrollers*. Englewood Cliffs, NJ: Prentice Hall.

1.10 MacKenzie, S. (1999) *The 8051 Microcontroller*, 3rd edn. Englewood Cliffs, NJ: Prentice Hall.

1.11 Bursky, D. (ed.) (1978) *Microprocessor Data Manual*. Philadelphia, PA: Hayden.

EXERCISES

1.1   Find a non-technical friend and describe to him or her what an embedded system is and what its common characteristics are.

1.2   List five products or product sub-systems which could be embedded systems, choosing examples from the domestic, automotive, industrial or office environments. For each one, outline briefly the effects of unreliable performance.

1.3   For the products listed in Exercise 1.2, consider the importance of time in the operation of each. Is *fast* operation required? Does the system operate within strict time demands?

1.4   The Harvard memory structure gives some clear advantages over the conventional von Neumann structure. Can you think of any disadvantages? (Consider and expand on: system complexity, flexibility of memory utilisation, ease of accessing data tables in program memory, access to Stack.)

1.5   Three microprocessors, A, B and C, have maximum clock speeds respectively of 10 MHz, 24 MHz and 20 MHz. Processor A divides its clock by 4 to give one machine cycle, processor B by 12, and processor C by 8. A and B take two machine cycles to perform an 'add accumulator to immediate data' instruction, while C takes three cycles. Place the processors in order of the speed in which they can perform this instruction.

1.6   An 80C552 can execute an 'add register to accumulator' in one machine cycle. A PIC 16F84 can also perform this in one machine (instruction) cycle. If the PIC is running at a clock frequency of 10 MHz, at what speed should the 80C552 run in order for them to execute the instruction in the same time?

1.7 For the 16F84, the 80C51, the 68HC05 and the 68HC08, draw up a table showing all arithmetic and logical flags appearing in their 'Status Registers' (or equivalent), indicating which flag is implemented in which microcontroller. Which register has the potential to give the most information?

1.8 The now obsolete Intersil IM6100 microprocessor had a 12 bit data bus and a 12 bit address bus.
(a) How many words of data could it address?
(b) Assuming 12 bits in every memory location, how many bits of data could it address?
(c) With the data coded in two's complement, what range of numbers could it represent?

(Refer to Appendix A if you are initially unable to do this question.)

1.9 How long do the longest and the shortest instructions take to execute, for each of the 16F84, 80C552, and 68HC05, when each is operating at
(a) its fastest clock frequency?
(b) an external clock frequency of 2 MHz?

1.10 In the 16F84 Instruction Set (Appendix B), identify all instructions which operate on single bits.
(a) Which bits may be used as operands?
(b) What operations on the operand bits are possible?
(c) Describe clearly how operand bits are identified. How many bits within the instruction are needed for this?

From humble beginnings –
towards the minimum system

IN THIS CHAPTER

Chapter 1 has already mentioned the broad division of the embedded system into two integrated parts: the hardware and the software. Each of these areas has the potential to become complicated, and each therefore requires a good understanding of underlying principles and of the design processes involved.

Because of these looming complexities, this book introduces both hardware and software in a staged way. This chapter is intended to introduce the 'minimum system' from a hardware point of view. It parallels Chapter 3, which introduces the minimum system from the software point of view. Most of the topics in this chapter, which are introduced in their simplest and most digestible form, are developed in later chapters of the book.

The terminology 'minimum system' implies a simple microcontroller or microprocessor, surrounded with only the bare essentials adequate for it to operate in a meaningful way. Therefore in this chapter we will explore the essential features of the microcontroller itself, and the essential external circuit elements that accompany it. It is assumed that the reader already has knowledge of the microcontroller core. To this must now be added those distinguishing features of the microcontroller, the peripherals, and the means to communicate with them. The external elements will include simple user interface devices, power supply and clock oscillator. Examples in this chapter are based predominantly on the Microchip PIC 16F84, which because of its simplicity finds itself at the heart of many a minimum system.

While it would seem sensible to include a section on hardware testing, that is in fact difficult to do in the absence of any program. Therefore there is a section in Chapter 4 which covers introductory testing for both hardware and software.

The overall aim of the chapter is that, on its completion, the reader should have the confidence to embark on the hardware design of a simple microcontroller-based system. In more detail, the chapter aims:

- to describe how the microcontroller CPU communicates, through Special Function Registers, with its peripherals

- to describe simple Interrupt structures

- to introduce the principles and application of the Parallel Port, together with simple associated interfacing techniques

- to introduce the principles of the Counter/Timer, and simple applications

- to describe those essential external hardware items, the power supply and clock oscillator, which complete the minimum working environment

The chapter makes frequent reference to TTL (Transistor–Transistor Logic) and CMOS (Complementary Metal Oxide Semiconductor) Logic, and related concepts. Readers who wish to refresh themselves on these issues should refer to Ref. 1.1.

2.1   Interfacing with peripherals

We met the concept of the microcontroller peripheral in Chapter 1. In this chapter we will consider two important peripheral types. In order to do this, it is necessary to understand not only how the peripherals work, but also how they interface with the microcontroller CPU. This is the point we now start with.

2.1.1   Special Function Registers

Almost all microcontroller peripherals can be configured in software to operate in a number of different modes. Before they are put to use, therefore, certain control data must be sent to them to set them up (initialise them) in the required way. Once in use there is a need for data flow to or from the peripheral, and there may also be a need for further flow of control data. We shall see, for example, that bits in a parallel port can be set up as input or output, and that data can then be read from them or sent to them. If the peripheral has a need to demand urgent action from the microcontroller core, there may be one or more interrupt lines dedicated to it. Some parallel ports, for example, generate an interrupt whenever a change is detected on selected input lines.

These needs are commonly met by means of dedicated registers, called either *Special Function Registers* (SFRs, by Microchip or Intel), or *Control Registers* (by Motorola). We will use the terminology SFR when speaking

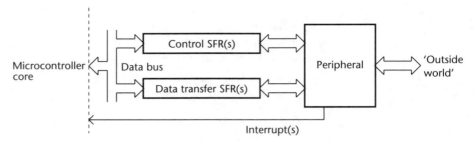

Figure 2.1 Generalised interface between peripheral and core.

generally. The generalised interface between a peripheral and the microcontroller core is shown in Fig. 2.1. One or more SFRs are used for control purposes, and one or more are used for transfer of data. In practice, one SFR may be used for both input and output, or be used for several different peripherals. Interrupt lines may also be included. The precise structure is of course dependent on the function of the peripheral and on the manufacturer. SFRs are usually 8-bit registers lying in the microcontroller memory map.

The outline memory structure of the PIC 16F84 appears as part of Fig. 1.11(a). The map of its RAM, which includes the microcontroller SFRs, is now shown in Fig. 2.2. We can see that there is an area of general-purpose static RAM (in the form of 68 general-purpose registers), as well as a block of SFRs. Some of these, like PCL and STATUS (which we have met already as Fig. 1.12), relate to the CPU and its control. Others relate to the microcontroller peripherals, i.e. ports A and B, and the Counter/Timer module. These will be described within the context of the peripherals as they are introduced.

The flow of control and information data between peripherals and CPU through the SFR structure gives the microcontroller manufacturer great scope to extend a microcontroller family. As long as there are uncommitted SFR locations in the memory map, a new peripheral can be easily added to the controller architecture and new SFRs defined to provide the necessary interaction.

2.2 Interrupts

A microcontroller feature essential for interaction with peripherals and the external world is the *interrupt*.

2.2.1 An interrupt review

An interrupt is an external input to the CPU which can be used to force the processor to provide a rapid response to external events. It thereby allows the system hardware to call for program response at instants which it determines, rather than the program. Interrupt sources may signal emergency situations – the power supply has been lost, or the temperature has

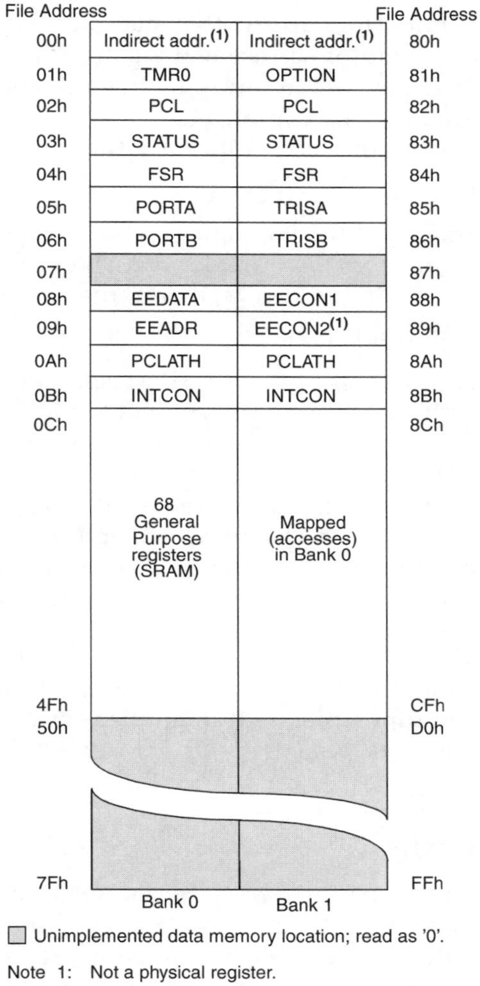

File Address			File Address
00h	Indirect addr.[1]	Indirect addr.[1]	80h
01h	TMR0	OPTION	81h
02h	PCL	PCL	82h
03h	STATUS	STATUS	83h
04h	FSR	FSR	84h
05h	PORTA	TRISA	85h
06h	PORTB	TRISB	86h
07h			87h
08h	EEDATA	EECON1	88h
09h	EEADR	EECON2[1]	89h
0Ah	PCLATH	PCLATH	8Ah
0Bh	INTCON	INTCON	8Bh
0Ch			8Ch
	68 General Purpose registers (SRAM)	Mapped (accesses) in Bank 0	
4Fh			CFh
50h			D0h
7Fh			FFh
	Bank 0	Bank 1	

☐ Unimplemented data memory location; read as '0'.

Note 1: Not a physical register.

SFR	Summary function
TMR0	Counter/Timer (TMR0) Register
PCL	Holds lower byte of program counter
STATUS	Holds CPU status flags, and other control bits
FSR	File Select Register; holds indirect file address
PORTA/B	Input/output bits for Port A/B, also containing external interrupt and Counter/Timer inputs
EEDATA	EEPROM data
EEADR	EEPROM address
PCLATH	Transfer buffer for higher bits, Program Counter
INTCON	Interrupt Control
OPTION	Contains control bits for Interrupt, Weak Pull-up, and Counter/Timer
TRISA/B	Port A/B Data Direction Register; sets data direction of Port A/B
EECON1	EEPROM Control 1
EECON2	EEPROM Control 2

Figure 2.2 16F84 register map. Reprinted with permission of the copyright owner, Microchip Technology Incorporated © 2001. All rights reserved.

exceeded a safe limit. Alternatively they may signal more routine, possibly periodic events – the motor has reached top dead centre, a keypad has been pressed, a data byte has been received by the serial port, or an analogue to digital conversion has been completed.

When it is activated the interrupt causes the processor to:

- complete the current instruction
- store the contents of the Program Counter (i.e. the address of the next instruction), and possibly other key variables (for example contents of Accumulator and Status Register), onto the stack
- jump to an interrupt service routine (ISR)

The ISR is written to make the processor respond in a suitable way to the interrupt. It is similar to a subroutine, except that it is initiated by the Interrupt, and it must be ended by a 'Return from Interrupt' instruction (retfie for the 16F84). On executing this last instruction the Program Counter value previously saved on the stack is retrieved and put back, any other variables stored on the stack are also retrieved, and the program resumes from the point where it had been interrupted. The starting address of an ISR is determined by the contents of a location in program memory called the *Interrupt Vector*.

When an ISR is being executed, we say that that interrupt is being *serviced*. An interrupt that has occurred, but which has not yet been serviced (in a multiple interrupt structure it might be waiting for another ISR to be completed), is said to be *pending*.

2.2.2 Interrupt hierarchy

Many microcontroller peripherals have the capability of generating interrupts. Timers can interrupt on overflow, and serial ports can interrupt when a new data byte has been received. A sophisticated controller therefore has many interrupt sources: a few from external sources through the processor pins, and many arising from its on-chip peripherals. In general, all can be enabled or disabled individually by setting or clearing *mask* bits in the relevant SFR. A *global interrupt enable* is usually available to enable or disable the overall interrupt capability. There may be a *non-maskable interrupt*, for events of highest priority, which can never be disabled.

In many controller architectures the occurrence of an interrupt sets an associated *flag*, usually a bit in an SFR. This is set even if the interrupt is disabled (and can be usefully accessed by the program). It *may* be a processor requirement to clear the flag in software during the course of the interrupt routine.

Each interrupt source is linked in the microcontroller hardware with an Interrupt Vector. In some cases each source has its own Interrupt Vector. In others (for example the 16F84), one Interrupt Vector is shared by several sources. The ISR then has to poll the interrupt sources (in general by checking the state of the interrupt flags) to determine which one was active.

R/W-0	R/W-0	R/W-0	R/W-0	R/W-0	R/W-0	R/W-0	R/W-x
GIE	EEIE	T0IE	INTE	RBIE	T0IF	INTF	RBIF

bit7   bit0

```
R = Readable bit
W = Writable bit
U = Unimplemented bit,
    read as '0'
- n = Value at POR reset
```

bit 7: **GIE:** Global Interrupt Enable bit
1 = Enables all un-masked interrupts
0 = Disables all interrupts

Note: For the operation of the interrupt structure, please refer to Section 8.5.

bit 6: **EEIE**: EE Write Complete Interrupt Enable bit
1 = Enables the EE write complete interrupt
0 = Disables the EE write complete interrupt

bit 5: **T0IE**: TMR0 Overflow Interrupt Enable bit
1 = Enables the TMR0 interrupt
0 = Disables the TMR0 interrupt

bit 4: **INTE**: RB0/INT Interrupt Enable bit
1 = Enables the RB0/INT interrupt
0 = Disables the RB0/INT interrupt

bit 3: **RBIE**: RB Port Change Interrupt Enable bit
1 = Enables the RB port change interrupt
0 = Disables the RB port change interrupt

bit 2: **T0IF**: TMR0 overflow interrupt flag bit
1 = TMR0 has overflowed (must be cleared in software)
0 = TMR0 did not overflow

bit 1: **INTF**: RB0/INT Interrupt Flag bit
1 = The RB0/INT interrupt occurred
0 = The RB0/INT interrupt did not occur

bit 0: **RBIF**: RB Port Change Interrupt Flag bit
1 = When at least one of the RB7:RB4 pins changed state (must be cleared in software)
0 = None of the RB7:RB4 pins have changed state

Figure 2.3 16F84 INTCON Register. Reprinted with permission of the copyright owner, Microchip Technology Incorporated © 2001. All rights reserved.

2.2.3 The 16F84 interrupt structure

The 16F84 has four interrupt sources: External Interrupt, Timer Overflow, Interrupt on Port Change, and EEPROM Write Complete. Only the first of these is generated externally. The other three are due to events internal to the processor; Timer Overflow and Interrupt on Port Change will be introduced later in this chapter. Each interrupt sets a flag when an interrupt occurs, and each has its individual interrupt mask. These flag and mask bits are located in the INTCON Register (Fig. 2.3), which controls interrupt activity. The interrupt logic structure is shown in Fig. 2.4. In every case, the interrupt flag value is ANDed with the interrupt mask bit. If both are high, and the global interrupt enable (GIE, also in the INTCON register) bit is high, then an interrupt is transmitted to the CPU.

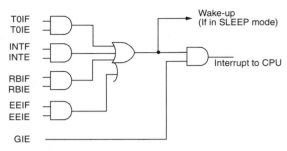

Figure 2.4 16F84 interrupt structure. Reprinted with permission of the copyright owner, Microchip Technology Incorporated © 2001. All rights reserved.

It is a processor requirement that interrupt flags are cleared by software within the ISR; otherwise the routine will be called again (repeatedly) every time it completes.

Simple programming with interrupts is covered in Chapter 3, and advanced interrupt structures and applications are dealt with in Chapter 8.

2.3 Parallel input/output (I/O) ports

2.3.1 The bidirectional port

A parallel I/O port is an interface through which digital data can be received into the microcontroller or output from it. Initially there may appear to be some confusion between a parallel data bus and a port. Are they not both used for bidirectional transfer of digital data? The difference is that the data bus is multi-tasking – it is shared by different devices for use at different times, for different data transfer functions. Any device connected to the data bus has the possibility to read data from it or to put data onto it. It follows that the device must be able to recognise (by interpreting the processor control lines) when it can connect to the bus and when it must not. Unlike the data bus, bits on a parallel port are committed full-time to a single task, and data on the port lines relate only to that function. Devices connected to the port do not need the sophistication of those connected to the data bus, and can be continuously active. Therefore I/O ports are essential as a primary channel of I/O data, with ports being easily connected to switches, keypads, light-emitting diodes and displays.

In the past, microprocessors used external ICs with names like Parallel Interface Adaptor (PIA) or Versatile Interface Adaptor (VIA) to interface between the data bus and digital input or output data. The PIA/VIA typically had two ports, each of which was a group of 8 bits. It could set ports to be input or output, or have some bits as input and some as output. It was memory-mapped, so by reading the right memory location it could determine the digital values being asserted at the port input; alternatively, by writing to the port it could set new values to output bits.

In the microcontroller world the features of the PIA/VIA are incorporated on-chip. Indeed, in the larger controllers there are many I/O ports (for example six in the 80C552), generally 8-bit. Usual features of each port are:

- simple memory-mapped access, generally as an SFR
- an ability to act either as input or output (i.e. 'bidirectional'), with the direction of each port bit being independently programmable
- an ability to read or set individual bits.

Common further features are:

- an ability to further modify pin characteristics, for example by enabling internal pull-up resistors
- an ability to drive LEDs or other small loads
- an ability to use a single pin for several functions (for example an interrupt or counter input)
- port pins of differing capabilities (for example some may have higher power drive capability, others with Schmitt trigger input)

2.3.2 The 16F84 ports

As shown in Fig. 1.11, the 16F84 has 13 I/O lines, configured as:

- *Port A*: five bidirectional bits, RA4 to RA0. All are TTL-compatible input with CMOS output, except for bit 4 (RA4), which is Schmitt trigger input and open drain output. RA4 also doubles as the Counter/Timer external clock input.
- *Port B*: eight bidirectional bits, RB7 to RB0. Bit 0 (RB0) also acts as the external interrupt input. Port B includes a weak pull-up and 'interrupt on change' capability, described below.

The I/O bits of each port are located in the PORTA/B SFRs (Fig. 2.2), and the data direction control bits are located in the SFR's TRISA/B.

The key to the usefulness and flexibility of the I/O port and pin lies in the 'pin-driver' circuit connected to the port pin and how it is controlled. As an example, the driver circuit for each of bits 0 to 3 of Port B is shown in Fig. 2.5. The I/O pin itself is seen at the top right of the diagram. The logic value of this pin can be set by the output buffer lying just to its left, *and* it may be read via the 'TTL Input Buffer'. The output buffer is, however, controlled by the 'TRIS Latch'. If its output is logic 0, then the output buffer is enabled and data on the 'Data Latch' is transferred to the I/O pin. The port bit is then acting in output mode. If the 'TRIS Latch' output is a logic 1, then the output buffer is put into a high-impedance mode. The pin can then not output data, but it can be used as an input. The 'TRIS Latch' therefore effectively controls data direction. The microcontroller program can set the values in the TRIS and Data latches, which are both connected to the data bus, by writing to the appropriate bits in the SFRs.

If the value of the port pin is to be read, then the 'RD Port' line is pulsed and the state of the I/O pin, via the 'TTL Input Buffer', is latched into the

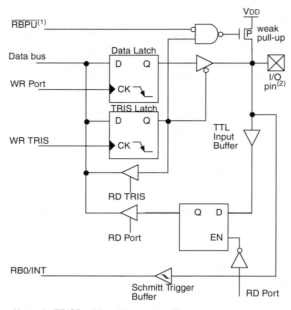

Figure 2.5 Pin driver circuit for bits 0–3 of the 16F84 PortB. Reprinted with permission of the copyright owner, Microchip Technology Incorporated © 2001. All rights reserved.

associated input latch (not labelled). The data is then transferred onto the bus, and thence to the CPU. This occurs whatever the state of the TRIS latch. It is worth noting that even if the port bit is set to output ('TRIS Latch' output at logic 0), it is the actual pin value that is read, rather than the value of the data latch output.

This circuit is replicated for each port bit, and similar ones are used for the Port A and other B bits. Each 'TRIS latch' forms one bit in the TRISB SFR, and each 'Data Latch' and input latch together form one bit in the PORTB SFR. The diagram also shows the direct input path of the external interrupt, connected only to pin 0.

2.3.3 Interrupt on change

Commonly a parallel port is connected to a user interface. The programmer can repeatedly poll the port to check for a change in input, but this is time-consuming for the program. Therefore some ports are designed to detect a *change* in input, and initiate an interrupt if a change is detected. An example is the 16F84 Port B pins 4 to 7. Each of these has a circuit similar to Fig. 2.5, except that there is an extra input latch which holds the value of the pin from the *previous* time the port was read. When a difference between the present and previous values occurs, an interrupt is generated. Figure 2.4

shows that this has the further advantage that it can be used to wake the controller up from 'sleep' (i.e. low power, at rest) mode.

2.3.4 Unassigned inputs and weak pull-ups

Serious problems occur when the input voltage of a CMOS gate is undefined. The voltage may drift midway between logic high and low and cause excessive (even destructive) current flow, or weird oscillations can take place. Some controllers with CMOS inputs therefore allow the option of a weak pull-up capability when a port bit is set to input. The purpose of this is to define the logic level of the input if it is unconnected. Its disadvantage is that current consumption is increased slightly when the input is taken to logic 0. All 16F84 Port B pins, when in input mode, can switch on a weak pull-up if bit $\overline{RBPU}$ (in the OPTION Register, another SFR) is set low. This can be seen in Fig. 2.5.

2.3.5 Electrical characteristics

When designing with microcontroller digital I/O it is important to have an understanding of its electrical characteristics. Logic voltage thresholds, both of the microcontroller and any interconnected devices, must be satisfied, and drive current capabilities may need to be evaluated.

The output of a logic gate or port pin can either *sink* current (i.e. current flows *into* the gate output when at logic 0), or it can *source* current (i.e. current flows *out* of the gate output when at logic 1). These output source and sink capabilities are an important aspect of the microcontroller's ability to drive simple loads.

The input of a logic gate or port pin requires the voltage to be *below* a certain maximum in order to be recognised as a a logic 0, or *above* a certain minimum to be recognised as a logic 1. As well as this, the input represents a loading effect to the driving device which needs to be understood. This load is not necessarily purely resistive. In TTL logic, for example, a gate input when pulled to logic 0 actually sources current to the device driving it.

The input characteristics of the 16F84 port pins (excluding the Schmitt trigger input) are shown in Table 2.1. These indicate TTL voltage level requirements but CMOS current characteristics.

The port pin output characteristics of the 16F84 are shown in Fig. 2.6. Figure 2.6(a) shows the output voltage for the output high condition with varying source current. It can be seen that the output voltage is 5 V for negligible output current, but falls to around 4.5 V when 7 mA is being

Table 2.1 PIC 16F84 port input characteristics (TTL buffer, 5 V supply).

Minimum input high voltage, V_{IH}	2.4 V
Maximum input low voltage, V_{IL}	0.8 V
Input leakage current, I_{IL}	±1 μA

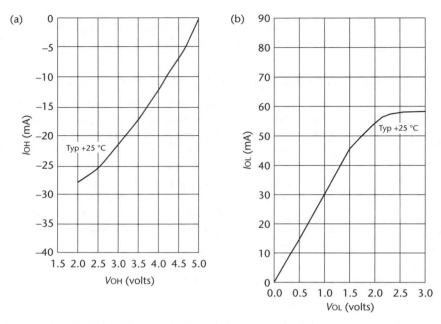

Figure 2.6 PIC 16F8X output characteristics; supply voltage = 5 V. (a) Output high; (b) output low. Reprinted with permission of the copyright owner, Microchip Technology Incorporated © 2001. All rights reserved.

sourced. It falls further, to around 2.5 V, when 25 mA is being sourced. Figure 2.6(b) shows the output low condition, with current being sunk by the output. It can be seen that this is 0 V for negligible sink current. The output voltage rises to 0.5 V for around 12 mA of sink current, and to 1.0 V when 30 mA is being sunk. Clearly, the output appears to have slightly better sink than source capability. For application of these curves, see Example 2.2 below.

2.3.6 The quasi-bidirectional port

We turn aside briefly here to consider another way of producing a bidirectional port, which does not need its data direction to be specified. Figure 2.7 illustrates the principle behind the 'quasi-bidirectional' port. This circuit can be used as an alternative to that of Fig. 2.5, and certainly appears simpler. Here the output stage is CMOS, but deliberately made asymmetrical by the inclusion of a low-value constant current source (typically around 100 μA), in the upper arm of the output. When data is written to the pin, it acts as a normal output, except that it has very limited ability to source current when its output is high. For this reason, when the output is high, it is easy for that high value to be overridden (i.e. to a logic 0) by another device connected to the I/O pin. The only requirement is that the other device must be able to absorb the small amount of current from the constant current source.

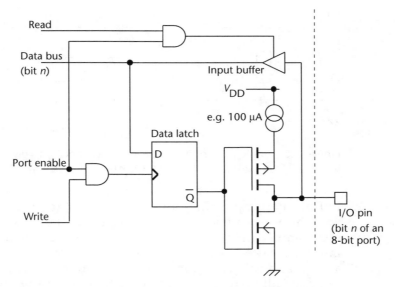

Figure 2.7 Pin driver circuit – quasi-bidirectional parallel port.

To use the pin as an input therefore, it must simply be set to output logic 1. When the logic value of the pin is read, it reads its actual value, rather than the '1' set in its data latch.

This type of parallel port is used in the Philips 80C51 microcontroller family and in related peripheral ICs. Its main advantage is that it does not require any sort of Data Direction Register. Some care must be taken in using it. Its asymmetric output means that it cannot drive simple loads like LEDs by sourcing current to them; it can only usefully sink current. An important characteristic is that generally outputs are set *high* on power-up (i.e. potentially to input mode). It is important therefore *not* to use logic high as a state which activates an important external device; otherwise at the instant of power-up, all such devices are immediately switched on! Application of this type of port is discussed in Example 2.3.

2.4 Simple interfacing

2.4.1 Switches

Switches come in many forms, and are the commonest way of arranging human (and some mechanical) interaction with a microcontroller. Push-button switches are used for simple control actions, while keypads and keyboards are used to enter alphanumeric information. Dual-inline and thumbwheel switches can be used to set up rarely changed system variables, like printer settings, while microswitches are used to detect mechanical displacement.

The simplest way of deriving a logic level from a single-pole, single-throw switch, for example a push-button, is shown in Fig. 2.8. Most commonly, as

(a)   (b)

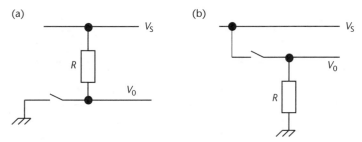

Figure 2.8 Single-pole, single-throw switch connected to give logic levels. (a) With pull-up resistor; (b) with pull-down resistor.

in Fig. 2.8(a), a pull-up resistor is connected to one contact of the switch, with the other contact connected to ground. If the switch is closed, then V_O is 0 V, and a current V_S/R flows to ground; if open, then V_O is equal to V_S. To minimise wasted current when the switch is closed, the value of R should be as high as possible. In some controllers (for example the 16F84 Port B) the internal pull-up resistor can be used for R. Attention should be paid to the effect of any current sourced from the input of logic gate or port input connected to V_O. Example 2.1 explores possible values for R. The circuit *can* be reconnected as Fig. 2.8(b). The characteristics of some logic families (for example true TTL) do, however, place restrictions on the use of this circuit, as the current sourced from the gate input affects the action of the pull-down resistor.

WORKED EXAMPLE   **2.1**

The circuit of Fig. 2.8(a) is to be implemented for a 16F84 port input, with power supply 5 V. The internal pull-up is not available. What are the constraints on the pull-up resistor value?

Solution: The input conditions of Table 2.1 must be satisfied. When the switch is open, the resistor R will pull the port pin high. Any leakage current which the pin input draws will flow through R, and cause a voltage drop RI_{IL}. R should not be so high that this voltage drop is excessive (in particular, the input high voltage must not drop below 2.4 V, i.e. a voltage drop of 2.6 V is theoretically tolerable). When the switch is closed, resistor R is effectively shorted to earth, and a current of V_S/R flows through it. R should not be so low that this current is excessive.

The values of 1 μA leakage current and maximum permissible 2.6 V voltage drop lead to a maximum value for R of 2.6 MΩ. To operate so close to the logic threshold is, however, living dangerously, particularly in the presence of any electromagnetic interference. Typically values from 10 to 100 kΩ are usually chosen. These define the logic 1 level safely, and, unless the application is extremely power-conscious, do not lead to excessive current when the switch is closed.

2.4.2 Light-emitting diodes

A light-emitting diode (LED) exploits the phenomenon that in certain semiconductor materials light is emitted as current flows across a forward-biased p–n junction (actually as electrons and holes combine). LEDs made of gallium arsenide (GaAs) emit light in the infrared, and if phosphorus is added in increasing proportions the light moves to visible red and ultimately to green. LEDs are widely available in red (most popular), green and yellow; as single devices and as arrays, bar graphs and alphanumeric displays.

As diodes, LEDs display the normal voltage–current relationship of a forward-biased diode (Fig. 2.9(a)). This means that, to a reasonable approximation, the voltage across an LED is constant if it is conducting. Note, however, that this forward voltage is considerably higher for GaAs than it is for silicon, while the reverse breakdown voltage is very low (usually around 5 V). When in conduction, a red LED has a forward voltage V_D across it of around 1.7 V, but green and yellow LEDs will have a V_D closer to, or just above, 2 V. Nor do the different colours give equal intensities for equal drive currents; red is the most efficient, which probably accounts for its greater popularity. For a single LED to be comfortably visible, it normally requires around 10 mA of current. Brighter ones may require up to 20 mA, but special low-power devices need as little as 1 or 2 mA.

An LED is easy to drive from a logic output, for example a microcontroller port, as long as the modest current requirements can be met. Depending on the capabilities of the port output they can be connected so that the output is sinking current (Fig. 2.9(b)), or sourcing current (Fig. 2.9(c)). TTL logic families can source very little current, and therefore have to be used in the configuration of Fig. 2.9(b). CMOS logic families, however, have symmetrical outputs, and can source or sink equally well. A current-limiting resistor must normally be included in series with the LED. Its value need not be calculated precisely, but account should be taken of the gate or port output voltage corresponding to the drive current required. This will be available in data sheets, and for the 16F84 has already been given in Fig. 2.6.

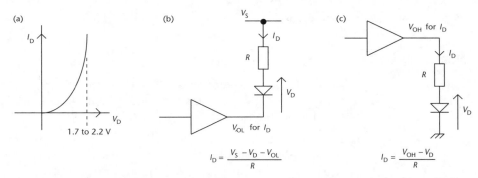

Figure 2.9 Driving LEDs from logic gates. (a) LED *V–I* characteristic; (b) gate output sinking current from load; (c) gate output sourcing current to load.

WORKED EXAMPLE 2.2

The circuit of Fig. 2.10, where the logic gate represents a 16F84 port pin output, is to be used to light either a red LED (gate output high, gate is sourcing current), or a green LED (gate output low, gate is sinking current). To balance light intensity levels the red LED requires approximately 10 mA of drive current, while the green requires approximately 14 mA. At these currents the LEDs both have 1.8 V of forward voltage. Calculate values for R_1 and R_2. The supply voltage is 5 V.

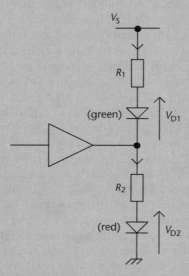

Figure 2.10 An example LED drive circuit.

Solution: For the red LED, we apply Fig. 2.9(c). From Fig. 2.6 we can see that for 10 mA of output current the output voltage is around 4.2 V. Therefore $R_2 = [(4.2 - 1.8)/10^{-2}] = 240\ \Omega$.

For the green LED, we apply Fig. 2.9(b). Figure 2.6 shows that for 14 mA current sink, the output voltage is around 0.5 V. Therefore $R_1 = [(5 - 1.8 - 0.5)/10^{-2}] = 270\ \Omega$.

WORKED EXAMPLE 2.3

An 8-bit quasi-bidirectional I/O port, with pin driver circuits of the form shown in Fig. 2.7, is to be connected to four switches and four LEDs. Draw a suitable circuit diagram.

Solution: Looking at the port outputs first, it should be remembered that weak pull-up outputs can only source a very limited amount of current, in this case around 100 µA. Therefore to switch the LEDs on, the port pins should be connected to sink current (assuming here that the CMOS output will be able to sink the necessary current, which is the case for standard LEDs). LEDs are switched off by setting their corresponding port bit high. The bits connected to the switch inputs should be programmed to logic 1. Pull-up resistors are unnecessary, as the weak pull-up takes over their function. The circuit appears as shown in Fig. 2.11. Given full data on the port I/O characteristics, LED series resistors can be calculated, as in Example 2.2.

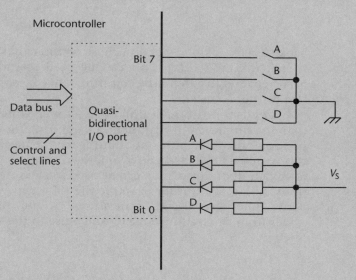

Figure 2.11 Quasi-bidirectional port interfaced to switches and LEDs.

2.5 Counters and timers

2.5.1 The digital counter reviewed

An important microcontroller peripheral, easily incorporated into the IC, is the digital counter. A counter is a series of bistables, connected together in such a way that its parallel output, formed from all the bistable outputs, indicates in binary form the number of logic pulses that have been received at its clock input. A general-purpose 8-bit counter is shown in block diagram form in Fig. 2.12(a). It has a clock input and an 8-bit output. It also has a *Clear* input, which resets the output to 0, and a *Load* input, which preloads the counter to the 8-bit value set on the *Preload Inputs*. A counter such as this, used in conjunction with the right sort of sensor and interface, gives us the capability to count pulses, objects or events.

An n-bit binary counter can count from 0 to $(2^n - 1)$. If it reaches this maximum number, and another clock pulse is then received, it *overflows*

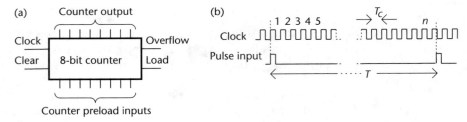

Figure 2.12   Counter/Timer fundamentals.

back to zero. Many general-purpose counters have an overflow output, as shown in Fig. 2.12(a). This can be used to indicate that overflow has occurred, and to extend the counting range by connecting it as clock input to a further counter.

An application of the counter is as a timer, which has greater importance than the simple counting function itself. This allows the microcontroller to perform one of its most important feats: the measurement of time. It does this with a stable clock source of known frequency. The period of the clock forms a unit of time which can be used to measure much greater time durations. In Fig. 2.12(b) for example the time between the two pulses of the 'Pulse Input' signal can be measured by counting the number of clock cycles from the leading edge of one pulse to the leading edge of the next. Then

$$T = nT_c \tag{2.1}$$

where T_c is the (known) clock period. Arising from this, the counter can both measure the time interval between external events *and* initiate events on a timed basis.

Microcontrollers normally incorporate at least one Counter, also configurable as a Timer (hence Counter/Timer, or C/T), as a peripheral. The basic microcontroller C/T takes some (but maybe not all) of the features shown in Fig. 2.12(a), and adds to them elements which make possible the timing function.

2.5.2   The 16F84 TIMER0 module

Figure 2.13 shows the features of the 16F84 C/T, called by Microchip the TIMER0 module. Also shown is the OPTION register, which controls it. The module is based around an 8-bit counter, represented by the 'TMR0 register', whose value can be read or preloaded through SFR TMR0 (Fig. 2.2). This SFR therefore acts as both the Counter Output and Counter Preload Inputs of Fig. 2.12(a). When the counter overflows, i.e. counts from FF_{16} to 00, its overflow output generates an interrupt signal, which enters the interrupt logic shown in Fig. 2.4.

The input clock signal to the counter is determined by the state of two bits in the OPTION register: T0CS and PSA. Each of these controls a two-input multiplexer. If T0CS = 0, then the internal instruction cycle signal is

(a)

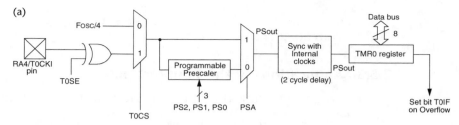

Note 1: Bits T0CS, T0SE, PS2, PS1, PS0 and PSA are located in the OPTION_REG register.
2: The prescaler is shared with the Watchdog Timer (Figure 6-6)

(b)

R/W-1	R/W-1	R/W-1	R/W-1	R/W-1	R/W-1	R/W-1	R/W-1
RBPU	INTEDG	T0CS	T0SE	PSA	PS2	PS1	PS0

bit7   bit0

R	= Readable bit
W	= Writable bit
U	= Unimplemented bit, read as '0'
- n	= Value at POR reset

bit 7:   **RBPU**: PORTB Pull-up Enable bit
1 = PORTB pull-ups are disabled
0 = PORTB pull-ups are enabled (by individual port latch values)

bit 6:   **INTEDG**: Interrupt Edge Select bit
1 = Interrupt on rising edge of RB0/INT pin
0 = Interrupt on falling edge of RB0/INT pin

bit 5:   **T0CS**: TMR0 Clock Source Select bit
1 = Transition on RA4/T0CKI pin
0 = Internal instruction cycle clock (CLKOUT)

bit 4:   **T0SE**: TMR0 Source Edge Select bit
1 = Increment on high-to-low transition on RA4/T0CKI pin
0 = Increment on low-to-high transition on RA4/T0CKI pin

bit 3:   **PSA**: Prescaler Assignment bit
1 = Prescaler assigned to the WDT
0 = Prescaler assigned to TMR0

bit 2-0: **PS2:PS0**: Prescaler Rate Select bits

Bit Value	TMR0 Rate	WDT Rate
000	1 : 2	1 : 1
001	1 : 4	1 : 2
010	1 : 8	1 : 4
011	1 : 16	1 : 8
100	1 : 32	1 : 16
101	1 : 64	1 : 32
110	1 : 128	1 : 64
111	1 : 256	1 : 128

Figure 2.13   The 16F84 TIMER0 module. (a) Block diagram; (b) OPTION Register.
Reprinted with permission of the copyright owner, Microchip Technology
Incorporated © 2001. All rights reserved.

selected (FOSC/4), and the module is in Timer mode. If T0CS = 1, then an
external input is selected, which enters the microcontroller via the RA4/
T0CKI pin. The module can then be used in a counting mode. The Exclusive
Or gate following the input acts as an edge select for this signal, controlled
by bit T0SE.

Downstream from the first multiplexer there is an optional prescaler,
which can introduce division of the incoming clock signal in binary powers
from 2 to 256. It is selected if bit PSA is set to 0, and bits PS2, PS1 and PS0

determine the division ratio. The prescaler can be allocated to the C/T *or* the Watchdog Timer (WDT; see below).

2.5.3 Applications of the Counter/Timer

We consider now three of the most important applications of the C/T.

2.5.3.1 *Event counting*

This is illustrated by Example 2.4.

WORKED EXAMPLE 2.4 _____

The external input of the 16F84 C/T is connected to an opto-sensor, which is used to count objects passing it on a conveyor belt. The objects are packed into a box in a 12 × 12 array, and when enough for one box has passed the sensor, the conveyor should be momentarily stopped. Suggest values that should be preset in the OPTION and TMR0 SFRs if the C/T Interrupt on Over-flow capability is to be used to stop the conveyor. Assume that the C/T can be preset by the programmer to any value desired.

Solution: Clearly the external input of the C/T is required, and the prescaler will not be used. Assuming that a rising edge on the C/T input is to cause a counter increment, the binary number to be placed in the OPTION SFR is shown in Fig. 2.14, where 'X' denotes a 'don't care' condi-tion (for the purposes only of this question). The Timer Overflow Flag T0IF in the INTCON register (Fig. 2.3) should be set to 0 as it will other-wise cause a spurious interrupt. The Timer will overflow as it rolls over from 255_{10} to 0, i.e. on the 256th count. Therefore the TMR0 SFR should be preset, before counting starts and on every overflow, with the number $256 - 144$, i.e. 112_{10}. To enable the interrupt, the GIE and T0IE bits of the INTCON register should be set.

| x | x | 1 | 0 | 1 | x | x | x | OPTION |

Figure 2.14 OPTION Register setting for Example 2.4.

2.5.3.2 *Time measurement*

Returning to the question of time measurement, illustrated in Fig. 2.12(b), we need to find a means of determining the number n. The counter could be cleared and disabled until the arrival of the first pulse, enabled thereafter, and stopped on the arrival of the second pulse. Then the value held in the counter would be the number n. Alternatively, the counter could be free-running, and values of its output recorded on the arrival of each of the pulses. The count *n* is then the difference between these two values. Provision must be made for the case when the counter overflows to zero during the period between the pulses. This is not an entirely simple problem, as either of these approaches could lead to timing errors. Time measurement is such a vitally important area that we return to it in a number of places in this book, and find that some very interesting hardware and software solutions have been developed to address these apparently simple problems. For now, we just consider a simple timing application, in the form of Example 2.5.

WORKED EXAMPLE   2.5 _____

The TIMER0 module of a 16F84 microcontroller is to be used to measure the period of a square wave whose frequency is known to be 10 kHz ± 25%. It is running with a clock oscillator frequency of 8 MHz. The microcontroller program will be written so that the counter starts running from zero on the rising edge of the measurand signal, and is stopped and then read on the next rising edge. How should the OPTION register be set, so that the measurement can be made without the counter overflowing during a measurement cycle?

Solution: The measurand frequency will range from 10 kHz – 25%, i.e. 7.5 kHz, to 10 kHz + 25%, i.e. 12.5 kHz. These frequencies translate to periods of 133 μs to 80 μs. The TIMER0 module can only count up to 255, so it would appear convenient to explore the possibility of forcing it to count with a clock period of 1 μs, which would then lead to counts in the range 80 to 133. The clock oscillator is running at 8 MHz, indicating an internal frequency of 2 MHz. Using the prescaler to divide by 2, this can be reduced to the 1 MHz required. The resulting OPTION register setting is shown in Fig. 2.15.

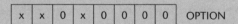

| x | x | 0 | x | 0 | 0 | 0 | 0 | OPTION |

Figure 2.15   OPTION Register setting for Example 2.5.

2.5.3.3   *Generating repetitive interrupts*

It should be evident that if a C/T is running continuously in Timer mode (i.e. it is clocked from the internal processor clock), then the C/T will repeatedly overflow, and a regular series of interrupts can be generated. There are many situations, even in simple systems, when activities have to be initiated in a repetitive, timed pattern, and this capability is therefore very useful. If the clock to the C/T is running at frequency f, the C/T is n-bit, and the prescaler is not used, then the interrupt frequency will be given by:

$$f_{int} = f/2^n \tag{2.2}$$

If the prescaler is in use, with division by 2^p, then the interrupt frequency will be:

$$f_{int} = f/(2^n \times 2^p) \tag{2.3}$$

Application of this is described in Chapter 3.

WORKED EXAMPLE **2.6**

The oscillator of a 16F84 runs at 4.194304 MHz. Using the Interrupt on Overflow to generate a regular series of interrupts, what frequencies can be achieved?

Solution: The internal instruction cycle frequency is the oscillator frequency divided by 4, i.e. 1.048576 MHz. The maximum interrupt frequency will occur if the prescaler is not used, i.e. Equation (2.2) is applied. Then

$$f_{int} = (1.048576 \times 10^6)/2^8$$

$$= 4.096 \text{ kHz}$$

A range of lower frequencies can be achieved by including the prescaler in its different settings, and applying equation (2.3). For example, with the prescaler set to divide by 2, f_{int} = 2.048kHz. With increasing values of prescaler setting, f_{int} takes the values 1.024 kHz, 512 Hz, 256 Hz, 128 Hz, down to 16 Hz for the maximum prescaler division of 256.

2.5.4 A Counter/Timer enhancement: the auto-reload

We consider one enhancement to the basic C/T structure here, as we will need it before Chapter 8, where other C/T enhancements are introduced. This one develops from the need, already mentioned, to have a repetitive interrupt on overflow. However, it takes note of the fact that the simple means of achieving this, just described, only allows binary division of the original clock frequency. Suppose in Example 2.6 one wanted an interrupt frequency of 1.5 kHz. Given the oscillator clock frequency in use, this would be impossible to achieve with this method.

Some C/Ts are therefore designed in such a way that the counter value is automatically reloaded to a predetermined value whenever it overflows. This reload value is held in a presettable SFR. As the overflow also generates an interrupt, this again provides a means of generating a regular series of interrupts at a frequency which the programmer can more precisely determine. If the C/T is n-bit, then the overflow occurs as the count goes from $(2^n - 1)$ to 0, i.e. the equivalent of it being incremented to 2^n. Therefore, if the reload value is R, it will count $(2^n - R)$ clock cycles before it overflows. With an input clock of frequency f Hz and period T, the Interrupt on Overflow period T_{int} will be

$$T_{int} = (2^n - R)T \tag{2.4}$$

and the interrupt frequency f_{int} will be

$$f_{int} = f/(2^n - R) \tag{2.5}$$

WORKED EXAMPLE   **2.7**

A 16-bit Counter/Timer is to be used in Auto-Reload Mode to generate an interrupt every 500 μs. Its input clock will have a frequency of 1.2 MHz. What should the reload value be?

Solution: Applying equation (2.4),

$$500 \ \mu s = (2^n - R)/f$$

$$R = 2^{16} - 1.2 \times 10^6 \times 500 \times 10^{-6}$$

$$= 65536 - 600$$

$$= 64936_{10}. \text{ This is the reload value.}$$

2.5.5 The Watchdog Timer

A common 'nightmare scenario' of any computer-based system is for the computer or processor to lock up and cease all interaction with the outside world.[1] Yet most embedded systems have a need for high reliability; total failure due to lock-up is just not unacceptable. Means must be therefore be found to restart a processor, should it enter this unhappy condition.

An uncompromising solution to the problem is the Watchdog Timer (WDT), which resets[2] the processor if it is ever allowed to overflow. It is up to the programmer to ensure that this overflow never ever happens in normal program operation. He does this by including periodic WDT resets throughout the program. *If*, however, the program crashes, the WDT overflows, the controller resets, and the program starts from its very beginning.

As with all safety features, it is important that the WDT is not enabled or disabled accidentally. An *enabled* WDT in a program which does not cater for it is almost as disastrous as a *disabled* WDT which is meant to be

---

1 We explore the causes of such lock-up in later chapters.
2 A reset forces the program to start again from its very beginning. The program counter is therefore set to its reset value.

protecting a vital system. Therefore most WDT systems employ a hardware or firmware solution to their enabling. Protection is also employed in the actual WDT reset, so that it cannot be reset by accident.

The novice designer is unlikely to want to use the WDT, and details on applying it are left to Chapter 8. You must, however, know of its existence, if only so that you can ensure that it is switched off in simple designs.

2.6 Power supply and reset

2.6.1 Power supply requirements

The system designer must supply power, at an appropriate voltage and with adequate current supply capability, to all circuit devices which require it. In a minimum system this may be little more than the microcontroller. In a more complex system there will be a number of different devices, each having its own supply requirement. The manufacturer of any device specifies the range of voltages within which it can operate. If the supply voltage falls below the specified value, the device is unlikely to operate correctly. If it exceeds the specification, the device may be damaged. With this in mind, the manufacturer provides a set of 'Absolute Maximum Ratings'. These are conditions beyond which the device should never be allowed to operate. It is important not to confuse these ratings with normal operating conditions. The manufacturer also usually indicates current consumption figures. These are not always easy to forecast, as they depend on a number of factors, including operating conditions and clock frequency.

It is a characteristic of logic circuits that current is consumed mainly in pulses, occurring at logic transitions. Power is supplied along wires and PCB track, which, however short, inevitably have some resistance and inductance associated with them (Fig. 2.16(a)). Therefore when the microcontroller takes a pulse of current from the supply, there is a danger that the supply voltage at its pins drops instantaneously below the minimum required supply voltage, represented by V_{min} in Fig. 2.16b). It is therefore essential to 'decouple' the power supply to the microcontroller (and to any other digital device) with a capacitor placed across the supply rails, close to the body of the IC. Current pulses are then drawn primarily from the capacitor, and the supply is stabilised. Manufacturers frequently recommend the value and type of capacitor to be used; 100 nF is common. Generally a larger reservoir capacitor (of the order of 100 µF) is also placed on a printed circuit board close to the point where power is connected.

2.6.2 Reset

The reset condition of a microprocessor is one in which the program counter is set to the start address of the program, defined by the *Reset Vector*. Usually certain internal registers, like SFRs, are at the same time set to a predetermined logical value. Frequently manufacturers specify in their

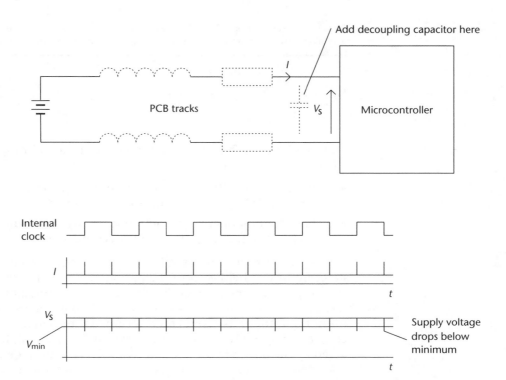

Figure 2.16 The need for supply decoupling. (a) Supply tracks equivalent circuit; (b) supply voltage and current without supply decoupling.

data the reset condition of some or all the SFRs. On leaving the reset condition, the program starts execution from its first instruction.

There may be several sources of reset for a processor. We have just seen that a WDT can be one of them. The commonest is an external input. In the PIC 16F84, the reset input is labelled MCLR. For normal operation it is held at logic high. If it is taken low, then the controller enters the reset condition just described. The input can be connected to a switch, applying the circuit of Fig. 2.8(a). The manufacturer's maximum recommended value for R is 40 kΩ. This allows a user to force a reset if program operation fails.

2.6.3 Power-up

The time of power-up is a potentially difficult and dangerous one for any embedded system. As the supply rises and stabilises (which itself takes a finite time), some parts of the system will be ready for action before others. What action, for example, will actuators take if they are powered and ready before their controlling peripherals? Even the clock oscillator takes time to start and stabilise.

A 'time-line' representing microcontroller power-up is shown in Fig. 2.17. The controller should not start program operation until the supply

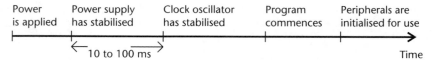

Figure 2.17 Stages in power-up.

and oscillator are stable, and it then needs time to initialise its peripherals. Clearly it is essential:

- that the controller 'wakes up' with its SFRs set in a safe state, i.e. with peripherals inactive, and interrupts disabled
- that program operation commences only when the power supply and clock oscillator are fully established and stable

Several techniques are used to meet these requirements. Traditionally the Reset input was held active for a period of time intended to be longer than the anticipated system stabilisation time. A simple power-on reset circuit recommended for 80C51 devices is shown in Fig. 2.18. This holds the reset input high (the active state) for a period of time determined by the *R–C* time constant. Other processors incorporate an internal digital delay timer, triggered by the power being applied. These have a similar effect to the external *R–C* network, holding the processor in reset for a period of time. An example of this is the 16F84, which has an internal *Power-up Timer*, driven from a dedicated internal *R–C* oscillator. This holds the microcontroller in reset for a nominal period of 72 ms after power-up has been detected. It is only necessary to use an external *R–C* network if it is anticipated that power-up time will be longer than this.

These features go some way towards ensuring successful power-up. However certain modes of microcontroller operation must be in place from the instant of power-up, independent of any initialisation that the microcontroller program will undertake. A good example is the WDT. As a device which is meant to oversee correct program operation, we cannot let it be dependent on that program! This requirement is approached in one of two ways. The more traditional one is to commit external controller pins

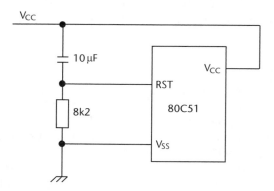

Figure 2.18 Power-on reset for the 80C51.

R-u	R-u	R-u	R-u	R-u	R-u	R/P-u	R-u	R-u	R-u	R-u	R-u	R-u	R-u
CP	CP	CP	CP	CP	CP	DP	CP	CP	CP	PWRTE	WDTE	FOSC1	FOSC0

bit13   bit0

R = Readable bit
P = Programmable bit
- n = Value at POR reset
u = unchanged

bit 13:8 **CP**: Program Memory Code Protection bit
1 = Code protection off
0 = Program memory is code protected

bit 7 **DP**: Data Memory Code Protection bit
1 = Code protection off
0 = Data memory is code protected

bit 6:4 **CP**: Program Memory Code Protection bit
1 = Code protection off
0 = Program memory is code protected

bit 3 **PWRTE**: Power-up Timer Enable bit
1 = Power-up timer is disabled
0 = Power-up timer is enabled

bit 2 **WDTE**: Watchdog Timer Enable bit
1 = WDT enabled
0 = WDT disabled

bit 1:0 **FOSC1:FOSC0**: Oscillator Selection bits
11 = RC oscillator
10 = HS oscillator
01 = XT oscillator
00 = LP oscillator

Figure 2.19 16F84 configuration word. Reprinted with permission of the copyright owner, Microchip Technology Incorporated © 2001. All rights reserved.

for configuration. The 80C552, for example, has an external pin ($\overline{\text{EW}}$) dedicated to the WDT. If tied low, the WDT is enabled, if tied high, it is disabled.

As an alternative to this, the 16F84 uses a *Configuration Word*, embedded in program memory. This is shown in Fig. 2.19. Once programmed, the contents of this determine the enabling or otherwise of the WDT, program code protection, and the power-up timer, and also determine the selection of oscillator type. The configuration word is programmed along with the program memory.

2.7 The clock oscillator

An essential part of the microcontroller system is the clock oscillator. Not only does the oscillator drive forward all processor activity, it is also the basis of any accurate time measurement or generation. The actual choice of oscillator frequency itself depends on several variables. The simplistic approach is to set it 'as fast as possible', remaining of course within the specified microcontroller limits. In some cases, however, particularly when precise timing values are needed, a highly predictable and stable clock frequency is required. An example of this is the 32.768 kHz crystal oscillator, the basis of the wrist watch. When this frequency is divided by 2^{15} it gives a one second timebase. Other very precise frequency values arise from the need to set up close synchronisation with other systems, e.g. through serial data links.

Clock oscillators can be based on resistor–capacitor networks or ceramic or crystal resonators. While the choice is normally crystal, this is the most expensive, and the designer should always be aware of the lower cost options that are available.

2.7.1 R–C oscillators

Many possible configurations exist for R–C oscillators (or, more correctly, multivibrators). We are interested in those which are simple and easily integrated onto an IC. Two possible configurations are shown in Fig. 2.20. In the first, a Schmitt inverter charges capacitor C through resistor R from its output. When the Schmitt trigger input threshold is passed the output goes low, and the capacitor proceeds to discharge until the negative-going input threshold is passed, when the output again goes high. In Fig. 2.20(b) the capacitor is charged from the supply rail. When the Schmitt threshold is passed the output goes high and switches on a transistor, which rapidly discharges the capacitor. The Schmitt trigger output goes low again, and the capacitor recommences charging. In either case, the action described continues indefinitely.

Circuits such as these represent the cheapest form of clock oscillator available, and both are widely used as IC clock generators. If, as is usually the case, external components are to be used, then the circuit of Fig. 2.20(b) has the advantage, as only one IC interconnection pin is required. Operating frequencies depend on the values of R and C and the input thresholds of the Schmitt trigger. All of these have significant temperature dependence. Their main disadvantage therefore, which in many situations proves

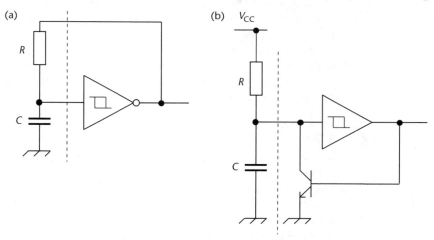

Figure 2.20 *R*–*C* multivibrator circuits.

to be decisive, is that their precise frequency of oscillation is unpredictable, and that it drifts with temperature and time.

It is possible to integrate every component of an R–C oscillator within an IC if no user setting of the frequency is required. The 16F84, for example, has two internal R–C oscillators, for Watchdog and Power-up Timers, whose frequencies are predetermined by the IC design.

2.7.2 The quartz crystal

A quartz crystal oscillator is based on a very thin slice of crystal. The crystal is piezoelectric, which means it has the property that if it is subjected to mechanical strain then a voltage appears across opposite surfaces, and if a voltage is applied to it, then it experiences mechanical strain, i.e. it distorts slightly. The crystal is cut into shape, polished and mounted so that it can vibrate mechanically, the frequency depending on its size and thickness. The resonant frequency of vibration is very stable and predictable, with very high 'Q factor' (of the order of 10 000). If electrical terminals are deposited on opposite surfaces of the crystal, then due to its piezoelectric property the vibration can be induced electrically, by applying a sinusoidal voltage at the appropriate frequency.

Crystals are grown in laboratory conditions, and then cut to the physical dimensions which set the resonant frequency. Different shapes, for example disc or bar, are possible. Each of these has its own mode of vibration. Low-frequency crystals often use a bar shape, which leads to a flexure mode. The actual angle of the cut (referred to by codes such as XY, DT and AT) within the crystal also influences crystal characteristics, in particular temperature dependence. In the frequency range 1 to 20 MHz, AT-cut crystals offer a good compromise between temperature stability and frequency accuracy, and are normally chosen.

Electrically, the crystal behaves like the equivalent circuit shown in Fig. 2.21(a). R, L and C_s are notional components, representing the crystal itself, with L being a function of the crystal mass and C_s a function of its stiffness. C_p represents the electrode capacitance. Some typical values are given, taken from Ref. 2.1. The components R, L and C_s combine together to create a series resonant circuit, whose reactance in Fig. 2.21(b) is seen to pass through zero at the resonant frequency f_s. The point f_a is called the anti-resonant frequency.

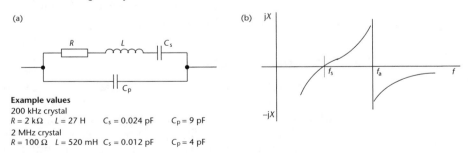

Example values
200 kHz crystal
$R = 2\,k\Omega$ $L = 27\,H$ $C_s = 0.024\,pF$ $C_p = 9\,pF$
2 MHz crystal
$R = 100\,\Omega$ $L = 520\,mH$ $C_s = 0.012\,pF$ $C_p = 4\,pF$

Figure 2.21   (a) Crystal equivalent circuit; (b) crystal reactance vs. frequency.

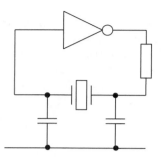

Figure 2.22   Crystal oscillator circuit.

The requirements for oscillation in a feedback oscillator are for the phase shift to be an integer multiple of 360°, and the loop gain (i.e. amplifier gain × feedback fraction) to be greater than 1. If the crystal is put into a circuit satisfying these requirements, and it is the only frequency-dependent element, then oscillations will take place at its resonant frequency.

A practical and commonly used circuit is shown in Fig. 2.22. Here the amplifier is a CMOS logic inverter, which effectively produces a phase shift of 180°. The pi network of crystal plus load capacitors provides a further 180° of phase shift. The resistor is required in certain applications to reduce the drive level to the crystal, which is damaged if driven with too high a voltage. A high-valued resistor is also sometimes placed directly across the inverter, to bias it into its linear region of operation.

In almost any microcontroller circuit the logic gate of Fig. 2.22 is integrated on-chip, with its two terminals available for external connection of the crystal and associated components. While formulae for calculating the components are available, generally suppliers give recommended values, which are adequate for standard applications. As load capacitances are small and operating frequencies high, it is very important that the crystal circuit is designed to be compact. Long PCB tracks will result in additional stray capacitance and inductance which may result in a crystal failing to oscillate, or oscillating at a spurious frequency.

Demanding crystal applications, for example designs incorporating significant temperature or power supply variations, very low power consumption, or high-frequency stability, need a detailed understanding of the design principles. Reference should then be made to application notes such as Refs. 2.2 or 2.3.

2.7.3 The ceramic resonator

Ceramic resonators use other materials which can display piezoelectric properties, for example barium titanate. They are moulded into shape while hot, and then cooled in the presence of a strong electric field, which aligns the molecular electric dipoles and leaves a piezoelectric-like effect. They then have similar properties to quartz crystals, but are smaller and cheaper (as the manufacturing process is simpler).

Ceramic resonators have lower Q factors than quartz, and much higher temperature dependence (for example 3000 ppm (parts per million)/°C, compared with 10–20 ppm/°C for AT-cut crystals). The lower Q factor *can* be used to advantage, as frequency is more susceptible to adjustment.

2.7.4 The 16F84 oscillator

The 16F84 has two pins for oscillator connection, labelled OSC1/CLKIN and OSC2/CLKOUT. If an external clock source is to be used, then it can simply be connected to the first of these. Otherwise, the PIC can be configured in one of four modes to achieve an oscillator. The choice is up to the user, and is programmed into the Configuration Word. The modes are itemised in Fig. 2.19; each uses one of the techniques just described. If used in *R–C* mode, then only the OSC1 pin is used, and the circuit is that of Fig. 2.20(b), with the *R* and *C* being added externally. For LP, XT and HS modes, the circuit is that of Fig. 2.22, with the logic gate being internal to the device. LP mode is used for frequencies up to around 200 kHz, XT for frequencies from around 100 kHz to 4 MHz, and HS for from 4 to 10 MHz.

2.8 Some practical build tips

It is hoped that the reader will soon be planning to design and build a small embedded system. While this is not a book on electronics, many students make such a system their first electronic design and build project, and then face many of the problems of the inexperienced electronics builder. Therefore a small number of essential build tips are given:

- Aim to do your hardware prototyping on PCB or strip board. There are a range of prototyping systems available, which allow the user to make interconnection by sliding wires and components into a connector grid. These may be adequate for very simple circuits, but are more trouble than they are worth at this level of construction. Connections are unreliable, it is easy for a wire to slip out, and the resulting mass of wiring can create interference problems.
- Design the PCB or stripboard layout with ease of test in mind. Aim for it to be compact, without becoming excessively small. Ensure test and connection points are easily accessible.
- Keep earth and power supply tracks short and wide, with good power supply decoupling.
- Do not solder the microcontroller to the PCB. Put it in an IC socket, preferably of the 'zero insertion force' (ZIF) type.
- Keep crystal tracks short, with the crystal close to the microcontroller IC body.
- Design in a couple of diagnostic LEDs.
- Build neatly and well. Any wired interconnections should be short and tidy. Learn the art of soldering properly if you have not yet mastered it.

SUMMARY

1. The microcontroller normally configures peripherals, and exchanges data with them, through memory-mapped Control or Special Function Registers.

2. Most peripherals have the capability to generate interrupts. Microcontrollers therefore tend to have complex interrupt structures.

3. The peripheral of perhaps most fundamental importance is the parallel port, which forms a highly flexible channel for I/O data.

4. Another important peripheral is the Counter/Timer, which forms the basis of most timed activities in the microcontroller world.

5. The hardware designer must recognise and meet the power supply requirements of the system under development, and also be aware of the power-up characteristics of the system.

6. Selection of the clock oscillator has important implications on the cost and performance of the overall system.

REFERENCES

2.1 *Crystal Product Data Book* (1994) IQD Ltd, Crewkerne, UK.
2.2 *Microcontroller Oscillator Design Guide*. Application Note AN588. Microchip Technology Inc.
2.3 *Microcontroller Oscillator Circuit Design Considerations*. Application Note AN1706/D/1997. Motorola Inc.

EXERCISES

('+' indicates a simple review question)

+**2.1**   The INTCON Register of a PIC 16F84 is set as shown in Fig. 2.23(a).
(a) Determine which interrupts are enabled.
(b) An interrupt occurs, and the INTCON register is found to have changed to Fig. 2.23(b). Which interrupt source has called?
(c) Which bit must the user change before the end of the ISR, and why?

(a)   (b)

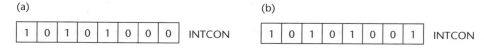

| 1 | 0 | 1 | 0 | 1 | 0 | 0 | 0 | INTCON

| 1 | 0 | 1 | 0 | 1 | 0 | 0 | 1 | INTCON

Figure 2.23   INTCON Register settings for Exercise 2.1.

+**2.2**   Explain the difference between a data bus and a parallel port.

2.3   Four bits of Port B of a 16F84 are to be used to drive LEDs, and four are to be used to receive inputs from push button switches. For these the interrupt on change option is to be used. The LEDs each require 15 mA when 'on', and their forward voltage is then 1.9 V. Two should be on when their associated port bits are at logic 1, and two should be on when they are at logic 0. Draw a diagram showing how the switches and LEDs may be connected, indicate the value of any resistors used, and show how all relevant SFR bits should be set. The microcontroller is supplied from 5 V.

2.4 In its device summary for the 16F84, the manufacturer states that the maximum port output current sink capability is 25 mA per pin and source capability is 20 mA. For a 5 V supply, what is the output voltage for each of these values, and what power is dissipated in the microcontroller due to this current in each case?

2.5 In applying the circuit of Fig. 2.10, the port output (represented by the logic gate) is accidentally disconnected from the rest of the circuit. Describe and evaluate what happens in the circuit. Take resistor values, and any other data needed, from Example 2.2.

2.6 The Counter/Timer of a 16F84 is to be used to measure the speed of an object falling through a column of water. When the object passes an upper sensor the controller starts the counter running from zero. When it passes the lower sensor, it stops the counter and reads its value. The microcontroller is clocked from a 1.024 MHz crystal oscillator, and the counter is required to increment every 1ms. How should the OPTION register be set?

2.7 A 16F84 microcontroller with an oscillator clock source of 8 MHz requires a regular interrupt every 128 μs.

(a) Explain how this requirement can be met, to the best accuracy possible.

(b) How should the requirement be met if the interrupt is to occur every 2048 μs?

(c) How should the requirement be met if the interrupt is to occur every 500 μs?

Assume no other interrupts are enabled.

2.8 A small embedded system is powered from a 5 V supply. The supply itself cannot source more than 75 mA. The total system decoupling capacitance is 220 μF. Estimate the time required on power-up for the system voltage to reach 5 V. Assume the supply is otherwise ideal, and that the current drawn by the system is very small.

2.9 A microcontroller, with a clock frequency of 1 MHz, is to be used for a time measurement application, which must be accurate to within ±0.1% over an operating temperature range of 12 °C to 30 °C. The designer can choose between a crystal clock source of stability 100 ppm (parts per million)/°C, a ceramic resonator of stability 800 ppm/°C, or an R–C network of stability 0.4%/°C. Assuming each clock source is set initially to give the correct frequency at the mid-range temperature, what frequency range will each clock source give for this temperature range? Which sources, if any, are acceptable for this application?

2.10 Using the 16F84, complete the *hardware* design of an electronic lock, to the following outline specification:

Electronic lock

The lock will activate if the correct code is entered. The code is set by six toggle or Dual Inline (DIL) switches, which must be set to the correct binary combination. When the user has set the binary code, he or she presses an 'Enter' push button.

If the code is correct, then the lock is activated for a period of 10 seconds. This should be indicated by an LED requiring 20 mA current, controlled by a port line, activated by a logic low output. If the code is incorrect, then a 'wrong code' indication is given, by an LED being illuminated for 5 seconds. This LED requires 10 mA. The lock will then not accept another input for 30 seconds.

Preliminary programming

IN THIS CHAPTER

This chapter introduces programming for the minimum system, much in the same way that Chapter 2 introduced hardware design for the minimum system. To do this it makes use of the programming language most directly associated with the structure and instruction set of the microcontroller, Assembly Language. As with Chapter 2, most examples relate to the 16F84. Most programming examples moreover are taken from the Case Study of Appendix C. It will therefore be useful to read that in parallel with this chapter.

   The chapter's aims are:

- to describe the knowledge required before programming can start

- to introduce the characteristics of Assembler programming, and to describe the features of well-designed programs

- to consider how data is handled in an Assembler program

- to describe the tools used to develop and debug an Assembler program

- to describe simple test and commission techniques (both hardware and software) for the minimum system

3.1 An introduction to Assembler

Whether the programmer writes in Pascal, Fortran, C or Assembler, the program that a microcontroller actually executes is a list of binary codes, known as machine code, stored in program memory. An example of machine code, for a 16F84 controller, is shown in Fig. 3.1. It is a brew of instructions drawn directly from the controller's instruction set (Appendix B), together with associated data and address information. Clearly, it would be a miserable task for the human programmer to work directly with the machine code.[1] Not only must all the instruction codes be worked out and listed, but also a host of memory address and data information needs to be in place.

```
10 1000 0001 0000
01 0110 1000 0011
11 0000 0000 1000
00 0000 1000 0110
11 0000 0000 0000
00 0000 1000 0101
01 0010 1000 0011
....
```

Figure 3.1 Part of a 16F84 machine code listing – but what does it all mean?

To assist programmers in reaching their final goal of a machine code listing, many programming languages have been devised. These allow the programmer to write in a human-friendly language, and then automatically convert the program into machine code, ready for download to program memory. These programming languages are very attractive, but a potential problem is that they tend to distance us from the detail of the hardware system that we are controlling, and in the world of embedded systems this can be a significant disadvantage.

A compromise, which retains direct usage of the instruction set, is with Assembler programming. Using this technique, the program is written using instruction *mnemonics* defined by the manufacturer. Mnemonics are short descriptive names given to each instruction, and are meant to be easy for the human programmer to understand. The 16F84 mnemonics can be seen in Appendix B. A program written in mnemonics can then be converted to machine code using a purpose-designed computer program, itself called an Assembler.[2] Given the use of a computer anyway for this translation process, it is easy to give it all the other routine and repetitive tasks of program generation, such as keeping track of memory addressing.

From this stage on, we will recognise the fact that we are effectively always dealing with *two* computer systems. The first is the *Host Computer*, sometimes also called the *Development Computer*, used to undertake all program development. It is usually a personal computer. The second is the *Target System*, which is the embedded system for which development is being undertaken.

---

1 In the early days of microprocessors, that was what was done, but only for a short time!
2 The terminology 'Assembler' is strictly correct only if the computer doing the assembling is developing code for itself. If it is developing code for another machine or processor it is called a *Cross-Assembler*. This is the usual case for the embedded system, as it is not normally designed with the facilities needed (keyboard, screen and so on) to allow easy code development.

3.2 Prerequisites for programming

Assembler programming makes direct use of the processor instruction set, and deals in a very direct way with the system hardware. Before programming can begin, therefore, it is important to have a clear picture of the following:

- the programming model, including knowledge of all CPU registers
- the instruction set, including the function of each instruction, how it addresses memory, and how the instruction affects the Status or Condition Code register
- the system memory map
- on-chip peripherals, their SFRs, and how they are initialised and used
- interrupt sources and their hierarchy

Some of these items have already been covered in the preceding chapters. We will now spend some time in closer examination of the others.

3.2.1 Memory maps

Chapter 1 described the broad option of memory map design, choosing between the conventional von Neumann and Harvard structures. We saw that the 16F84 has a distributed memory pattern, as is evident from Fig. 1.11(a). This shows separate areas for program and for data, in keeping with its Harvard memory model. In reality there is further subdivision, leading to no fewer than *four* distinct memory areas: program, EEPROM, stack and RAM, each with its own independent addressing structure. The program memory map, addressable by the Program Counter, is shown in Fig. 3.2. The *Reset Vector*, which is the program start address, is fixed in hardware as memory location 0000. There is only one *Interrupt Vector*, which is fixed in hardware as location 0004 (labelled 'Peripheral Interrupt Vector'). This is the starting address of any interrupt routine.

By contrast, the 68HC05 and '08 give excellent examples of the conventional von Neumann structure. Program memory, data memory, EEPROM, stack and peripheral control registers *all* lie within a single memory map, and all can be addressed by the Program Counter. This is illustrated in Fig. 3.3, which shows the memory map of the 68HC05B6. The reset vector is located at memory locations $1FFE and $1FFF (the $ symbol indicates a hexadecimal quantity). Unlike the 16F84, this *contains* the program starting address, but is not the starting address itself. Similarly, the 68HC05 interrupt vectors are memory locations which *contain* the start address of the interrupt routines, but are not the start addresses themselves. They are seen in Fig. 3.3, at memory locations $1FF2 to $1FFD.

In any one processor, memory can be addressed from the program in a number of different ways; these *addressing modes* arise from the architecture of the processor itself. They also have a fundamental influence on the format of the instruction. CISC processors tend to have a sophisticated range of addressing modes. A good example is the 68HC05, whose

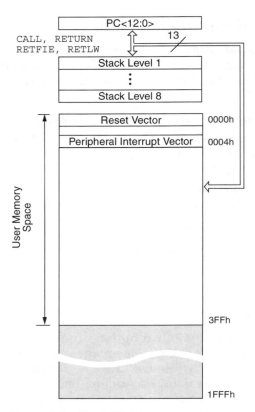

Figure 3.2 The 16F84 program memory map and stack. Reprinted with permission of the copyright owner, Microchip Technology Incorporated © 2001. All rights reserved.

addressing modes are summarised in Appendix D. In contrast, the addressing modes available to the 16F84 are fewer (as is to be expected from its RISC structure) and simpler. We turn now to consider these in greater detail. Because they are all intertwined, we will consider more or less at the same time the instruction format, and finally the instruction functions.

3.2.2 16F84 instruction format and addressing modes

It is now time to get a clear understanding of the 16F84 instruction set, which appears in Appendix B. Table B.1 and Fig. B.1 should be studied together, as the figure shows the format of the instructions appearing in the table, and also explains some of the abbreviations. It is worth taking time to sort out clearly the information given. The first table column gives the mnemonic of the instruction, together with the codes (**b**, **d**, **f**, and **k**) for the address or operand data which *must* accompany the instruction. The function of these codes is detailed in Fig. B.1. The second column, *Description*, states briefly what each instruction does. The third column indicates the number of instruction cycles that the instruction takes to execute. As the controller has a RISC-like structure, it comes as no surprise to see that all of

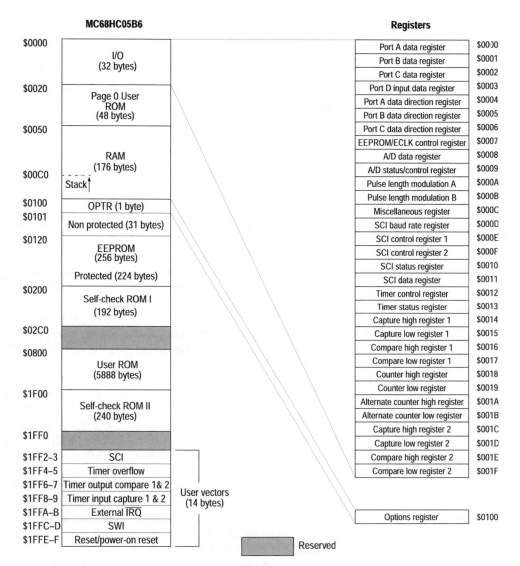

Figure 3.3 Memory map of the 68HC05B6. (Copyright of Motorola, Inc. Used by permission.)

them are one cycle only, except for those instructions which cause a program branch. The next column shows the actual 14-bit binary instruction codes. This contains both the instruction code and the address or data information which the programmer will have supplied. The next column indicates which flags in the Status Register are affected by the instruction.

Almost every 16F84 instruction is formatted according to one of the four patterns shown in Fig. B.1. Each pattern reflects a different mode of addressing. In each case the instruction word is made up of the instruction information itself, represented by the 'OPCODE' field. To this is added address or data information, according to the addressing mode in use. Three[3] possible operand locations may be specified. These are:

- a file register address in RAM, represented in the figure by **f**

- literal data, i.e. constant data embedded into the instruction, represented by **k**

- the working register (accumulator) **W**, implicitly specified within certain op codes

Further addressing information is contained within the **d** bit, which specifies a destination for the result of certain instructions, and the **b** bits. These represent a 3-bit address, applied to identify one bit within an 8-bit file register.

The instruction set table is roughly grouped according to the main addressing mode used, which are now described.

Direct register addressing In this addressing mode the operand address, which lies within the register map of Fig. 2.2, is specified 'directly' within the instruction word. The instruction format follows the first pattern of Fig. B.1, and includes most (but not all) of the instructions in Table B.1 labelled 'byte-oriented file register operations'. This mode is the one most commonly used for access to the SFRs and the general-purpose RAM.

Because the register map is banked, it is sometimes necessary to specify bank information as well as address information. The way that the addressing information accesses the memory is shown in generalised form in Fig. 3.4. The RAM area of the 16XXX series PIC microcontrollers is structured in a number of banks. The bank which is actually selected for access depends on the setting of the RP0 and RP1 bits in the STATUS register (Fig. 1.12). These must be preset by the programmer to select the bank desired. Within the bank, the actual location addressed depends on the file register address specified within the instruction code. The 16F84 has only *two* memory banks, so only the RP0 bit is applied. Moreover, in the 16F84 it is *only* the SFRs that are banked, and for ease of access some of these are mirrored in both banks (for example STATUS).

As an example, consider the first instruction in Table B.1, addwf. Looking at the '14-bit op code' of Table B.1 we can see that the instruction information is in its six most significant bits. The value of the **d** bit determines whether the result of the operation is placed in the **W** register or the file register. The seven least significant bits form the file address itself. This conforms with the instruction format of Fig. B.1.

Bit addressing In this addressing mode single bits within register files (but not within the **W** register) can be accessed. The instruction follows the second pattern of Fig. B.1, and includes both the 7-bit file identifier **f** used

---

3 The exception to this is certain control instructions, like clrwdt, where the operand is none of the three options listed.

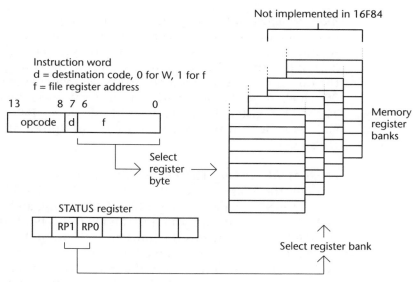

Figure 3.4 Register bank addressing, 16XXX.

above, as well as a 3-bit bit identifier **b**. Only the four most significant bits of the instruction word contain instruction information.

Literal addressing   In this mode an 8-bit data value, called the *literal* and symbolised in Appendix B by **k**, is embedded in the instruction code. The instruction follows the third pattern of Fig. B.1.

An example of this mode is the instruction addlw. This adds to the **W** register the literal value embedded into the instruction. The instruction information again occupies the six most significant bits, while the literal value is placed in the eight least significant bits.

Unconditional branching: Call *and* Goto   These two instructions cause unconditional branches to occur within a program, i.e. under every circumstance the branch takes place. The pattern is the fourth of Fig. B.1. The target address of the branch, an 11-bit literal number, is specified within the instruction. Only three bits are available for instruction information.

Inherent addressing   In inherent addressing, the operand address is implicit within the instruction itself. No address therefore needs to be specified, and the instruction code occupies the whole 14 bits. No pattern is specified in Fig. B.1. Examples include clrw and clrwdt.

Conditional skipping   It is essential for any microprocessor to make conditional program branches. The 16F84 does this with just four skip instructions, btfsc, btfss, decfsz, and incfsz. These do not on their own constitute a distinct addressing mode. They can however be combined with call and goto instructions to form conditional branching functions. They will be considered further in Section 3.2.3.

Indirect register addressing In this addressing mode the programmer does not explicitly specify the operand address. Instead, an SFR is made available within which an address can be held. This is the File Select Register (FSR, Fig 2.2). When this addressing mode is applied, the contents of FSR are used as the operand address. The mode is invoked whenever register file location 00 (labelled INDF) is specified in an instruction as the file register address. Then the register file actually addressed is *not* INDF, but the one whose address is held in the FSR. INDF itself is not physically implemented as a register.

EEPROM addressing The EEPROM area can only be addressed indirectly, through the SFRs. Figure 2.2 shows four SFRs (EEDATA, EECON1, EEADR and EECON2) that are used for EEPROM reading and writing. The register EEADR holds the memory address, and EEDATA holds the data. The register EECON1, which controls the transfer, is shown in Fig. 3.5. This shows that there are register bits to enable the write process, to initiate read or write cycles, to flag the completion of a write cycle, and to flag an incomplete write operation. The location EECON2 is required in the write sequence to

U	U	U	R/W-0	R/W-x	R/W-0	R/S-0	R/S-x
—	—	—	EEIF	WRERR	WREN	WR	RD

bit7   bit0

R   = Readable bit
W = Writable bit
S   = Settable bit
U   = Unimplemented bit, read as '0'
- n = Value at POR reset

bit 7:5   **Unimplemented:** Read as '0'

bit 4   **EEIF**: EEPROM Write Operation Interrupt Flag bit
1 = The write operation completed (must be cleared in software)
0 = The write operation is not complete or has not been started

bit 3   **WRERR**: EEPROM Error Flag bit
1 = A write operation is prematurely terminated
(any $\overline{\text{MCLR}}$ reset or any WDT reset during normal operation)
0 = The write operation completed

bit 2   **WREN**: EEPROM Write Enable bit
1 = Allows write cycles
0 = Inhibits write to the data EEPROM

bit 1   **WR**: Write Control bit
1 = initiates a write cycle. (The bit is cleared by hardware once write is complete. The WR bit can only be set (not cleared) in software.
0 = Write cycle to the data EEPROM is complete

bit 0   **RD**: Read Control bit
1 = Initiates an EEPROM read (read takes one cycle. RD is cleared in hardware. The RD bit can only be set (not cleared) in software).
0 = Does not initiate an EEPROM read

Figure 3.5 16F84 EECON1 Register. Reprinted with permission of the copyright owner, Microchip Technology Incorporated © 2001. All rights reserved.

enhance the security of the write operation; it is not physically implemented as a register.

Other addressing techniques   There are other means of addressing data with the 16F84, which rely on certain of the instructions and the controller architecture, but do not constitute addressing modes as such. These include the *Computed Goto* and *Table Reads.* These are described in the appropriate sections later in this chapter.

3.2.3   Exploring the 16F84 instruction set by function

The 16F84 instruction set is shown in reconfigured form in Table 3.1, with instructions now grouped according to the categories introduced in Section 1.2.2. There are apparently only three data transfer instructions: movwf (move **W** to **f**), which moves the contents of the **W** register to the specified file register, movf (move **f**), which does the reverse (as long as **d** is set to 0), and movlw (move literal to **W**), which moves a byte of constant data held within the instruction into the **W** register. It is not possible in one instruction to move a byte of literal data straight into a file register. To do this a movlw instruction must be followed by a movwf. It should be noted that any file register instruction which makes use of the **d** destination bit contains the possibility of a data transfer. Depending on the state of the **d** bit, the result may be placed in a file register or **W** register.

The Arithmetic and Logical instructions reflect very closely the basic capabilities of the ALU, as shown in Table 1.1. The arithmetic operations of

Table 3.1   PIC 16F84 instruction set, grouped by function

	Data transfer	Arithmetic	Logical	Branch	Other

Byte-oriented	movf f,d
	movwf f
		addwf f,d
		subwf f,d
		decf f,d
		incf f,d
			andwf f,d
			iorwf f,d
			xorwf f,d
			clrf f
			clrw
			comf f,d
			rlf f,d
			rrf f,d
			swapf f,d
				decfsz f,d
				incfsz f,d	

Bit-oriented			bcf f,b bsf f,b	btfsc f,b btfss f,b	
Literal	movlw k	addlw k sublw k	andlw k iorlw k xorlw k	call k retlw k goto k	
Control and other				return retfie	clrwdt nop sleep

add and subtract can be performed between **W** register and file register, or between **W** register and literal value. Increment and decrement can be performed on file registers only.

The logical range of instructions seems the most extensive, with the possibility to implement 8-bit AND, OR and Exclusive OR operations between either file register and **W** register, or **W** register and literal value. In every case the operation is performed between corresponding bits of the two operand bytes. Clearing, complementing and rotating of file register contents are also possible. Logical operations are not possible between single bits. These can however be set and cleared, with the bcf (bit clear f) and bsf (bit set f) instructions.

There are four conditional 'branch' instructions. In each case these cause the following instruction to be skipped if the tested condition is 'true'. The instruction decfsz (decrement f, skip if zero) therefore decrements the specified file register, and causes the next instruction to be skipped if the result of the decrement is zero. The instruction incfsz acts in a similar way on increment instructions. The instructions btfsc (bit test f, skip if clear), and btfss (bit test f, skip if set), allow single bits to be tested, and lead to the following instruction being skipped, if the bit was found to be cleared or set respectively. This is extremely important for testing single bits in SFRs, as well as input bits on ports. The other branch instructions are unconditional: goto causes a branch to any specified point in the program, and call is used to call a subroutine, which must be ended with either return or retlw. Interrupt routines must be terminated with retfie.

3.3 Writing in Assembler

An assembler is the computer program which converts the text file written in Assembly language into machine code. There are now very many assemblers available, both commercially and as freeware. Some, like the Microchip MPASM™ (Ref. 3.1), are developed by the microcontroller manufacturer to support their particular product. Others, like the IAR Embedded Workbench (Ref. 3.2), are developed by 'third party' software houses. A common practice is to offer the Assembler as part of an Integrated Development Environment (IDE), which integrates a number of related program development tools (for example Assembler, simulator, C compiler).

3.3.1 Source code format

Despite differences in detail, the main features of all Assemblers are the same. This section describes them. To make full use of any particular Assembler, the reader will of course have to study its accompanying manual or Help information.

Assembler code is developed in a text file, usually case-insensitive, whose format is clearly defined. There are four possible fields on a line: the label, instruction mnemonic, operand(s) and comment. This is illustrated in the

```
;************************************************
;Configure System
;************************************************
;set port bits
start   bsf     status,5    ;select bank 1
        movlw   08
        movwf   trisb       ;all portb bits output
                            ;except 3
        movlw   00
        movwf   trisa       ;all porta bits output
        bcf     status,5    ;select bank 0
label               operand

        mnemonic            comment
```

Figure 3.6 Assembler format.

16F84 program fragment of Fig. 3.6, taken from the Case Study of Appendix C.

Where a label (for example start) is used, it must commence in the leftmost position of the line. Labels are essential for every line which is to be the target of a program branch or jump. The following field is the instruction mnemonic (for example bsf, movlw), which must start in the second or subsequent spaces, and be separated from any label by at least one space. This is followed by the operand, which must again be separated from the instruction mnemonic by at least one space. Multiple operands are separated by commas. A comment, preceded by a semicolon, may then follow. Alternatively; a whole line or lines may be used for adding comments. This is seen in the first few lines of the example, where the program section heading is made up of comment lines.

Constant data may be entered in binary, octal, decimal or hexadecimal. The radix is usually identified by a leading or trailing letter. For example, in MPASM the radix is specified as in Table 3.2. However, the assembler also has a default radix, which can be changed. Therefore if this radix is always applied, it is unnecessary to specify the radix within the program text. This

Table 3.2 Radix specification in MPASM.

Radix	Format	Example
Binary	B'<digits>'	B'00111001'
Octal	O'<digits>'	O'777'
Decimal	D'<digits>'	D'100'
Hexadecimal	H'<digits>'	H'9F'
or		0x9F
ASCII	'<character>'	'c'
or	A'<character>'	A'c'

is the approach used in Fig. 3.6. Hexadecimal numbers starting with A to F must be preceded by a 0, otherwise the assembler will attempt to interpret them as labels.

In other assemblers, the radix is specified by a trailing alphabetic code. A number preceded by the dollar sign $ is also widely used to represent a hexadecimal quantity. For example, following the sequence of bases used in Table 3.2,

$$11111011B = 373Q = 251D = 0FBH = \$0FB.$$

A number to be treated as immediate data (as opposed to an address), is frequently preceded with a hash: #. This is not, however, used in MPASM.

3.3.2 Assembler file structure

Used in simplest form, assembler program development follows the pattern shown in Fig. 3.7. The program is written in the format just described, called the *source* file, frequently given the extension .asm. If there are no errors in the source file, then the process of assembly produces the executable machine code as a Hex file. It also produces a *List* file, which is a listing of the original program, with the hex code alongside each instruction line. If errors have been detected, these may be listed in an Error file, with extension .err. The Hex file is not then produced.

The List file corresponding to the example of Fig. 3.6 is shown in Fig. 3.8. The leftmost field is now the program memory address, followed by the machine code stored at that address. The third simply shows the line number. For example we can see that the instruction labelled start is located at memory location 0010_{16}, and is encoded as 1683_{16}. With a little care, it is possible to confirm in a listing like this that the instruction code given does indeed correspond to the mnemonic and operand entered by the programmer.

The process of assembly is normally done in two passes. A feature of writing in assembly language is that labels can be used before they are defined, so the first pass identifies all labels and allocates them values. In the next pass it applies the label values to the code generated.

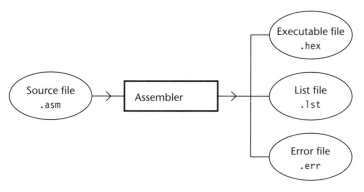

Figure 3.7 File structure of simple Assembler program development.

```
;**************************************************
                  00052 ;Configure System
                  00053
;**************************************************
                  00054 ;set port bits
0010 1683         00055 start  bsf    status,5  ;select bank 1
0011 3008         00056        movlw  08
0012 0086         00057        movwf  trisb     ;all portb bits
output
                  00058                          ;except 3
0013 3000         00059        movlw  00
0014 0085         00060        movwf  trisa     ;all porta bits
output
0015 1283         00061        bcf    status,5  ;select bank 0
                  00062 ;set up display
```

Figure 3.8 List file fragment for example of Fig. 3.6.

The model of Fig. 3.7 is suitable for simple programs. Usually, however, programmers do not develop every new program from square one. Instead they make considerable use of program modules which have either been developed previously or are available in the public domain or commercially (for example supplied with the Assembler or by the microcontroller manu-facturer). The development model then followed is shown in Fig. 3.9. Now source files are assembled into intermediate files called object files. These have the characteristic that they are *relocatable*, i.e. their position in program memory is not yet defined. They are combined by a program called

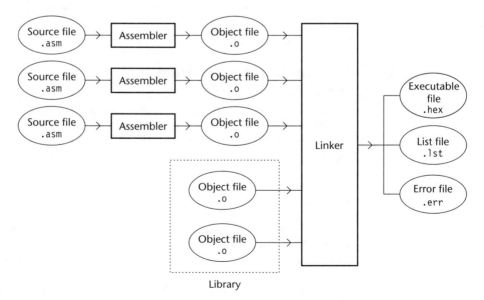

Figure 3.9 File structure of advanced Assembler development.

Table 3.3 Example MPASM directives.

CONFIG	Sets processor configuration bits.
END	Defines the end of the source program.
EQU	Assigns an integer value to a label.
INCLUDE	Includes an additional source file.
LIST	Listing options. This includes selection of processor type and radix.
MACRO	Defines the start of a macro.
ENDM	Ends a macro definition.
ORG	Sets the program origin, i.e. it specifies the program counter value at a point or points in the program listing.
RADIX	Specifies the default radix to hexadecimal, decimal, or octal. The default is hexadecimal.

a *Linker*, which has the capability of combining newly assembled programs, as well as object files held in a library, into the final executable code. This development environment gives the programmer the flexibility to build up programs from a range of modules, and to add new modules to the library.

3.3.3 Assembler directives

The usefulness of an assembler is greatly enhanced by a number of 'Assembler Directives', which are embedded in the source code. These have the appearance of instruction mnemonics, but are recognised by the assembler as instructions to which it must respond. Example directives from MPASM are shown in Table 3.3. These directives will be illustrated in the following section.

3.4 Developing the program

One of the criticisms directed against Assembler programming is that it is too easy to develop programs with no structure, which are thereby difficult to debug, maintain or develop. There is some justification in this criticism, for reasons discussed in Chaper 7. When writing in Assembler it is therefore especially important to pay attention to the need for structure.

3.4.1 Flow diagrams

In almost any engineering activity, we use diagrams to express our intentions. A good engineering diagram or drawing allows designers to express and develop their views, *and* it encourages easy transition from the stage of drawing to the stage of production. Program design is no different; we need a diagramming method which allows us to lay down the structure of the program, and then allows that program to be written in an error-free way.

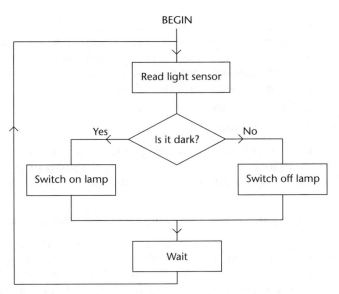

Figure 3.10 Simple flow diagram example.

The traditional way of diagramming computer programs has been by flow diagram.

A flow diagram example, for a program which measures ambient light level and switches a lamp on if it is dark, is shown in Fig. 3.10. The ambient light level is first measured, and a program branch is then made depending on the light level. After a pause the program restarts. It is possible to develop just about any microcontroller flow diagrams using just the two symbols used in the example.

3.4.2 Program flow

The processor always follows instructions sequentially, unless disturbed by an interrupt, or unless an instruction causes a program branch to take place. As even the simple diagram of Fig. 3.10 suggests, we can expect any program to have branches and loops in it. The microprocessor instructions which give programs this capability are:

- Unconditional branches. These are sometimes also called Jump instructions. An example is the 16F84 goto.

- Conditional branches. Here a certain condition is tested (for example the state of a bit in a register or memory location) and a branch takes place if the condition is met. Many such instructions are available in the 80C51 and 68HC05 controllers, but the 16F84 has none. The target address of the branch must be specified in some way in the instruction (the 68HC05 does this with relative addressing – Appendix D).

- Conditional skip. Here a certain condition is tested and the instruction which follows is skipped if the condition is met. Examples are the 16F84 btfsc and btfss instructions. These test a specified bit in a specified memory location, and skip if it is clear or set respectively. While the versatility of this branch is less, the instruction format itself is simpler than the conditional branch, as there is no need to specify the target address of the branch (which is always the next instruction but one).

- Subroutine calls and returns.
- Interrupt returns.

Most embedded systems programs are designed to run continuously, so are likely to be based on a major loop. If a program is to execute once only, then it can be terminated in a continuous loop on the last instruction.

3.4.2.1 Developing the 16F84 branching capability: the 'Computed Goto'

When a branch instruction is implemented in any microprocessor, the address of the instruction to where the branch is taking place is moved into its Program Counter. While the 16F84 does not have the extensive branching capability of the 68HC05, it does have some flexibility in such instructions, using the 'Computed Goto' technique. Because the lower eight bits of the program counter are directly accessible, as SFR PCL (Fig. 2.2), they can be written to in one instruction. The subsequent instruction is then taken from this new, modified address. Unless the goto thus implemented is short, attention must also be paid to the higher bits of the program counter. These are accessible, under certain constraints (detailed in Ref. 3.4), through SFR PCLATH. An example of the Computed Goto appears in Section 3.6.2.1.

3.4.3 Starting the program

The opening section of the program of Appendix C shows the main features needed to start a program, and is reproduced in Fig. 3.11. The opening title block and program definition are created using comment lines. Reference information is then given on essential hardware features, such as clock frequency and port pin allocations. The microcontroller type is identified with a LIST directive. The program listing then gives labels to some of the on-chip RAM, using the EQU directive. These are names and locations determined by the programmer, all falling within the area of 'general-purpose registers' of Fig. 2.2. In a similar way the SFR addresses are then defined. Note that the Assembler does not 'know' these, even though they are defined by the processor design. The section defining SFR addresses *could* be (and often is) replaced with an INCLUDE statement, which allows a standard register listing to be included in the assembly process. Such files are provided as part of the MPLAB package.

The 16F84 Reset Vector, located at memory address 0000, is specified by use of the ORG directive. This location is then followed shortly after by the Interrupt Vector, so the first act of the program must be to jump beyond the interrupt vector to a memory location where the main program can start, which is defined by further use of the ORG directive.

```
;***********************************************************
;Program to run Panel Meter ADC.
;Binary to BCD, and display driver routines not included.
;Program operates in unipolar mode, ie negative input
;voltages are not recognised.
;ADC6                                            2.11.98
;***********************************************************
;
;HARDWARE ALLOCATION
;Clock frequency = 10MHz. Clock mode = HS.
;WDT disabled.
;
;RB0 = E (Display)                    RB4 = D4 (Display)
(etc)

                LIST    P=16F84
;specify SFRs
indf    equ     00
timer   equ     01
status  equ     03
fsr     equ     04
(etc)

;name memory locations
sarhi   equ     10   ;successive approx. reg.
sarlo   equ     11
counter equ     12   ;used in DAC output routine
(etc)
;
;Reset Vector
        org 00
        goto start
;
;Interrupt Vector
        org 04
        goto loop0
;
        org 0010
;
;*********************************************
;Programme Starts - Configure System
;*********************************************
;set port bits
start   bsf     status,5    ;select register bank 1
        movlw   08
        movwf   trisb       ;all portb bits output
                                    ;except 3

        movlw   00
        movwf   trisa       ;all porta bits output
        movlw   07
        movwf   option      ;sets OPTION register
        bcf     status,5    ;select bank 0 for later
                                    ;memory transfers
        bcf     intcon,2    ;clear pending interrupts
        bsf     intcon,5    ;enable Timer Overflow int.
        bsf     intcon,7    ;set global interrupt enable
wait    goto    wait        ;wait for next Interrupt
;
```

Figure 3.11 Opening program section.

The program itself starts with initialisation of the peripherals. Bit 5 (RP0) of the Status Register (Fig. 1.12) is set to 1 to select Bank 1 of the RAM. The number 08_{16} (0000 1000_2) is moved to the **W** register, and thence to SFR TRISB, which sets all port B bits to output, except bit 3. A similar procedure is then followed for Port A. The Counter/Timer is then set up; this is described in Section 3.5.3.

Having set up the peripherals and interrupts, this particular program enters a wait loop. All program activity is contained within its interrupt routine, and the subroutines that that calls.

An Assembler program should be terminated with the END directive. It should be clearly understood that this directive gives information to the assembler; it is not passed on to the processor as some form of instruction to halt operation.

3.4.4 Subroutines

A subroutine is a group of instructions, together performing some identifiable function, which are called from within the main flow of the program. The usual application is when a section of program is to be used more than once, for example if two numbers are to be multiplied together. However, any group of instructions which together perform an identifiable function can be made into a subroutine. It is much easier to read a program listing which is mainly a series of subroutine calls than it would be to read the whole sequence of instructions.

Once developed and tested, it is easy to transfer routines from one program to the next, and to build up a library of subroutines suitable for a particular processor and application. Most manufacturers are also good at providing free subroutine libraries, for example for mathematical functions.

Subroutines are one means to realising a well-structured modular program in Assembler, and hence are one of the keys to controlling program complexity. To this end, each subroutine should have a limited and clearly defined task, and be kept reasonably short.

A subroutine is called by a 'jump to subroutine' instruction. This transfers the address specified as the subroutine start address to the Program Counter, and saves the present contents of the Program Counter onto the Stack. The subroutine is terminated by a 'return from subroutine' instruction, which loads the Program Counter from the stack. The 16F84 subroutine jump instruction is call, and the subroutine return is return. An alternative is retlw, whose use we shall see in Section 3.6.2.1. Subroutine calls and returns must always be used to balance each other. It would be unworkable for example to try to call a 16F84 subroutine with a goto instruction and then end it with a return – the processor would load a non-existent return address from the Stack, with disastrous consequences.

A subroutine is likely to require certain data input, and will generate some data output. In so doing it *may* modify certain other memory locations or registers, and *may* call other subroutines. It is therefore necessary to identify:

- what input variables are needed, and how they are passed
- what registers or memory locations (if any) are altered during the course of the routine,
- what outputs are generated, and how they are passed
- what other subroutines are called from within this routine

To control the way subroutines alter memory locations, two conventions have been proposed (Ref. 3.3), either of which can be applied.

1. Subroutines leave *all* CPU registers, except the status register, *exactly* as they find them.

2. The subroutine may change *any* CPU register; it is up to the calling program to preserve any registers it needs before calling the routine.

The first of these probably leads to more reliable programming. If a register is to be used in a routine, then its original contents must be saved at the start of the subroutine and restored at the end. It does, however, lead to the overhead of saving these registers, even when it is not necessary. The second requires the programmer to identify any register contents which are vulnerable, and then save these before the routine is called.

In a similar way, it is important to exercise consistency in the passing of parameters. These may be passed in CPU registers; for example the accumulator, dedicated memory locations, or on the stack. If the latter (not possible with the 16F84 hardware stack), then it is important to ensure that the return address of the subroutine is not lost in the process of retrieving other data from the stack.

One subroutine can be called from inside another; this is called a 'nested subroutine'. Clearly this places certain demands on the stack size. The 16F84 in particular has limited stack capacity, which restricts the number of nested routines possible.

Information on all the above points should be given in the Subroutine Header; for example Fig. 3.12.

```
;****************************************************************
;Subroutine to multiply 2 16-bit numbers. The numbers are passed
;in DEC4-3 and DEC2-1, and returned in BIN4-1.
;No other memory locations altered. No other SRs called.
;****************************************************************
```

Figure 3.12 A subroutine header.

3.4.5 Interrupt service routines (ISRs)

The ability of a program to respond rapidly to external events is provided by the processor's interrupt capability, the purpose and mechanism of which were discussed in Chapter 2. The ISR start address is defined by the interrupt vector. In the listing of Fig. 3.11, it can be seen that the instruction goto loop0 is placed at this location. The ISR proper starts at a location labelled loop0 later in the program.

The actual execution of the ISR is superficially similar to a subroutine. The big difference, however, is that the subroutine is called when the programmer wants, whereas the interrupt can appear at an entirely unpredictable time. The interrupt routine must therefore be more or less 'invisible' to the main program, except of course for the actions it is designed to introduce. This means that interrupt routines should be kept short, and should make no change whatsoever to the CPU registers (the change of one flag in the Status Register could have disastrous consequences). It is therefore up to the programmer to check whether 'context saving', i.e. saving of all key registers, is done automatically by the processor (as it is in the 68HC05), or whether it is up to her to save whatever is necessary (as in the 16F84 or 80C552).

On completion of the routine the programmer should check whether she should clear the calling Interrupt Flag – this is a requirement with the 16F84. The ISR must then be terminated with a 'Return from Interrupt' instruction. Just as a 16F84 subroutine *must* be called with a call instruction and terminated with a return or retlw, so a 16F84 ISR *must* be terminated with a retfie instruction.

If two or more interrupts are enabled simultaneously, the processor must have a means of determining priority between them. Interrupt prioritisation is discussed in Chapter 8.

3.4.6 Macros

One directive listed in Table 3.3, but not illustrated in the case study, is the macro. This gives a capability which helps to get around the minimalist nature of the instruction set. Groups of instructions can be put together to define a 'macro instruction'. This can then be written into the assembler program just like any other instruction. To the user this makes a macro similar to a subroutine. The significant difference is that when the program is assembled the instructions forming the macro are embedded into the program every time the macro is used. Using macros makes the actual program residing in memory longer than if subroutines had been used, but it executes more quickly. This is because 'jump to subroutine' and 'return from subroutine' instructions are eliminated.

3.4.7 Laying out the program

A well laid-out program is likely to have the features of Fig. 3.13. The Heading should contain:

- program title
- target application
- name of programmer
- date of last update

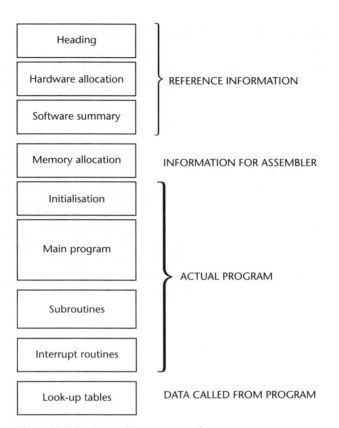

Figure 3.13   Assembler program layout.

It is very important that this reference information should be kept clear and up to date. The Hardware allocation specifies:

- how I/O pins are connected
- allocation of on-chip peripherals
- clock frequency (for timing purposes)
- use of interrupts
- allocation and operating conditions of any other microcontroller hardware, for example the Watchdog Timer

The Software summary should outline the program operation and hierarchy, possibly including listing of all subroutines, and indicating which routines each calls.

Comments within the program should be applied liberally, describing both individual program steps and blocks. This aids the thought process, and makes it easy for a programmer to come back to her work after an interruption (of hours, days or months). It also allows another programmer to take over the work.

3.5 Generating time delays and intervals

A very common need in the microcontroller environment is to perform activity on a timed basis. While this is considered systematically in Chapter 8, we look here at some introductory timing techniques.

3.5.1 Software-generated delays

The simplest way to generate time delays is by setting up a timing loop, such as is shown in Fig. 3.14(a). Here a counter (which can just be a memory location) is loaded with a predetermined number and decremented within a loop until its value is zero. 'Padding', i.e. a number of time-wasting operations, can be included in the loop to extend the time duration. As the time taken for each instruction is known, the number originally loaded into the counter can be calculated to give the desired duration.

The case study of Appendix C includes a subroutine delay1, reproduced as Fig. 3.15(a), which follows this flow diagram. The system clock frequency in this example is 10 MHz, and with four clock cycles per instruction cycle

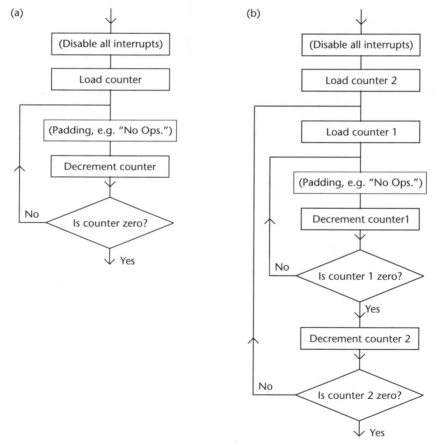

Figure 3.14 Timing loops: (a) for short durations; (b) for longer durations.

```
(a)      ;introduces delay of 500us
         delay1  movlw    0fa
                 movwf    delcntr1
         del1    nop                    ;(1 cycle)
                 nop                    ;(1 cycle)
                 decfsz delcntr1,1      ;(1 cycle)
                 goto del1              ;(2 cycles)
                 return


(b)      ;introduces 100ms delay
         delay100 movlw    0c8          ;200 decimal
                  movwf    delcntr2
         del3     call delay1
                  decfsz  delcntr2,1
                  goto del3
                  return
```

Figure 3.15 Software timing loops with the PIC 16F84. (a) Short software delay; (b) extended delay by subroutine call.

each instruction takes 400 ns. The duration (in instruction cycles) of each instruction within the delay loop is indicated in the figure. There are a total of 5, which together take 2 μs to execute. To achieve the nominal delay time of 500 μs, the loop must be iterated 250_{10} times, or $0FA_{16}$. This is the number loaded into the memory location delcntr1, used as the delay counter. In fact, the total duration of this subroutine will be slightly longer than the nominal value just calculated, as the call, return and counter loading instructions have not been taken into account.

Longer delays can be introduced in one of several ways. Given a subroutine of known delay duration, this can be called repeatedly from another routine. An example of this is shown in Fig. 3.15(b). Here a subroutine delay100 calls the routine delay1 200 times, leading to a nominal total delay of 100 ms. The approach as shown is not entirely accurate, as the time taken by the looping within the delay100 subroutine has not been accounted for. Nevertheless this is a convenient way of creating approximate delays. Alternatively, precise delays can be achieved by designing the routines to accommodate these delays (this is the basis of Exercise 3.6). Extended delays within a single routine can be developed using flow diagrams of the form shown in Fig. 3.14(b).

3.5.2 Hardware-generated delays

The disadvantages of the delay loop are clear: the processor is tied up entirely with the timing function for all of its duration, and to ensure reliable delay generation all interrupts must be disabled. An improvement on the delay loop is to use a Counter/Timer (C/T) with an 'Interrupt on Overflow' capability, as described in Chapter 2. The C/T is preloaded with a suitable number, and when it overflows an interrupt is generated to which the controller can respond. While the Timer is running the controller is freed to perform other operations.

WORKED EXAMPLE **3.1**

Devise a means of replacing the 16F84 500 µs software generated delay of Fig. 3.15(a) with one based on the C/T module. The oscillator clock frequency remains at 10 MHz.

Solution: The internal instruction cycle time is 400 ns. It would take (256×400) ns = 102.4 µs for the C/T module to overflow, if counting from 0. This is inadequate for the time delay required. With a prescaler division of 8, however, it would take 819.2 µs, with the C/T incremented every 3.2 µs. With this increment rate, dividing 500 µs by 3.2 µs, leads to 156.25 cycles being required to generate the required delay. As the C/T overflows on the 256th count, it should be preloaded with $(256 - 156)$, i.e. 100_{10} (64_{16}). The OPTION register should be set to XX0X 0010. If the timer overflow interrupt is enabled, an interrupt will occur 500 µs (nominal) after the C/T has been loaded. A program fragment which implements this is given in Fig. 3.16.

```
bsf     status,5    ;select register bank 1
movlw   02
movwf   option      ;sets OPTION register, Timer mode
                    ;prescaler on, divide by 8
bcf     status,5    ;select bank 0 for later memory transfers
movlw   64          ;preload Timer with required value
movwf   timer       ;Timer will overflow in 500us
bcf     intcon,2    ;clear any pending overflow interrupt
bsf     intcon,5    ;enable Timer Overflow int.
bsf     intcon,7    ;set global interrupt enable
...
```

Figure 3.16 A 16F84 Counter/Timer delay.

3.5.3 Initiating repetitive time-based activity

There are many situations, even in simple systems, when activities have to be initiated in a repetitive, timed pattern. This may be necessary to generate required system outputs, or it may be a means of timing software execution, using techniques described in Chapter 8. Such repetitive interrupts can be instigated by the C/T Interrupt on Overflow. For precise timing it may still not possible to tolerate other interrupts, as these may delay response to the Timer-generated interrupt. The principle was introduced in Section 2.5.3.3.

As an example, the case study data conversion takes place in a repetitive pattern, with an approximate conversion period of 25 ms required. A regular interrupt was required to time these conversions. With the 2.5 MHz clock frequency applied directly to the Timer input, it would overflow every (256×400) ns, i.e. every 102.4 µs. This can be extended by any of the prescaler division rates. In this example the prescaler is set to 1:256, so the interrupt occurs every (256×102.4) µs, i.e. every 26.2 ms. The last seven or eight lines of the program fragment of Fig. 3.11 show the OPTION register being set to select this prescale value (07_{16}, i.e. $0000\ 0111_2$ is loaded into the OPTION register), and then the appropriate interrupt being enabled. All further activity of this particular program now occurs in the ISR instigated by this interrupt.

If the binary powers of division offered by the prescaler had not given the required interrupt rate, then it would be possible to adjust the interrupt rate by preloading the C/T to a suitable value at the beginning of every interrupt routine, in an manner similar to Example 3.1.

3.6 Data handling

Developing program structure with flow diagrams allows us to picture the structure of the program, but gives little information about how data is treated. Yet in most cases an essential function of the microcontroller is as a processor of data. Therefore we need to think independently about the flow of data in our system – where it is coming from, how it is processed and stored, and to where it is going.

3.6.1 Data types

Figure 3.17 illustrates informally the different types of data likely to be found in a system. Sources of data include sensors, human interface devices like keypads, and data links from other systems, for example through an RS232 serial link. Incoming data may need to be buffered. Once processing is under way, it may be necessary to call upon constant data, embedded in the program. Temporary data is likely to be generated. This may take the form of *global variables*, valid at all times during operation of the program, and *local variables*, valid only for specific program sections, for example during the course of one subroutine. Data is then output to a variety of targets. These include actuators and displays, as well as remote recipients, again possibly by a serial links.

Data storage needs to be planned for any program. In a simple application, each variable may be allocated its own labelled memory space, for example a RAM location. In more advanced applications, where memory

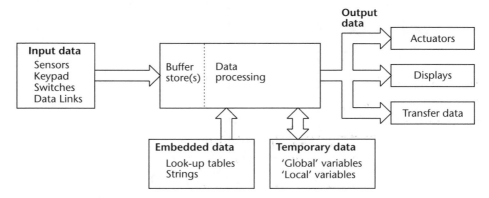

Figure 3.17 Data flow in the embedded system.

demands are heavier, it may be that only global variables are permanently allocated their own memory location. Local variables are then just held in a temporary location, which may include the stack.

In many cases quantities of data must be handled as blocks or tables, or as a continuous incoming flow. To allow us to cope with these situations, a number of simple data structures have been developed which allow the data to be held and processed in an orderly manner. Proper use of these optimises both processor time, and memory utilisation, and leads to reliable programming.

3.6.2 Tables and strings

It is often necessary to embed data constants in program memory. Two different ways of doing this are the Table and the String. A Table is used for a block of data. All data elements are of identical type, and the block is of known length. Figure 3.18(a) illustrates the general principle. In many processors (for example the 80C51 and 68HC05) such a table is accessed by indexed addressing (Appendix D). The Index Register is loaded with the address of the first memory location, and data is accessed using the Index Register as pointer. The Index Register can then be incremented to point to subsequent values in the table until the address of the end location is reached.

In the example of Fig. 3.19, a look-up table has been created using a certain cross-assembler for which DFB is the *Define Byte* directive. It contains sine wave values from 0° to 90°, each value being represented by 2 bytes.

Sometimes a program needs to access a sequence of data elements where the sequence is of unknown length. These are known as Strings. Because the length of the sequence is unknown, its end is identified by a predetermined terminator character or code. Figure 3.18(b) illustrates the general principle of a String.

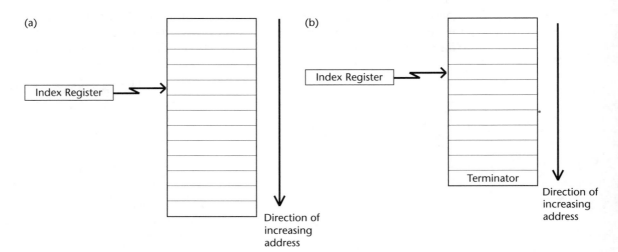

Figure 3.18 Storage of constants in program memory. (a) Table; (b) String.

```
SINTBLE   DFB  00,00,03,5AH,06,0B2H,0AH,09H,0DH,5CH  ;0 to 8 degs
          DFB  10H,0ACH,13H,0F6H,17H,39H,1AH,76H,1DH,0AAH; 10 to 18 degs
          DFB  20H,0D5H,23H,0F6H,27H,0CH,2AH,15H,2DH,12H ;20 to 28 degs
          DFB  30H,00H,32H,0DFH,35H,0AFH,38H,6DH,3BH,1AH ;30 to 38 degs
          DFB  3DH,0B5H,40H,36H,42H,0B0H,45H,0EH,47H,58H ;40 to 48 degs
          DFB  49H,8AH,4BH,0A6H,4DH,0AAH,4FH,96H,51H,6AH ;50 to 58 degs
          DFB  53H,23H,54H,0C3H,56H,49H,57H,0B3H,59H,02H ;60 to 68 degs
          DFB  5AH,36H,5BH,4DH,5CH,48H,5DH,26H,5DH,0E7H ;70 to 78 degs
          DFB  5EH,86H,5FH,11H,5FH,79H,5FH,0C4H,5FH,0F1H,60H,00;80 to 90
                                                          ;degs
```

Figure 3.19   Look-up table (80C552).

```
TITLE       DFB  "1kV Power Supply",0FFH
START       DFB  "# to set op",0FFH
DEFINE2     DFB  "Enter 4 Dgts",0FFH
AGAIN       DFB  "# to restart",0FFH
EXCESS      DFB  "Set Lower Volts",0FFH
TIMEOUT     DFB  "Timed Out",0FFH
CMNDS       DFB  28H,0EH,06H,01H,0FFH
```

Figure 3.20   Data strings (80C552).

A common application of Strings is for the storage of text messages. In the example in Fig. 3.20, written for the same assembler as the previous example, a set of messages is stored. These are to be transferred at appropriate moments to a display. Because of the inverted commas the assembler recognises the data as text information and assembles it into ASCII codes. Access can again be by Indexed Addressing. The terminator $0FF_{16}$ is used to show the end of the string, and is identified by the calling routine. Another terminator often used is 04_{16}, which is the ASCII EOT (end of transmission) character. If only ASCII data is being stored, it is possible to avoid the use of a terminator altogether. ASCII uses only 7 bits, so the 8th (sign) bit can be left at zero for all elements in the string except for the last, when it is set to 1. The program must of course restore this bit to 0 once it has played its purpose.

3.6.2.1   16XXX table reads

It is rather more difficult to embed data into the program of a machine with Harvard memory structure, as the data must somehow be transferred from the program memory domain to the data memory domain. The 16XXX family does not have an indexed addressing mode, but can access constant data within a program by using the retlw instruction. This is actually a subroutine return instruction, but the literal data accompanying the instruction is placed in the W register, and the return is made with that data in place. A table is made up of a subroutine containing a series of these instructions, each one having as operand one of the table constants. Each retlw instruction executes a subroutine return; the one which is actually accessed is determined at the start of the subroutine by applying the computed goto technique described earlier in this chapter.

```
int    movf  sample_no,0
       call  table
       movwf ccpr1l        ;output new pwm value
       decfsz sample_no
       goto  intout
       movlw 14            ;reload counter with 20d
       movwf sample_no
       bcf   portb,7
intout bcf   intcon,t0if
       retfie
;
table addwf pcl
       retlw 7f
       retlw 0a6
       retlw 0ca
       retlw 0e6
       retlw 0f8
       retlw 0ff
       retlw 0f8
       retlw 0e6
       retlw 0ca
       retlw 0a6
       retlw 7f
       retlw 57
       retlw 33
       retlw 17
       retlw 5
       retlw 1
       retlw 5
       retlw 17
       retlw 33
       retlw 57
```

Figure 3.21 Implementing a look-up table (16C74).

In the example of Fig. 3.21, values corresponding to a sine wave are held within the subroutine labelled table. An interrupt routine labelled int moves the contents of a counter labelled sample_no into the W register, and then calls table. At the start of table, the W register contents are added to the program counter contents, and a jump forward, dependent on the value of the W register, is then automatically made. This is an example of the computed goto. Whichever retlw instruction is then reached transfers its constant number into the W register, and the subroutine is ended. Precautions in practice must be made so that the table does not cross page boundaries. These are detailed in Ref. 3.4, which also gives further details on the computed goto.

3.6.3 The stack

It is assumed that the concept of a stack, implemented in hardware, is familiar to the reader. In formal terms it is known as a LIFO (Last In, First Out) structure. Stacks are used to hold return addresses for subroutines and ISRs. In many processors they also act as a temporary data store, allowing data to be 'pushed' onto them, or 'popped' off. This is true of the 68HC05 (whose stack lies in the general-purpose RAM area, and can be seen in

Fig. 3.3) and 80C51 microcontrollers. It is not the case with the 16F84. If a data stack is required for the 16F84, or its hardware stack needs to be extended, then it must be implemented in software (Ref. 3.5). Software stacks may also be implemented on other processors, for example when more than one stack is required.

3.6.4 Queues and buffers

Frequently, incoming data must be stored temporarily before it can be processed. Two means of doing this are the Queue and the Circular Buffer.

A queue is like a line at a supermarket checkout. It is of variable length, and the person who has waited the longest is the next to receive attention. In a data structure it has the form shown in Fig. 3.22(a). In formal terms this is known as a FIFO (First In, First Out) structure: new data is added at one end, and the oldest data is taken from the other. Two pointers are needed, one to indicate where the next new piece of data is to be placed (the 'put' pointer), and the other indicating from where the next piece of data is to be taken (the 'get' pointer). Each time a 'get' or 'put' takes place, the associated pointer is adjusted to point to the next location. For example, in Fig. 3.22(a) the most recent 'put' has been Data $n + 5$. The pointer has been advanced to point to the next empty location. The next 'get' will be of Data n. When this has been read, it can be overwritten.

When implementing a queue, it is important to include control software to track the queue length and limit pointer values. The maximum permissible queue length should be predetermined, depending on the available memory and anticipated quantity of data. Control should be then be exercised to ensure that the queue is not allowed to under- or overflow.

A circular buffer is a simpler structure, and assumes only that the most recent data is required. It is of fixed length, and new data automatically overwrites the oldest. One way of implementing it is shown in Fig. 3.22(b). Here the circular buffer has eight locations. On every write operation, the

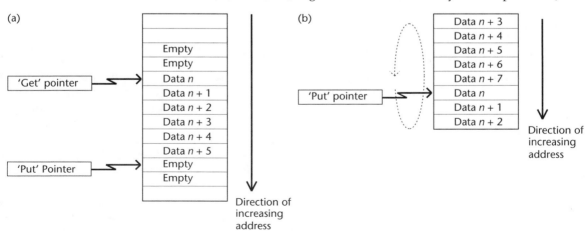

Figure 3.22 Temporary storage of incoming data. (a) Queue; (b) circular buffer.

next memory location is written to, overwriting the oldest piece of data in the buffer. In this case the most recent piece of data written is Data $n + 7$, and the next to be overwritten is Data n. Addressing for buffers of binary power length (for example, 8, 16, 32) is easy to implement. For example, if an Index Register is used as the pointer, incremented on every write operation, and all but the three least significant bits are masked out, then the register will point repetitively at eight sequential memory locations, exactly as shown in Fig. 3.22(b). An application of the circular buffer, to average an incoming sampled signal, is described in Chapter 11.

3.7 Commissioning the minimum system

When the hardware of the minimum system has been designed and built, and a simple program written, the program *may* be downloaded into the hardware and the system switched on. The designer is then in that unenviable situation – she has untested software attempting to run in untested hardware. When it doesn't work (as it is bound not to do) it is difficult to know where to start looking for the problem. The difficulties are linked to two things. First, the system is largely invisible; the registers and peripherals inside the microcontroller simply cannot be accessed to find out what is going on. Second, program execution is happening so fast that a fault condition may come and go before there is time anyway to detect it.

Fortunately there are an assortment of tools and techniques which exist to tame the embedded system, to allow the designer to access the apparently inaccessible, and to view, or slow down, the galloping sequence of events that the program is causing. These tools are introduced at different stages in the book, beginning with the simple ones in this chapter. Let us start by *not* putting untested software into untested hardware. Let us test the software before it even leaves the host computer, so that we know that it is working by the time it is matched up with the hardware. We do this by applying the Instruction Set Simulator.

3.7.1 The Instruction Set Simulator

The *Instruction Set Simulator*, usually just called the *Simulator*, is a software package running on the host computer, usually as part of the IDE. It can simulate the action of a particular processor, and attempt to execute any program written for that processor. A good simulator replicates all aspects of the processor, including the peripherals. As it does so it keeps a record of the contents of all internal registers, including SFRs, and internal memory locations. The user can select which of these she wants to see, displayed on the host computer screen.

The simulator can generally run in one of at least three modes. The simplest is *Run*, in which normal program execution takes place. The screen of the host computer, with its display of register contents, memory and so on, is updated when the program is halted. *Single Step* allows the user to step through the program one instruction at a time, with the host computer screen updated at every step. *Animate* is a form of automated single step. The program runs continuously at a slow speed, with the host computer screen again being updated on every instruction.

One of the most powerful features of the simulator is its ability to run to a pre-defined point, called a *breakpoint*, and then stop. Associated with breakpoints is the ability to keep a *Trace*, i.e. a record of past events. This record is held in a *Trace Buffer*, and may include instructions executed, and register and stack contents. The simplest breakpoint definition is on a match with the program counter; the program is made to execute until it reaches a certain predetermined instruction, when it halts. Breakpoints may also be available on stack overflow, or when the Trace Buffer is full. Another diagnostic tool is the *Pass Counter*, which allows the user to force a halt only after the program has executed a particular instruction a specified number of times.

The simulator does not interface with the real world, so all inputs must be simulated. A number of techniques are used to achieve this. Most simulators allow inputs of the simulated processor to be controlled in one way or another from the host computer's keyboard keys. This allows input changes to be introduced at will by the user, generally when running in Single-Step or Animate Mode. Alternatively, simulators allow a stimulus file to be generated. This allows the user to define input changes to be made after a specified number of instruction (machine) cycles have elapsed. Note that the simulator is generally unable to replicate the effects of genuinely asynchronous input, for example if an input changes for a duration of less than one instruction cycle. Changes introduced by the stimulus file are synchronised with the instruction cycle, and even keyboard-generated input changes are usually sampled only on a new instruction cycle. Finally, it is usually possible to 'connect' simulated clock signals to specified inputs, at signal periods based on instruction cycle times. This allows C/T systems to be exercised.

Every program instruction that the simulator executes involves the host computer in making numerous operations of its own. The speed of execution of the simulated program is therefore dependent on the speed of the host computer, and is usually far slower than the intended program speed of the target system. Execution times of the order of 1 ms per instruction are not uncommon, i.e. a factor of one thousand or more slower than the commonly used target system clock frequencies. Therefore the simulator is of little direct use in testing the timing characteristics of the program. It does, however, normally have the facility to allow timing measurements to be made. This is based on its ability to count the number of program cycles required to execute a particular section of program code, from which the actual program execution time can be calculated, given knowledge of the target system clock speed.

Any of the simulator modes of execution can present a problem in certain situations, one of which is a delay loop. For example, if a subroutine was written to introduce a delay of 1 s, and the simulator is running 1000 times slower than the target system speed, then that will work out as a delay of over 16 minutes, not something one really wants to wait through. Far less would one want to single step through a routine like this! Therefore another option usually offered is to *step over* subroutine calls.

The simulator should be viewed as an almost indispensable tool in testing program sections as the program is being developed, and then to test the completed program. If this is done, then the designer should be able to download the program to the target system with a high level of confidence.

3.7.2 Program-related faults

Whether testing with simulator or in the hardware, it is worth checking for the faults listed below, which are common in Assembler programs.

General
- *Simple typographical errors*, leading to an Assembler program which is grammatically correct but functionally faulty. This is very easily done with mnemonics. A small error turns addwf to andwf for example, with quite different outcomes!

Program flow errors
- *Unintended infinite loops*, where the program is unable to break out of a loop, either because it is written that way, or because it is waiting for an 'impossible' event or condition
- *Incorrect branching*, for example wrong condition tested, or branching to wrong target address
- *Incorrect program termination*, where the program continues execution into unprogrammed program memory
- *Incorrect subroutine and interrupt routine call or termination*
- *Incorrect use of reset and interrupt vectors*
- *Failure to clear interrupt flags after the interrupt has been serviced*

Numerical errors
- *Variables not preset*
- *Numbers over-range*; for example addition of 8-bit numbers with results greater than 8-bit

Errors with peripherals
- *Incorrect or absent initialisation*

Errors with memory
- *Errors with labelling*
- *Single RAM location used for two or more different conflicting functions*
- *Stack overflow*, for example from too many nested subroutines;
- *Memory map misunderstandings*, for example accessing absent RAM

3.7.3 Program download: connecting hardware and software

Having made good use of the simulator, you should feel ready to download the program to the hardware. The principles of this process are described in Chapter 4. The practice will depend very much on the microcontroller in use, its memory technology and the associated development tools. For PIC devices, a popular tool is the PICSTART™, driven from MPLAB, which allows program download to any PIC device. If using a PIC, it is important to ensure that the configuration word is correctly set up when programming takes place. MPLAB will prompt for this just before program download. Most simple applications will disable the Watchdog Timer, enable the Power-up Timer, and select the oscillator according to the configuration used.

When testing the embedded system, it is essential to have correct documentation to hand. This includes both an up-to-date program listing (the .lst file is the best for this, it gives both source and machine code together) *and* the hardware circuit diagram. The circuit diagram is needed even when the program is being tested on the simulator, at least once I/O connections are being simulated.

3.7.4 Hardware commissioning

When the hardware is ready for test, the preliminary tests listed below may be found to be useful for the first-time builder.

- Before power-up, using the continuity tester on a digital voltmeter, ensure that earth and power are connected to every point they should be, and that they are not short-circuited together.

- After power-up, with an oscilloscope, ensure that power has reached every point that it should, that the voltage is correct, and that the supply is relatively free of ripple or spikes. While some ripple or spiking is to be expected on the supply, it should never cause the voltage to exceed the voltage limits defined by the manufacturer.

- If after the above checks, there is no apparent activity, check with an oscilloscope for the presence of a clock signal, at the right frequency, at the microcontroller oscillator pins. A ×10 'scope probe (or greater) should be used, as a normal ×1 probe will have a capacitive loading effect which may upset the delicate oscillator circuit. The waveform, if present at all, will not be a clean square wave, and will appear different between the two oscillator pins. With the 16F84 for example, a clearer signal should be seen at pin OSC2 than at OSC1.

- Ensure that input signals are present and reaching the right microcontroller pins.

- Ensure that all microcontroller pins are at a defined voltage, and not 'floating'.

Given a program which has tested successfully in the simulator, and hardware that has passed the simple tests above, it should be possible to evaluate system performance in every mode possible. More advanced commissioning procedures are introduced in later chapters.

SUMMARY

1. Assembler programming makes direct use of the processor instruction set in a moderately user-friendly way; it leads to compact code, which executes fast.

2. It is easy to write Assembler programs which lack structure, and are therefore difficult to debug, develop and maintain; special care should be taken to ensure that a program is well-structured, with plentiful comments.

3. A range of tools and techniques exist to enhance the productivity of the Assembler programmer.

4. Testing and commission (whether of hardware or software) should make appropriate use of all diagnostic tools available. The process requires good documentation, and testing of sub-systems or modules should be undertaken whenever possible.

REFERENCES

3.1 *MPASM User's Guide*. Microchip Technology Inc. Document no. DS33014F, 1997.

3.2 *8051 Windows Workbench Interface Guide*. IAR Systems. Part number EW8051-2. 1996.

3.3 Peatman, J. B. (1988) *Design with Microcontrollers*. London: McGraw-Hill.

3.4 *Implementing a Table Read*. Application Note AN556. Microchip Technology Inc.

3.5 *Software Stack Management*. Application Note AN527. Microchip Technology Inc.

EXERCISES

3.1 Write out in assembler format the program fragment represented in Fig. 3.1. Identify by label any SFRs addressed.

3.2 A man gets up each morning and checks the weather. If it is cold he lights a fire in the stove. If it is not raining he waters the vegetable patch. If it is raining and cold he leaves the horse in the stable, otherwise he moves it into the field. Then he has breakfast. Draw a flow diagram of his early morning activities.

3.3 The ISR of the Case Study of Appendix C loops back to the instruction labelled loop1 a certain number of times. Explain clearly what factors control the looping, and determine how many times the loop is executed.

3.4 Using 16F84 instructions, write a subroutine in Assembler which adds four 8-bit numbers, and then takes their average by dividing the sum by four. The division is achieved by shifting the sum right. The numbers are initially stored in memory locations val1, val2, val3, val4, and the average should be stored in the location labelled average.

3.5 In ASCII code the numbers 0 to 9 are encoded as 30_{16}, 31_{16}, 32_{16} up to 39_{16}. Four ASCII-coded numbers are held in 16F84 RAM memory locations 20_{16} to 23_{16}. It is wished to convert them to packed binary coded decimal (see Appendix A), storing the converted data to memory locations 28_{16} and 29_{16}. Write a well-commented section of assembler program for the 16F84 to do this. The original data may be overwritten.

3.6 The subroutine delay1 in the case study of Appendix C, shown also in Fig. 3.15(a), introduces a nominal delay of 500 μs. Calculate the delay it actually introduces, taking into account the time taken by every instruction in the routine, including the subroutine call and return. What number should be loaded into delcntr1 to make the generated delay as close as possible to the nominal?

3.7 Write, in the form of an Assembler source file, a 16F84 program fragment which introduces a time delay of 1.5 ms into a program. The clock frequency of the processor is such that each instruction cycle takes 4 μs to execute.

3.8 A Digital to Analogue converter (DAC) is a device which produces an analogue output whose value is directly proportional to a digital input applied to it. A certain DAC produces 0 V output for digital input of 00, and its maximum output for input of $0FF_{16}$. It is connected to Port B of a 16F84, and converts instantaneously its digital input data. Write a 16F84 program, including initialisation, which causes the DAC to output continuously a ramped waveform, doing this by repeatedly incremented the value of Port B from 0 to $0FF_{16}$. Determine the output frequency of the ramp waveform. The clock frequency of the processor is such that each instruction cycle takes 2 μs.

3.9 A microcontroller system is to generate a sine wave, taking values from a look-up table and transferring them to a DAC. Negative values must be converted to two's complement. The table contains values from 0° to 90°, in increments of 2°. Draw a flow diagram showing how the values from the table should be accessed and manipulated in order to produce the required output.

3.10 A 16F84 is driven from a clock oscillator source of 4 MHz. Using its Counter/Timer module, explain how a regular interrupt can be achieved, to the best accuracy possible, for the intervals:

(a) every 256 μs
(b) every 1.024 ms
(c) every 4.096 ms
(d) every 64 μs
(e) every 1.00 ms
(f) every 10 ms

Give SFR settings, and write any lines of Assembler code essential for this timing activity.

3.11 (a) Within the context of Microcontroller Counter/Timers, explain the meaning of the terms: *Prescaler* and *Interrupt on Overflow*.

(b) Draw a labelled block diagram of a typical microcontroller Counter/Timer, showing clearly the principal features. Indicate those features which can be configured by the microcontroller program.

(c) A microcontroller is required, among other tasks, to generate a variable frequency sinusoidal signal. It does this by taking successive values from a look-up table in memory and transferring them to a Digital to Analogue Converter (DAC). The time delay between successive samples is proportional to an 8-bit value held in a memory location labelled PERIOD; this value may from time to time be changed. It is, however, always less than 10 ms. The sine wave frequency must be held stable whatever other tasks the controller may also be doing. Describe the software/hardware strategy that could be implemented to achieve this requirement.

**Cambridge University Engineering Department.
Module D8 Examination January 1999. Part Question.**

3.12 Complete a flow diagram for the electronic lock of Exercise 2.10. In so doing, consider carefully whether the timing requirements should be met by program loops, or application of the C/T module. Write the program in Assembler format. Build and test your completed design.

3.13 A PIC16F84 microcontroller, running from a clock frequency of 10 MHz, is to be used as a variable mark/space ratio pulse generator. The repetition rate of the pulse stream must remain constant at all times, and should be approximately 10 kHz. The mark/space (M/S) ratio is determined by a BCD (Binary Coded Decimal) thumbwheel switch, connected to the microcontroller, which can be set to any value from 1 to 9. A switch setting of 1 should give an M/S ratio of 1:9, a setting of 2 should give an M/S ratio of 2:8 etc. If the switch setting is changed, the mark/space ratio should change, without interruption of pulse repetition.

Using the data given, indicate how you would use the resources of the microcontroller to meet this requirement, including in your answer the following information:

(a) a diagram of your hardware implementation
(b) a description of how the timing requirements of the problem are to be met
(c) the setting of OPTION and INTCON registers, with explanation
(d) a flow diagram of the proposed program

Include only information which is relevant to the given problem. Details of the thumbwheel switch are given in Fig. 3.23.

**Cambridge University Engineering Department.
Module D8 Exam. January 1997.**

Thumbwheel
switch

Switch setting	Bit* 3 2 1 0
1	0 0 0 1
2	0 0 1 0
3	0 0 1 1
4	0 1 0 0
5	0 1 0 1
6	0 1 1 0
7	0 1 1 1
8	1 0 0 0
9	1 0 0 1

bit 0

bit 1
                C
bit 2

bit 3

* '1' signifies switch closure

Figure 3.23   Thumbwheel switch details (Exercise 3.13).

3.14 'Concert A' has a frequency of 440 Hz. Describe how a PIC 16F84, running with an oscillator frequency of 4 MHz, can be used to output an accurate square wave of this frequency:

(a) using a software time loop, listing the loop itself, and showing clearly how the timing has been achieved.

(b) using its C/T module, describing exactly how it is set up and applied.

3.15 Undertake the designs below, using the 16F84 or microcontroller of your choice, completing both hardware and software.

(a) *Electronic dice.* The output pattern is made up of LEDs as shown in Fig. 3.24(a). To get a new pattern, the push-button must be pressed. A random number generator can be derived using the C/T running continuously. Consider carefully whether the strategy you adopt gives true randomness.

(b) *Electronic ping-pong.* The layout of this game is shown in Fig. 3.24(b). The two 'paddles' are represented by two push-buttons. The ball is initially 'out of play', with that LED illuminated. When one paddle is pressed, the 'ball' moves away from it, by each LED in the line being lit for a certain duration in turn. Experiment with what that duration should be. The person at the other end of play must press the paddle during the time duration when the LED at that end is illuminated in order to return the ball. Otherwise the ball goes 'out of play'. Play continues in this way.

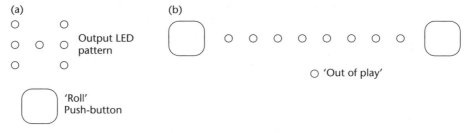

Figure 3.24   (a) Electronic dice; (b) electronic ping-pong.

Memory matters

IN THIS CHAPTER

The introduction to Chapter 1 described how the computer revolution came about, due to an explosive growth in processing power. Yet this revolution could not have taken place without a parallel growth in the ability to save information electronically. Dramatic changes in memory technology have appeared at every level of computing. The personal computer has its hard disk, floppy drive, CD-ROM[1] and megabytes of RAM.[2] Even the most modest computer today applies a host of memory technologies and tricks to store its program and look after its data. These changes have been reflected in changes at the microcontroller level. A microprocessor system of the 1980s would have had a selection of memory devices outside the microcontroller, dedicated to serving different memory needs and accessed through a maze of address decoding logic. A modern microcontroller will find most, if not all, of these memory blocks integrated onto the IC.

Unlike conventional microprocessor systems, small-scale embedded systems don't usually need to manipulate large data files, or have long and complex programs. This is important for the simple reason that a small memory can be physically small in size, and hence requires less space on the microcontroller chip and fewer address lines. In a small system like this, every byte of memory is important. While microcontroller circuits use less memory than larger computer systems, they do so with very great efficiency and inventiveness. It is the technologies and techniques applied to achieve this that we now explore.

1 Compact-Disc Read-Only Memory
2 Random Access Memory

The purpose of this chapter is to describe the current state of memory technology, and how it is applied to microcontrollers and the embedded system. Specific chapter aims are:

- to describe the structure of semiconductor memory, and review memory technologies which are important to the microcontroller designer
- to show the relative advantages of each technology
- to describe how memory technologies are applied in the embedded system
- to describe example memory devices used external to the microcontroller

The level of investigation of memory technology will be adequate to allow the reader to select the most appropriate memory type, and design with it. More detailed knowledge can be obtained from sources such as Ref. 1.3. The definitive source of information on memories is almost certainly Ref. 4.1.

4.1 A memory overview

An electronic memory is any device which can electronically store a number. The *ideal* memory has the following characteristics:

- read and write cycles are performed electrically, in negligible time, i.e. *fast*
- a large amount of information is held in a small volume/space, i.e. *dense*
- data is retained indefinitely, even when external power is removed, i.e. *non-volatile*
- negligible power consumption, with convenient supply voltage requirements
- high reliability and durability

In most cases it is only a subset of these desirable characteristics that are required for any one application. A mains-powered PC has a need for high-speed, high-density memory, but is not very power conscious. A handheld data logger, on the other hand, may require a comparatively small quantity of memory, but with very low power consumption and long-term data retention.

A variety of memory technologies have been developed to meet differing needs in the memory market. Each tries to optimise one or more of the characteristics listed above, generally at the expense of the others. A single system is likely to apply several memory technologies for differing applications. These technologies include both solid state memory (i.e. integrated circuits), and a range of magnetic storage devices, like floppy and hard disks. For the purpose of small-scale embedded systems, we consider only solid state devices.

Traditionally, solid state memory has been divided into two main types, called *RAM* (Random Access Memory) and *ROM* (Read-Only Memory). As its name implies, it is only possible to read data from a ROM – you cannot write to it. The process of writing the ROM contents takes place either at manufacture or by a distinct programming process. ROM also has the characteristic of being non-volatile, meaning that it retains its contents when the power supply is switched off.

It is possible to both read from and write to RAM, and it is therefore sometimes also called *Read–Write Memory*. The RAM name implies that, unlike certain other memory types, you can access any memory location with equal ease and speed; this is, however, also true of ROM! RAM is usually volatile, meaning its contents are lost on power-down, and for this reason it can only be used for holding temporary data. The old distinctions between RAM and ROM have however become increasingly blurred, with several recent memory types being both non-volatile and read–write.

The main memory types are illustrated in Fig. 4.1, which shows where some of the most significant memory technologies fall. Those important in the microcontroller world are described in greater detail in the following pages.

Memory devices themselves are generally classified according to their technology, as shown, their size (the number of bits or bytes they can store), the internal configuration, and the speed of access. Internally, the bits may be accessed singly, or in groups of four, eight or more. Memory for small-scale microcontroller systems almost invariably uses 8-bit locations.

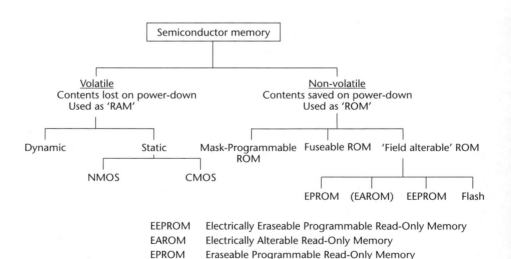

EEPROM	Electrically Eraseable Programmable Read-Only Memory
EAROM	Electrically Alterable Read-Only Memory
EPROM	Eraseable Programmable Read-Only Memory

Figure 4.1 The memory family tree.

4.1.1 The memory array

Any solid state memory device is made up of a large number of memory cells, each of which holds one bit of information. A number of fundamentally different methods are used to achieve storage in a memory cell. These are:

- a bistable (flip-flop) element (for example CMOS static RAM), *or*
- electrical charge stored in a capacitor (for example dynamic RAM), *or*
- electrical charge trapped on a conductor (for example EPROM, EEPROM, Flash), *or*
- the presence or absence of an electrical connection between two conductors (for example PROM)

A memory cell is made up of the actual memory element, plus whatever extra interfaces are needed to enable the selection and read–write process to take place, all of which are dependent on the memory technology used.

A solid state memory is made up of a large number of cells, fabricated into an array. The cells are interconnected by a network of lines, which provide a means of accessing individual cells, or groups of cells. The way this is commonly done is illustrated in Fig. 4.2, which shows the random access organisation. The cells are arranged in an array of 2^p rows and 2^q columns. An n-bit address bus, where $n = p + q$, is connected so that p address bits are decoded into 2^p lines (called *word lines*) and used to select one row at a time.

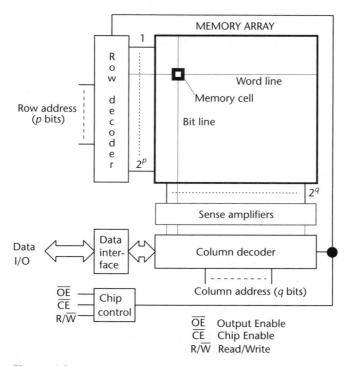

Figure 4.2 A memory array.

Data from the selected row is connected to the column lines (also called *bit lines*). This data is not at standard logic levels, and *sense amplifiers* are used to convert it to the required voltage. The desired bits are selected by the column decoder, which decodes q address bits to select one bit line, or set of lines. Each address combination therefore selects one unique memory cell or group, whose value(s) can be read, or written to in the case of RAM. A memory with n address lines will be able to address 2^n cells, or groups of cells.

This type of memory structure is used directly in memories which allow parallel data and address bus connection. For those with serial interconnection, the necessary circuitry will be required to convert serial address and data information to parallel.

WORKED EXAMPLE   4.1

The HM6264 Static RAM, detailed later in this chapter, is an 8 Kbyte (8096 bytes) device. As $8K = 2^{13}$, this indicates that 13 address lines are needed. The total memory size indicates that there are 8096×8 memory cells, i.e. 64 768. The manufacturer's data shows that the actual memory array is 256 $\times$ 256. Eight address lines are dedicated internally to the Row Decoder, and are decoded to select one of 2^8 (256) rows. Five address lines are connected to the column decoder. They can select 2^5 (32) different cell groups per row, each of 2^3 bits (i.e. each is a byte).

To be able to select and apply different memories we need a good understanding of their characteristics, in terms of speed, power, density and so on. These characteristics are primarily defined by the characteristics of the individual cell. If we can understand the cell, we will have made good progress towards understanding the overall memory. Therefore we look now at cell design for a range of important technologies.

4.2 Memory technologies 1: volatile memory

4.2.1 Dynamic RAM (DRAM)

The first form of RAM that we look at uses the 'memory' capability of a capacitor. An ideal capacitor when charged will hold its charge indefinitely, and it should be possible to read the voltage across it when wanted. This is the basis of DRAM. Here, a memory cell (Fig. 4.3) is made up of a single storage capacitor and a switching transistor. If the capacitor is charged it represents one logic state; if discharged it represents the other. If the cell transistor is switched on by the word line, then the voltage stored on the capacitor is connected to the bit line. This makes for a very simple, and hence small, device, as both capacitor and transistor can be integrated onto

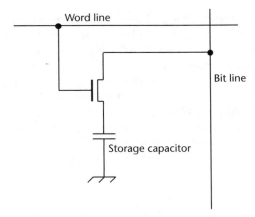

Figure 4.3 A dynamic RAM cell.

the IC. High-density memories can therefore be made at comparatively low cost.

The disadvantage of the DRAM is that the capacitor charge leaks, and the cell has to be 'refreshed', i.e. the memory state rewritten, around every 10–100 ms. This need for refresh makes DRAM cumbersome to use in small systems, and it is therefore not generally found in the microcontroller environment. It is, however, important to be aware of its characteristics, as it is the benchmark for high-density RAM.

4.2.2 Static RAM (SRAM)

CMOS SRAM cells are bistable (flip-flop) circuits, made up of two logic inverters connected 'back to back'. They are usually made up of either four transistors and two resistors (the '4T-2R' pattern) as shown in Fig. 4.4(a), or of six transistors (the '6T' pattern), as shown in Fig. 4.4(b). In either case, two of the transistors are used to connect the memory state to the bit lines, and both the bit value and its inverse are connected to the sense amplifiers.

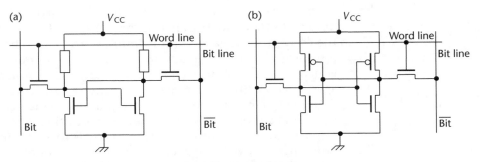

Figure 4.4 CMOS static RAM cells. (a) 4T-2R; (b) 6T.

The 4T-2R pattern allows a slightly denser memory. However, as one of the cell transistors is always switched on it consumes steady current. Moreover, as the pull-up devices are resistors, it is slightly more susceptible to interference. The 6T model gives very low power consumption, as with all CMOS logic devices the current consumption is negligible, apart from when switching. As the pull-up device is active, it is also more stable.

CMOS static RAM is widely used in the microcontroller world. It is often found embedded into controllers, where its low power consumption can be an important advantage. Its low density means that the amount of memory contained on-chip is comparatively low, for example 128 or 256 bytes.

4.3 Memory technologies 2: non-volatile memory

A semiconductor ROM is a device whose memory value is programmed into it towards the end of the manufacturing process. This is normally done at a stage when the silicon wafer is coated with a layer of aluminium. The aluminium is then selectively etched away, using a mask to determine the pattern that remains. This pattern determines the final interconnections, and hence the final memory contents. Such a memory is called a *Mask Programmed ROM*, and is very low cost when made in large quantities. It requires a long lead time to design and manufacture, and of course allows no modification whatsoever.

More versatile than the mask programmed ROM is the Programmable ROM (PROM). This can be programmed by the user, but only once! A typical arrangement uses polysilicon fuses which form interconnections in the memory array. They can be blown out at a programming stage by a high pulse of current, destroying the connection. Once again there is no chance of change once the programming has been completed, but at least the lead time is shorter.

Most embedded systems these days are in a state of more or less continuous change, where the ability to update program code is of paramount importance. For these requirements the old ROM and PROM technologies would be hopelessly restrictive. There are fortunately a number of very interesting alternative memory technologies available. These are sometimes called 'Field Alterable ROMs', and include the EPROM, EEPROM and Flash.

The three memory types just listed are all based on *floating gate* MOS technology. In this, a MOS transistor has an extra floating gate located between the conventional gate and the drain–source path. The floating gate has no electrical connection made to it, and is surrounded by silicon dioxide, which is an excellent insulator. By carefully controlling the terminal voltages, it is possible to charge electrically the floating gate. Two processes, described below, are used for this charging. Once charged, the gate will remain so almost indefinitely, unless some other action is taken to give the trapped electrons sufficient energy to recross the insulating layer. The differences between the memory types which follow lie in variations in the cell and array structure, leading to different charge transfer methods.

4.3.1 EPROM

An idealised diagram of an EPROM cell is shown in Fig. 4.5(a). Charge can be transferred to the floating gate if the drain and gate terminals are raised to a high potential (voltages around 25 V are used), while source and substrate are held at ground. A significant drain current then flows, and some electrons gain sufficient energy for them to cross the insulator and gather on the floating gate. This is called *hot electron injection*. It is a self-limiting process, as once charge builds up on the floating gate, it begins to repel further accumulation. The charge is trapped once the programming voltage is removed. This charge then alters the threshold voltage as seen by the control gate, and using normal supply voltages the transistor can no longer be turned on.

The way an EPROM memory cell is connected into an array is shown in Fig. 4.5(b). Word lines are connected to all the gates of a row of cells. If a cell is not programmed (no trapped charge), then the transistor turns on in the normal way when the word line is activated, and the associated column line is pulled low. If the floating gate of a cell *is* programmed, however, the transistor *cannot* be turned on, and the line remains high.

Once programmed, leakage from the floating gate is extremely low. EPROMs retain their programmed condition for at least 10 years; some estimates have put their data retention as high as 100 years.

If the EPROM memory is exposed to high-intensity ultraviolet light, then the electrons on the floating gate gain enough energy to move to the control gate or substrate. To allow this to take place, the memory has to be packaged in ceramic with a quartz window. The device can be erased in this way in the space of 10–30 minutes, depending on the light intensity. This effect actually takes place with any light of wavelength less than around 4000 Å, and as sunlight and fluorescent light contain components within this region, they also cause slow erasure. It is therefore good practice to cover over the quartz window when the EPROM has been programmed. A commonly used light source for EPROM erasing has a wavelength of 2537 Å.

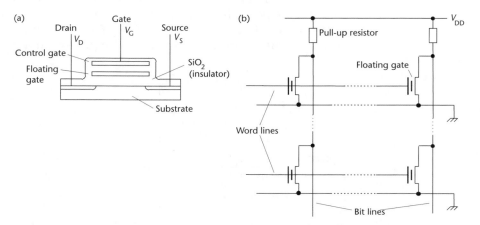

Figure 4.5 EPROM memory. (a) The memory cell; (b) part of a memory array.

To summarise, EPROM is a high-density memory, as each cell is effectively one transistor only. The technology is robust and well established. An EPROM can be programmed in single bits or bytes (and of course in a long sequence of bytes), with a comparatively high-energy programming process. Erasure is slow, and erasure of only selected parts of the memory is not possible. EPROMs have achieved very wide usage in the microcontroller world, and many microcontrollers are available with integral EPROM. The need for ceramic packaging increases their cost, especially when integrated onto a microcontroller. The need to remove them from the circuit to erase and reprogram is a further disadvantage to their use.

4.3.2 One time programmable (OTP)

A special category of the EPROM is the *One Time Programmable* memory. These are EPROMs mounted in plastic packages, with no quartz window. They can thus be programmed once, but never erased. OTP memories meet the need for a memory which is low-cost and easily programmed. They are suitable for production use, and are widely found embedded into microcontrollers.

4.3.3 EEPROM

Another mechanism for transferring electrons through the SiO_2 insulator in a floating gate structure is known as *Fowler–Nordheim Tunnelling (FNT)*. This requires that the insulating layer through which the charge transfer takes place is very thin, of the order of 10 nm or less, as opposed to the 100 nm which is tolerated by hot electron injection. As FNT takes much less current than hot electron injection, it is easier to include on-chip voltage converters to develop the supply voltages required. FNT is an effect which can be made reversible for the structures used, so that memory cells can be both programmed and erased electrically. This is the basis of the present-day EEPROM.

An example EEPROM byte is shown in Fig. 4.6. To make programming and erasure possible, each cell requires two transistors, one of which is the floating gate device, the other is for bit-select. A further transistor, shared by all cells programmed simultaneously (for example all those in one byte),

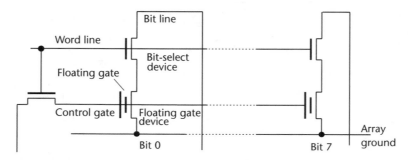

Figure 4.6 An EEPROM byte.

is required to switch the control gates. A single byte of an EEPROM programmable in bytes is therefore made up of 17 transistors. The overall Array Ground is also switchable.

To erase a byte, the common control gates are driven to the programming voltage V_{PP} (around 20 V), and the Array Ground is connected to 0 V. The high voltage between control gate and Array Ground causes electron tunnelling to the floating gate. This acquires a net *negative* charge, which it retains. During any subsequent read operation, the floating gate transistor cannot be turned on, and the cell will read as a logic 1.

To program a cell from 1 to 0, the control gate is connected to 0 V, and the row select and bit lines to V_{PP}. The Array Ground is not driven. Now a large voltage appears across the gate to drain of the floating gate transistor, and electrons tunnel from floating gate to drain region. The floating gate now has a net *positive* charge, and the transistor will conduct during a read operation.

To read an EEPROM cell, the word line is taken to the normal supply voltage V_{DD}, which enables the bit-select transistor. Any floating gate device which has been programmed, i.e. now has a positive charge on its gate, will conduct due to that charge. This pulls the bit-line low. An unprogrammed floating gate device will not conduct, and the bit-line will remain high.

An EEPROM memory has on-chip DC–DC converters to generate the extra voltage supplies needed. The programming action requires the programming voltage to be maintained for a significant time period (i.e. some milliseconds). Programming and erase cycles tend to take different time durations, and the time can vary from cell to cell and depend on cell age (as charge can become trapped in the insulating oxide layers). Therefore it is usually necessary to test and flag successful writes.

Because EEPROM programming and erasure takes place electrically, there is a risk of accidental programming on the event of power change (power-up, brown-out, power glitch) or controller program crash. Extensive precautions are therefore normally designed into EEPROM devices to avoid this. In the PIC 16F84, for example, the EEPROM write procedure requires that the codes 55_{16} and $0AA_{16}$ are written in turn to the EECON2 SFR before a memory write can take place.

It is important to note that, unlike many semiconductors, EEPROMs display a wear-out effect, and are likely to fail after extended use. Two common failure mechanisms are *charge trapping* and *tunnel oxide breakdown*. The first of these is a cumulative effect, in which charge becomes trapped in the oxide layers isolating the floating gate. This extends the time required to program and erase until these processes can no longer be achieved in the allotted time. Tunnel oxide breakdown is due to tiny defects in the oxide which distort the electric field and cause large localised gradients. Current is then not evenly distributed, leading to stresses in the high current areas, and ultimate breakdown when the insulating layer becomes short circuited.

For the reasons just described manufacturers usually quote an *endurance* figure for an EEPROM memory, expressed as the maximum number of permissible erase/write cycles. This figure applies to each memory location;

an endurance rating of 10 000 indicates that *every* location in the memory can be subjected to that many erase/write cycles. If an application uses a small number of EEPROM locations intensively, it is worthwhile using a larger memory than is apparently necessary and rotating memory location usage across it. Read cycles create negligible stress, and are not subject to any limit. A useful review of this topic is given in Ref. 4.2.

EEPROM is a very popular memory type in the microcontroller world, appearing both as memory embedded into a controller or as physically small serial devices.

4.3.4 Flash

Of the memory technologies surveyed here, Flash is the most recent, and the one developing the fastest. It represents a synthesis of EPROM and EEPROM, in that like EEPROM it is electrically programmable and erasable, but like EPROM it only needs one transistor per cell. *Unlike* EEPROM, it is not byte erasable – erasure takes place simultaneously for a whole block of memory. Because it is a new and evolving technology, there are different versions of Flash available (Refs. 4.3 and 4.4), together with significant differences between 'first generation' and 'second generation' memories. These differences appear not only in memory size, but also in supply voltage and current requirements and the sophistication of the internal 'memory management' capability.

Perhaps the most common Flash cell is known as ETOX[3] (EPROM Tunnel Oxide), developed by Intel. The cell, shown in Fig. 4.7(a), is very similar to the EPROM cell, except that it is not symmetrical, with the source diffusion being greater that the drain. Cells are connected into an array in the same pattern as EPROM, i.e. the interconnection shown in Fig. 4.5(b) is applied. All the Source terminals for one block of memory are interconnected.

Programming with this cell is by Hot Electron Injection, and erasure by FNT. Example bias conditions for read, write and erase are summarised in the table of Fig. 4.7(b). The necessary conditions for programming are achieved by biasing the bit line (connected to the Drain) to the required level, and pulsing the word line (connected to the Gate).

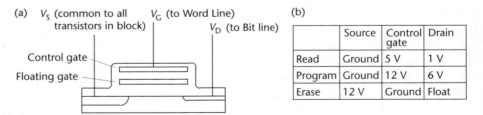

	Source	Control gate	Drain
Read	Ground	5 V	1 V
Program	Ground	12 V	6 V
Erase	12 V	Ground	Float

Figure 4.7 Flash memory. (a) The memory cell; (b) operating bias conditions.

---

3 ETOX is a trademark of Intel.

The distinguishing feature of Flash memory is its erase capability. To achieve this, all transistors are initially programmed to place them in the same condition. The (common) source connection is then pulsed high, with Control Gate grounded and Drains floating. This causes FNT to take place from the Floating Gate to the Source. Different cells will have different erase performance, and after each erase pulse the whole block is read to test for a successful erase. If not successful, another pulse is applied.

Because of the danger of accidental programming/erasing, as with EEPROM, it is necessary to take significant precautions to ensure that only 'legal' operations are allowed. Some of these data protection features, applied to the Atmel AT29C010A Flash memory, are described in Section 4.5.4.

Flash memory has similar density to EPROM. At the time of writing, ICs are readily available with capacity up to 16 Mbit, and this continues to increase. Programming and erase algorithms are generally handled by an on-chip microcontroller. Such memories are fabricated as a group of blocks or sectors, each of which can be independently erased.

Microcontroller manufacturers are increasingly including Flash program memory on-chip as a replacement for EPROM. Because of the simpler packaging requirements, this turns out to be much cheaper than the EPROM equivalent, and is even approaching OTP in price. It is therefore predicted that Flash will replace EPROM and even OTP altogether. It is also expected to replace EEPROM, except for those situations where the byte-programmability of EEPROM remains indispensable.

4.4 Microcontroller memory implementation

Having reviewed the memory technologies, this section considers some of the issues involved in their application.

4.4.1 Memory implementation – on-chip or off?

Broadly speaking, there are three major options for locating memory in a microcontroller-based system:

1. On-chip memory, if present at all, is inadequate to complete any system; the user *must* add external memory.

2. Some on-chip memory is provided, adequate to complete a minimum system; address and data buses are bonded out (i.e. they are connected to external pins); the user *can* add external memory.

3. On-chip memory is provided, adequate to complete a viable system. Address and data buses are not 'bonded out'. The user *has no chance* to extend the on-chip memory within the memory map.

Of our example microcontroller families, the 16F84 and 68HC05/08 are self-sufficient in memory, and fall into the third option. The Philips 8XC552 has versions in both options 1 and 2.

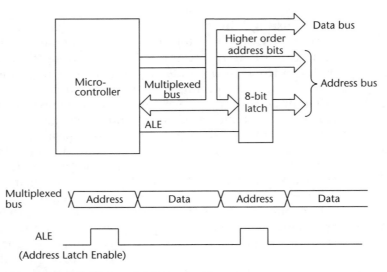

Figure 4.8 Multiplexed data and address buses.

The first option is typical of many older processors. There may be a little static RAM on the chip, but no program memory (which is more costly and complex to incorporate). The user must decode[4] the address bus, and design a memory map appropriate to her needs.

The second option implies the inclusion of program memory on-chip. Such devices have been available for more than 20 years, but have been until recently comparatively costly. It represents a flexible option, as it allows the controller to be used as a single chip, but also allows expansion if needed. The burden of bonding out address and data bus pins is, however, high; it inevitably increases the IC pin count, and hence its size. To make some reduction in the pin count, it is common to find that buses themselves are multiplexed, i.e. one set of IC pins is shared between two buses. For example the lower 8 bits of the address bus may be multiplexed with the data bus. The multiplexed bus is used for each of these functions in turn; a control line from the controller indicates which function is active. In the example of Fig. 4.8, a line called 'Address Latch Enable' (ALE) is used to identify when address bits appear on the bus, and latch them into an external latch. This approach of course does nothing to speed up a processor, indeed it exacerbates the 'von Neumann bottleneck'.

The third option above is preferred for the minimum system. It is technically quite feasible with present-day technology to integrate all the memory types so far described onto the microcontroller IC, and the most recent devices take full advantage of this opportunity. With no buses bonded out, the chip and the surrounding circuitry can be extremely small

---

4 It is assumed that address decoding forms part of the prerequisite knowledge for this book. A summary is however given in Appendix 4.

and compact. The very compactness of the device now, however, carries its own challenges. It remains necessary to program and then verify the memory, and pins must be made available for this. Also, when it comes to test, the inaccessibility of the buses makes the task that much more difficult.

Even when external connection cannot be made to the buses, it is still possible to provide memory expansion, though not within the system memory map. The practical way of doing this in any system which is not too speed conscious is by means of serial memory. While parallel data and address buses provide fast and efficient external memory access, they carry the major disadvantage of requiring a large number of microcontroller pins, as well as all the interconnection needed between controller and memory. Serial memory devices require very few pins for data interconnection, and are now very popular for applications where minimum physical size, rather than top speed, is the prime concern. Many suitable memory devices are now available, with both data and address information transmitted serially. The full details behind this will be described in Chapter 6. At the end of this chapter we will, however, see one example of a serial memory: the Xicor X24165.

4.4.2 Program memory

Let us pause for a moment to consider what characteristics we actually want from our program memory, bearing in mind the market forces that are driving development. There is an overwhelming need to get the product to market. While the (supposedly) finishing touches are being made to the program, hardware production may have started. Despite the best efforts of the program developers, residual bugs are still being found, and even when the product goes on the market, some further bugs appear. Then, within months, a new program release is made.

This leads to the inescapable conclusion that a high level of flexibility, in the ability to program and reprogram easily, is a most desirable attribute of program memory. The ability to delay programming memory to the last possible moment, to reprogram it even then, and to offer program updates in the field, are all very attractive capabilities.[5]

The historical progression for program memory has been essentially from Mask Programmable ROM and PROM, to EPROM, to Flash. ROM offers low unit cost when used in high numbers, but its lead time requires the program to be finalised well in advance of production, and allows no chance of last-minute revisions or fixes in the field.

EPROM goes some way towards meeting the demand for flexibility, and it has found very wide use in microcontroller systems. This is both as a

---

5 The most spectacular example of reprogramming 'in the field' was with the Mars Pathfinder space mission in July 1997. When the system crashed, *on Mars*, the software developers on Earth were able to identify the problem and transmit a program fix up to Mars! It worked.

Table 4.1 Microcontroller price comparisons for different program memory technologies.

80C552	ROMless, plastic	£14.08
87C552-4A68	8k OTP, 68 pin PLCC, plastic	£36.70
87C552-4k68	8k EPROM, 68 pin CLCC, ceramic	£206.29 (actual)
MC68HC705C8ACP	7744 bytes OTP 40 pin, plastic	£8.89
MC68HC705C8ACS	7744 bytes EPROM 40 pin, ceramic	£16.40
16C71	1K×14bit OTP, plastic	£3.54
16C71	1K×14bit EPROM, ceramic	£10.84

standalone device (see the 27C128 EPROM in Section 4.6) and embedded on the controller. As standalone memories their price fell over many years, despite the cost implications of the specialised packaging. It is comparatively easy in production to program them with an EPROM programmer and then insert them into the target circuit. Reprogramming, including in the field, can be done with a simple memory IC swap, with the replaced device being erased for subsequent reuse. EPROM can also be integrated onto the microcontroller IC. In this case the whole IC is packaged in ceramic, and the price goes up dramatically. Sample microcontroller prices for comparison purposes, quoted for unit quantities by a well-known UK supplier, are shown in Table 4.1. For purely production purposes, where no further change is anticipated, EPROM in the form of OTP is used.

As EPROMs more or less have to be removed from the embedded system for erasing, they do not permit genuine *in situ* reprogramming. This is where Flash (and for a while, EEPROM) comes into its own. Both Flash and EEPROM can be erased and programmed without removing the IC containing the memory from the system. Therefore it becomes possible to manufacture the hardware, insert all the ICs, and *then* program the memory, while it is in the end product.

Flash, however, goes much further than being just a technical improvement on EPROM. Because it is so easy to reprogram in-circuit, this can now be done remotely. For an Internet-linked device, for example, the manufacturer can download system firmware updates, maybe charging for it at the same time, without the device being removed from its working environment. Because of the importance of this type of memory programming, we look at it in further detail in the following section, using the 16F84 as an example.

4.4.3 16F84 Memory programming

The 16F84 on-chip non-volatile memory comprises Flash program memory, configuration word, identification data and EEPROM data storage. *All* of these can be programmed serially. The last-named can of course also be written to as part of normal microcontroller operation.

Table 4.2 16F84: programming memory map.

Address (Hex.)	Function
0000–03FF	Program memory
2000–2003	Identification (ID)
2004–2006	Reserved
2007	Configuration

The serial programming mode uses pin RB6 as clock, RB7 for bidirectional data and $\overline{\text{MCLR}}$ to signal entry to programming mode. The memory mapping for the purposes of programming is shown in Table 4.2.

A summary of the operating conditions for programming is shown in Table 4.3 (Refs. 4.5 and 4.6). It should be noted that the supply voltage requirement is more stringent than for normal operating conditions, the supply current is greater, and a higher voltage is required on the $\overline{\text{MCLR}}$ pin. It is therefore possible, if programming in-circuit, that the normal supply voltage will be inadequate for the task.

The program/verify mode is entered by holding pins RB6 and RB7 low, while raising $\overline{\text{MCLR}}$ from a logic low value to its program mode value of 13 V ± 1 V.

A typical circuit applied for in-circuit programming is shown in Fig. 4.9. As the $\overline{\text{MCLR}}$ pin is taken well above V_{DD}, it must be isolated from the rest of the circuit, which is the function of the diode. Circuit connections to RB7 and RB6 must not load the programmer, and again it may be necessary to isolate. For full programming verification, it must be possible to control the value of V_{DD}, so that correct programming can be tested at the extremes of operating voltage. For example, if a battery-powered system is to operate from 3 V to 4.2 V, it is necessary to check correct programming at these two extremes, while still maintaining the conditions of Table 4.3 when programming. A programmer which does not have this capability is categorised as being for prototype purposes only, and should not be used for production programming.

The programming action is controlled by the use of a small number of 6-bit command words, transmitted on the serial link, which instruct the processor to take the following actions:

Table 4.3 Selected 16F84 program/verify characteristics.

Supply voltage for programming	5 V ± 10%
Supply current for programming	50 mA max.
Supply voltage for verification	Maximum and minimum supply voltage for part
Voltage on $\overline{\text{MCLR}}$ for test mode entry	13 V ± 1 V
Supply current to $\overline{\text{MCLR}}$	200 µA (max.)
Erase cycle time	10 ms
Program cycle time	10 ms

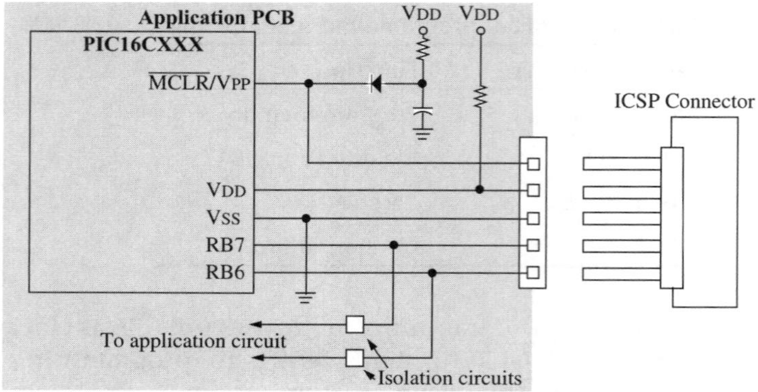

Figure 4.9 Connection for 16XXX in-circuit serial programming. Reprinted with permission of the copyright owner, Microchip Technology Incorporated © 2001. All rights reserved.

- load data for program memory
- load configuration data
- load data for EEPROM memory
- read data from program memory
- read data from EEPROM memory
- begin programming
- increment Program Counter address
- bulk erase program memory
- bulk erase EEPROM data memory

A new command must be issued for each word which is to be read in or out. The first three commands listed must each be followed by the appropriate data word, clocked in through the serial link. The next two commands cause the contents of the current memory location to be read back *out* of the microcontroller. The last four cause the actions stated. The current memory location is determined by the Program Counter, which is reset to 0 on power-up.

Each word is programmed and then verified in turn. A typical *program and verify* cycle is shown in Fig. 4.10. If the data output from the read operation is not the same as the data transferred for programming, the action is aborted. Even if this occurs for one location, the Program Counter can be incremented, and attempts to program other locations can still be made.

It can be seen that, although attractive, in-circuit serial programming for the 16F84 is not without some cost. There are added power supply requirements, as well as some possible increase in circuit complexity.

4.5 External memory devices

We close the chapter by looking at a selection of standalone memory devices, chosen to give examples of certain popular types. It is useful to compare the characteristics of each of these devices with their memory cell

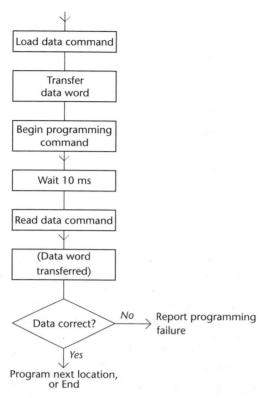

Figure 4.10 Program and verify cycle.

characteristics, as described earlier. To actually make use of any of them the reader must of course refer to the manufacturer's data sheets for full details. Where device characteristics are given, these are typical values. Performance is very dependent on operating conditions, and different versions may offer somewhat different speed and power capabilities from those given. Dual Inline plastic package outlines are shown in every case; each memory is, however, also available in one or more other (more compact) packages.

4.5.1 EPROM: the AMD (Advanced Micro Devices) 27C128

The 27XXX series of EPROMs is one of the most long-lived and successful of memories. It appeared in the very late 1970s with devices of small capacity, and multiple power supplies (the 2708 had a capacity of 1 Kbyte, and +12 V, +5 V and –5 V supply rails). The family soon appeared in single 5 V supply and in CMOS versions. It continued growing, with numerous manufacturers offering parts, of ever-increasing capacity. The pinout pattern used by the family has been adopted by many other memory types.

Memories like the 27C128 (Fig. 4.11), shown here to illustrate the principles of the family, have provided data storage for embedded systems for many years. They will continue to do so for a number of years to come,

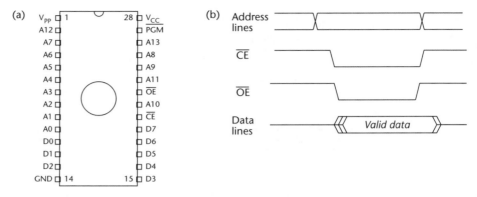

Figure 4.11   The 27C128 EPROM. (a) Pinout; (b) read cycle.

although they are increasingly being superseded by Flash memories. The family is extremely well established, low in cost, and has current consumption figures which make it a possibility for power-conscious applications. Within the family AMD now offer EPROMs from the 27C64 (8 Kbyte) to the 27C040 (4096 Kbyte). Other manufacturers continue to offer the smaller devices (for example the 27C32). While these smaller EPROM memories now offer little price advantage, their power consumption is less and their use is worth considering for this reason.

Device highlights (Refs. 4.7, 4.8)

Size	16 Kbyte
Power supply	5 V ± 10%
Access time	In range 45 ns to 200 ns
Supply current (active, switching frequency = 10 MHz)	25 mA typ.
Supply current (standby, CMOS level inputs)	100 µA typ.
V_{CC} Supply current, Program (V_{CC} = 6.25 V)	50 mA
V_{PP} Supply current, Program (V_{PP} = 12.75 V)	30 mA

The read cycle of this memory is straightforward, and is shown in Fig. 4.11(b). To program, the V_{CC} and V_{PP} pins are raised to the Program voltages shown in the Device Highlights table. Address and Data values are presented to the parallel inputs, and the $\overline{PGM}$ pin is then pulsed low for 100 µs. With the address lines unchanged, a memory read then takes place, i.e. following Fig. 4.11(b). If the data stored is found to be valid, programming proceeds to the next address. If not, another programming cycle takes place. The memory is deemed to be faulty if the data is still not correctly stored after 25 program cycles. Programming is flexible; single bytes may be programmed, or any selection of bytes, in any order. As the erased state of the memory is that all bits are at a logic 1, it is even possible to program single bits (all other bits in the byte being set to 1).

Erasure takes place by exposing the memory to a UV source, typically of wavelength 2537 Å and intensity 12 000 µW/cm^2, for a period of 15 to 20 minutes. The whole memory is erased.

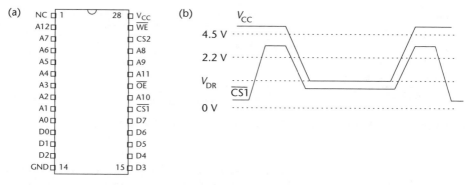

Figure 4.12 The HM6264 CMOS static RAM. (a) Pinout; (b) Low-power data retention.

4.5.2 Static RAM: the Hitachi HM6264

This is an 8 Kbyte static RAM, whose pinout is shown in Fig. 4.12(a). Similar devices are offered by a number of manufacturers, including Toshiba and Cypress. The pinout is very similar to the 2764 EPROM, giving flexibility in large systems (a user may choose to fill a memory location with EPROM or with RAM). The memory has the potential for very low power consumption when in standby, yet is moderately power-hungry when active.[6] A valuable feature of this type of memory is its data retention mode, illustrated in the timing diagram of Fig. 4.12b). With the chip deselected, the supply voltage can be dropped to a value specified by V_{DR}. In this state the memory cannot be accessed, but data will be retained indefinitely, at a very low power consumption indeed. This is the basis of battery-backed RAM systems.

Device highlights (Ref. 4.9)

Power supply	5 V ± 10%
Average operating power, typical	15 mW/MHz
Operating supply current, typical, $\overline{CS1}$ ≤ 0.8 V	7 mA
Standby supply current at 25 °C	2 µA
Access time	100/120/150 ns versions available
Minimum V_{CC} for data retention (V_{DR})	2 V
Fully static operation	

4.5.3 EEPROM: Xicor X24165

It has already been mentioned that the need to reduce physical sizes in many embedded systems has led to a move away from parallel data interconnection.

---

6 The average operating power gives an indication of power consumption when the memory is being continuously accessed at a certain frequency. The operating supply current gives the DC current consumption for the period when the memory is selected. Certain constraints on these figures appear in the manufacturer's data.

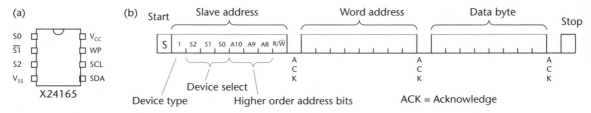

Figure 4.13   Xicor X24165 EEPROM. (a) Pinout; (b) byte write cycle.

The Xicor X24165 is an example of a serial EEPROM memory device. Its pinout is given in Fig. 4.13(a). The most striking feature of this device, when compared to all other memories in this section, is its very small size. Making use of serial data communication reduces the 19 pins that would be required for data transfer (8 data and 11 address) to 2. The trade-off is the greatly increased read and write times required.

Device highlights (Ref. 4.10)

Size	2 Kbyte	Power supply	2.7 V to 5.5 V
Active read current	< 1 mA	Write current	< 3 mA
Standby current	< 1 μA	Typical write cycle time	5 ms
Max. clock speed	100 kHz	Endurance	100 000 cycles

The X24165 uses a serial protocol similar to the Inter-Integrated Circuit (I^2C) bus described in Chapter 6. The data link uses a serial data line, SDA, and a clock line, SCL. It has three memory select bits, S0, $\overline{S1}$ and S2, which can be hard-wired high or low, or they can be put under microprocessor control. Eight units of this memory can therefore be connected to the same bus, each with a different address.

To illustrate the serial data communication, a 'Byte Write Cycle' is shown in Fig. 4.13(b). Transfer is initiated by a 'start condition', defined by the SDA line falling low while SCL is high. The following slave address byte contains both the IC identifier and the 3 highest address bits, as well as the R/W bit. If the memory recognises the start condition *and* its slave address it will acknowledge by pulling the SDA line low. It will subsequently acknowledge every byte received if Write has been selected. The second byte transmitted contains the Word Address, which is followed by a data byte. A Stop condition (SDA rising while SCL remains high) causes the memory to start its own internal write cycle to the non-volatile memory. During this time the IC will not respond to any requests from the controller. The memory then enters its low-power mode. A Single Byte Read operation follows a similar pattern to this. In addition, repeated read or write operations to sequential locations can be undertaken, for which only the start address need be specified.

The X24165 contains a number of features which provide write protection. A Write Protect register is located at the highest address, and settings of this allow individual blocks of memory to be write-protected. One bit also enables the hardware write protect pin (pin 7 of the IC). Using this, it is possible to program the memory in-circuit, and then enable the write protect pin. Once this is done, selected blocks in the memory are permanently protected.

4.5.4 Flash: the Atmel AT29C010A

This 1 Mbit device is representative of current standalone parallel access Flash memory chips. Its pinout is shown in Fig. 4.14, and follows the pattern adopted by the 27XXX EPROM memory family described earlier. It is one of a family of CMOS Flash memories, whose size ranges from 32 Kbyte to 512 Kbyte and which are available in 3 V and 5 V versions.

Device highlights (Refs. 4.11, 4.12)

Size and organisation	128 Kbyte, arranged in 1024 128-byte sectors; includes two 8 Kbyte Boot Blocks
Programming voltage	5 V
Read access time	70 ns
Sector program cycle time	10 ms
Active current	50 mA
Standby current (CMOS level inputs)	100 µA
Endurance	> 10 000 cycles

Sector program operation
Buffered programming
Hardware and software data protection
Flag polling for detection of end of programming
Optional chip erase

Memory organisation is in 1024 sectors. Address bits A16 to A8 specify the sector, while the lower bits specify the byte. While reading from the memory is as simple as from a parallel EPROM or SRAM, writing presents some interesting challenges. Any sector can be reprogrammed, but in so doing, the whole sector must be rewritten. This means that even if only a single byte needs to be rewritten, the whole sector must be reloaded,

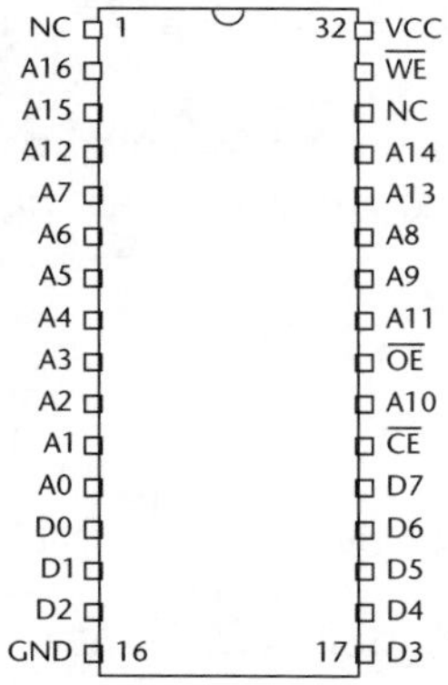

```
      NC ▢ 1       ᴗ    32 ▢ VCC
     A16 ▢                  ▢ WE
     A15 ▢                  ▢ NC
     A12 ▢                  ▢ A14
      A7 ▢                  ▢ A13
      A6 ▢                  ▢ A8
      A5 ▢                  ▢ A9
      A4 ▢                  ▢ A11
      A3 ▢                  ▢ OE
      A2 ▢                  ▢ A10
      A1 ▢                  ▢ CE
      A0 ▢                  ▢ D7
      D0 ▢                  ▢ D6
      D1 ▢                  ▢ D5
      D2 ▢                  ▢ D4
     GND ▢ 16          17 ▢ D3
```

Figure 4.14 Atmel AT29C010A 1 Mbit CMOS Flash memory.

otherwise any unspecified byte will have an indeterminate value. Data is loaded into a buffer memory using conventional memory write timing. However, if more than 150 µs elapses after a byte write cycle, the device enters a sector program mode and then programs the contents of its buffer memory into the selected sector (having first erased it). This process takes approximately 10 ms. During this time data bits 7 and 6 are used to indicate memory status, and can be polled to determine if the programming process is complete.

Extensive data protection features are included. In hardware the program function is inhibited if the supply voltage falls below 3.8 V. There is also a power-on delay of approximately 5 ms, during which the programming function is again inhibited. In software, it is possible to enable a protection mode whereby a certain sequence of bytes must be transmitted to the memory before programming is permitted.

The Boot Blocks are designed to provide a higher level of security than the remainder of the memory. There is a programming lockout feature for these blocks which, if enabled, inhibits any further programming of the block. Once activated (by sending a sequence of specific code bytes to the memory), the block can no longer be erased or programmed, and the chip erase feature is no longer available. The blocks are located at the top and bottom of the memory space, and so can easily be used on microcontroller/processor applications for program memory.

SUMMARY

1. Embedded systems make widespread use of CMOS SRAM, EPROM, EEPROM and Flash memory technologies. The particular characteristics of each need to be understood for their successful application.

2. All the memory types listed directly above can be readily integrated onto the microcontroller IC. With EPROM this raises the cost considerably, due to the packaging requirements.

3. Flash memory, with its high-density, electrical reprogrammability and non-volatility, is having a major impact on the world of memory. It is likely to replace EPROM and OTP completely, and push EEPROM into certain niche areas.

4. Where memory is not integrated onto the microcontroller IC, external memory devices can be applied. There is increasing use of serial memories for this purpose.

REFERENCES

4.1 Prince, B. (1996) *Semiconductor Memories: A Handbook of Design, Manufacture and Application*. Chichester: John Wiley & Sons.

4.2 *Serial EEPROM Endurance*. Application Note AN537. Microchip Technology Inc. Document DS00537B

4.3   Giridhar, B. V. (1996) Flash technology: challenges and opportunites. *Japanese Journal of Applied Physics*, **35**(1), 6347–51.

4.4   Pavan, P., Bez, R., Olivo, P. and Zanoni E. (1997) Flash memory cells – an overview. *Proceedings of the IEEE*, **85**(8).

4.5   *PIC 16F8XX EEPROM Memory Programming Specification*. Microchip Technology Inc. 1998. Document DS30262C.

4.6   *In-Circuit Serial Programming™. Mid-Range Reference Manual*, Section 28. Microchip Technology Inc. 1997. Document 31028A. http://www.microchip.com/.

4.7   *Am27C128 Data Sheet*. Advanced Micro Devices. May 1998. AMD Publication #11420. http://www.amd.com/.

4.8   *Programming AMD's CMOS EPROMs*. Advanced Micro Devices. January 1997. AMD Publication #19840. http://www.amd.com/.

4.9   *HM6264A Series Data Sheet*. Hitachi Semiconductors. http://www.hitachi.com/products/electronic/semiconductorcomponent/.

4.10 *X24165 Data Sheet*. Xicor Inc. 1996. http://www.xicor.com/.

4.11 *AT29C010A Data Sheet*. Atmel. http://www.atmel.com/.

4.12 *Atmel AT29 Flash Memories*. Application Note AN-1. http://www.atmel.com/.

EXERCISES

('+' indicates a simple review question)

+**4.1**   What are the different ways in which one bit of data can be stored electronically?

+**4.2**   (a) How many memory cells are there in a 16 Kbyte memory?
          (b) A memory contains 131 072 bits, organised into locations of one byte each. How many bytes does the memory have, and how many lines are needed to address it?

+**4.3**   Place the following memory technologies in order of cell size: Flash, CMOS Static RAM, EPROM and EEPROM.

+**4.4**   Explain why EPROM has been an extremely popular memory technology. What are its disadvantages?

+**4.5**   Flash memory is rapidly becoming the dominant memory technology. To what extent do you believe it will replace EPROM, CMOS Static RAM and EEPROM?

4.6   The floating gate of a certain Flash memory cell has 5 fC of charge deposited on it. Estimate the maximum average acceptable leakage rate of charge, in electrons per day, if the charge loss must be less than 10% in 10 years. (Electron charge is 1.6×10^{-19} C).

4.7   Assuming all the supply current of the HM6264 is consumed in the memory cells themselves when in standby, what is the maximum permissible current per cell in this mode?

4.8   A HM6264 is accessed once every 10 ms by a program. Initially the access took 1 ms, which was then reduced to 4 μs. What is the average power consumption of the memory device in each case? (Apply the

Operating Supply Current figure when the memory is selected, and the Standby Supply Current figure when it is not.)

4.9   A certain 40-pin microcontroller has an 8-bit data bus and 16-bit address bus. The lower eight bits of the address bus are multiplexed with the data bus. The manufacturer offers a new version of it in which all memory is now on-chip, and the address and data buses are no longer to be bonded out. All other functionality of the device (including all input/output connections) is to remain unchanged. How many pins would you expect the new device to have? State any assumptions made.

4.10   A microprocessor-based data logger reads sunlight intensity, air temperature, humidity and wind speed four times every hour, storing each variable as an 8-bit number. Once an hour it stores time data, which takes 4 bytes. If it is to run for one month without downloading data, what data storage memory size should it have? Advise on the type of memory that could be used.

4.11   Write a study of around 500 words on the requirements of program memory in a small-scale embedded system. Take note of both the characteristics of the memory itself, as well as the relative advantages of how it might be physically implemented in the system.

4.12   One byte per 8 ms is written to each of the Xicor X24165 and Hitachi HM6264. Data to the former is transmitted with a clock rate of 100 kHz; for the latter, the chip is enabled for 1.6 μs for each write. The memory is otherwise inactive. Using only the data given in the descriptions of the memories, estimate the worst-case power consumption of each device.

Analogue affairs

IN THIS CHAPTER

Although microcontrollers are essentially digital devices, they spend a good deal of their time interfacing with analogue signals. They must be able to convert input analogue signals, for example from a microphone or a temperature sensor, to digital form, and they must be able to convert digital signals to analogue form, for example if driving a loudspeaker or DC motor. These processes are broadly covered by the heading of 'data conversion'.

While conversion in both directions is necessary, it is conversion from analogue to digital form that is the more challenging task. It is also more widely applied, as once a signal is digitised it is usually preferable to keep it in that form. In many cases of analogue to digital conversion we find that the analogue signal must be conditioned in some way before digitisation can take place. The overall system for digitising a signal may therefore itself include several distinct processes. Taken together, it is called a data acquisition system. Some of the principles of the data acquisition system are sophisticated, and involve an advanced understanding of electronics and data processing. In the world of the microcontroller, most, if not all, of the data acquisition system may be embedded on the microcontroller. It nevertheless remains important to understand the principles, as it is easy otherwise to use even an apparently simple system below its optimum performance.

In this chapter we enter this very large field of data conversion. Important data conversion principles are introduced, and then applied to the microcontroller and the embedded system.

The aims of the chapter are:

- to describe means of deriving analogue outputs from a microcontroller system

- to review relevant elements of the data acquisition systems, and show how these should be specified

- to identify possible sources of error in the data acquisition system

- to review techniques of analogue to digital conversion relevant to the microcontroller environment

- to examine examples and applications of Analogue to Digital Converters integrated onto microcontrollers, or used in association with microcontrollers

5.1 Digital to analogue (D to A) conversion

A Digital to Analogue Converter (DAC) is a circuit whose analogue output depends on its digital input, and an associated reference voltage. This is represented in Fig. 5.1. The digital input may be presented in parallel or serial form, and the transfer of data and the conversion process may be controllable by certain digital control lines. Usually the reference voltage is fixed and the output directly proportional to the digital input, as shown in Equation (5.1). Here V_O is the output voltage, D is the n-bit digital input number, k is a converter constant (which may be unity), V_r is the voltage reference, and V_{os} is a possible offset voltage. The maximum value that D can take is $(2^n - 1)$, so the term $(D/2^n)$ always has a value less than 1. The term kV_r represents the output range.

$$V_O = \frac{D}{2^n} kV_r + V_{os} \qquad (5.1)$$

The smallest change in digital input that can occur is the change of one least significant bit (lsb). This leads to the smallest possible change in analogue output, which is referred to as the *resolution* of the converter. If D is an n-bit number, then the resolution is given by:

$$\text{resolution} = (1/2^n) \times kV_r \qquad (5.2)$$

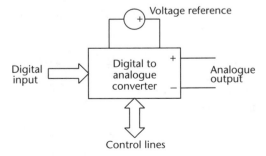

Figure 5.1 The Digital to Analogue Converter.

Resolution is in practice quoted in three ways – as a number of bits, as a percentage, and as a voltage. For example, an 8-bit D to A converter with an output range of 5 V may be stated as having an 8-bit resolution, or 0.39% (100%/256) resolution, or a resolution of 19.5 mV (5 V/256).

The D to A conversion process inevitably introduces errors, the greatest of which is often due to voltage reference instability. It is useful to note that if these are kept to less than one half of the resolution, then their effect can be treated as negligible. Greater detail on the principles of D to A conversion can be found in Ref. 5.1.

5.1.1 The standalone converter

There are a number of ways of designing DACs. The most common uses the 'R–2R ladder' (Refs. 1.1 or 5.1). This gives a fast response and good accuracy. Few microcontrollers have on-board DACs of this form, so to achieve the sort of performance they give the designer will often have to turn to an external device. There are many of these available, in highly integrated and easy to use form. A good example, with 12-bit resolution and low power consumption, is the Maxim MAX538, summarised in Fig. 5.2. This is easy to use, physically small, and requires only the addition of an external voltage reference. Its output voltage is very similar to Equation (5.1), and is given by:

$$V_{\mathrm{o}} = \frac{D}{2^{12}} V_{\mathrm{r}}$$

where V_{r} is the voltage applied at pin REFIN. The MAX539 is identical, except that it has a gain of 2. With a reference voltage of 2.048 V for example, the MAX538 resolution is 0.5 mV, and its maximum output value is 2.0475 V.

A similar DAC, of higher resolution, is the 16-bit DAC714, made by Burr Brown and used in the Case Study of Appendix C.

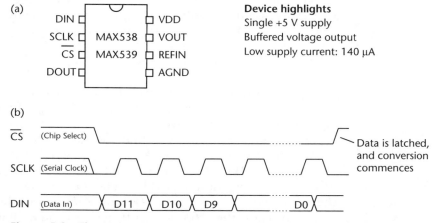

Figure 5.2 The MAX 538 – a serial DAC. (a) Pinout and device highlights; (b) data transfer timing diagram.

(a)

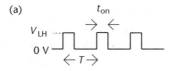

(b)

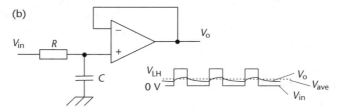

Figure 5.3 Using PWM for D to A conversion. (a) PWM signal; (b) filtering the PWM signal.

5.1.2 Pulse width modulation (PWM)

External DAC ICs such as the MAX538 provide good data conversion, at the limited extra expense of an IC and some serial interfacing. If the specification is not demanding, however, it is possible to achieve D to A conversion almost direct from a microcontroller digital output. This alternative technique is PWM.

PWM provides a means of controlling analogue voltages and currents with a digital waveform. Its commonest application is in the control of power flow, for example in motor drives or switched mode power supplies. It can, however, be viewed and used as a DAC, and it is in that guise that we consider it here.

Essentially, a PWM signal is a rectangular wave (for example that of Fig. 5.3(a)), with fixed period T, but variable 'on' time t_{on}. The logic high value is V_{LH}, and for simplicity the logic low value is assumed to be 0 V. If the rectangular wave of Fig. 5.3(a) is applied as V_{in} to the simple filter circuit of Fig. 5.3(b), then the voltage at V_o will appear as shown, with an average value of V_{ave}. This average value is given by:

$$V_{ave} = \frac{t_{on}}{T} \times V_{LH} \tag{5.3}$$

If t_{on} is changed (or modulated), then V_{ave} will change as well, according to Equation (5.3). As it is easy to generate in digital form the waveform of Fig. 5.3(a), this provides (with some limitations) a means of achieving D to A conversion. Note, however, that the bandwidth restriction imposed by the R–C filter will act on the varying V_{ave} signal as well.

Equation (5.3) is of the same form as Equation (5.1). We notice that the logic high voltage now acts as the reference voltage, and the resolution of the output voltage will be the resolution of the time duration t_{on}. For accurate conversion therefore, the duty cycle of the digital waveform must be accurately controllable, the filtering must be effective, and the logic 0 and 1 levels must be known and stable. As we would expect, given a microcontroller driven by a stable clock source, the first of these requirements is easy to meet. The question of filtering we will consider further below. If the requirement for stable logic levels proves troublesome, then a PWM signal can be separately buffered with a logic gate, supplied from a stable voltage (the 74HC logic family is useful for this application).

WORKED EXAMPLE   5.1

A designer wants a microcontroller-based PWM source to have a 10-bit resolution. The logic outputs are known to switch reliably between 0 V and the supply rail, which is to be 5 V. To what accuracy should the power supply be controlled in order to ensure that the 10-bit resolution is achieved?

Solution: For a 5 V range and 10-bit resolution, one lsb will be worth $(5/2^{10})$ V, i.e. 4.88 mV. Any variation introduced by power supply fluctuation should be less than half of this, i.e. 2.44 mV. The worst deviation from ideal which the power supply fluctuation can cause is when the PWM source is at maximum mark/space ratio. In percentage terms the supply must therefore be controlled to

$$\frac{2.44 \times 10^{-3}}{5} \times 100\% = 0.049\%$$

The permissibly supply voltage range is therefore 5 V ± 2.44 mV! This requirement is a sobering thought, and shows how much care will have to be taken if such resolutions are to be genuinely achieved.

Considering a little further a filter section of the type shown, the turn-over frequency is given by

$$f = 1/2\pi RC$$

and, by considering the exponential rise and fall times of the output waveform, the peak-to-peak output voltage ripple V_r can be shown to be

$$V_r = \frac{V_H[1 - \exp(-t_{on}/RC)][1 - \exp(-t_{off}/RC)]}{[1 - \exp(-T/RC)]} \tag{5.4}$$

where V_H is the voltage difference between logic 0 and logic 1, t_{off} is $(T - t_{on})$, and all other values are as shown in Fig. 5.3(b). Ripple is the worst when the mark/space ratio is 1:1, in which case

$$V_r = \frac{V_H[1 - \exp(-t_{on}/RC)]}{[1 + \exp(-t_{on}/RC)]} \tag{5.5}$$

The performance of a practical application using this circuit is described in Example 5.2.

5.1.3 Generating PWM signals in software

A PWM output can be generated in software by any microcontroller having digital output, based on the flow diagram of Fig. 5.4. Here two memory locations, PWM_Width and PWM_Prd, are preset with numbers proportional to the PWM pulse width and period respectively. The PWM output is initially set high, and a memory location called Counter is then incremented from zero, and compared with the two PWM registers. The output is cleared, and the cycle recommences, at the times indicated. Clearly, the duration of the cycle can be extended by inserting dummy instructions, or a delay loop. The programmer will need to find means of incorporating changes to the value of PWM_Width, at the start of every cycle, or as needed.

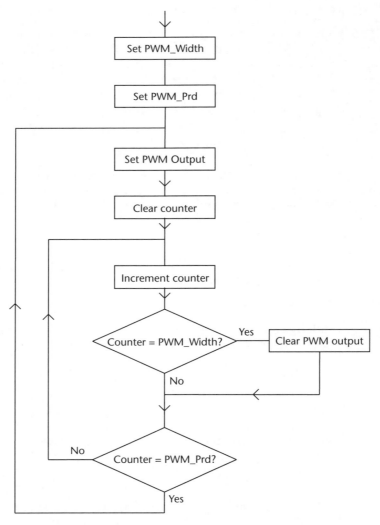

Figure 5.4 Generating PWM in software.

If T is the time taken to execute one cycle of the inner loop of Fig. 5.4, PWM_Prd holds a number N_P and PWM_Width holds a number N_W, then the PWM period will be $N_P T$ and pulse width will be $N_W T$. Equation (5.1) can then be rewritten:

$$V_{ave} = \frac{N_W T}{N_P T} V_{LH} = \frac{N_W}{N_P} V_{LH} \tag{5.6}$$

The pulse width resolution will be T, giving a voltage resolution of (V_{LH}/N_P).

This software-generated approach is just about adequate when the loading on the controller is not high. In most applications, however, there is a great advantage in passing the task of PWM generation to a hardware subsystem. This is offered by a number of microcontrollers.

5.1.4 Generating PWM signals in hardware

PWM generators can be made by introducing simple enhancements to the Counter/Timer structure. The internal form of a simple generator is shown in Fig. 5.5(a). A free-running counter is driven from the internal clock source via a presettable prescaler. A PWM register (normally a SFR) holds a value set in it by the program. This value is compared with the counter value using a digital comparator. The comparator output is logic 0 as long as the counter value is below that of the PWM register; when it is higher, the logic output is 1. This action is illustrated in the timing diagram. The duty cycle of the PWM output can be changed by changing the value of the PWM register, and the frequency by changing the prescaler value. The resolution of the PWM source is determined by the number of bits in the counter/comparator/PWM register sub-system.

In this structure the period of the PWM signal is given by:

$$T = 2^{n(PSC)} 2^{n(CNTR)} / f_{osc} \tag{5.7}$$

where $n(PSC)$ is the number of bits in the prescaler and $n(CNTR)$ is the number of Counter bits. The PWM period can only be set to binary multiples of the oscillator clock frequency, which limits system flexibility.

The on time is given by:

$$t_{on} = \{2^{n(CNTR)} - PWMR\} 2^{n(PSC)} / f_{osc} \tag{5.6}$$

where PWMR represents the contents of the PWM register.

This approach is similar to that of the 80C552 PWM source. The '552 actually has two (8-bit) PWM outputs, each derived from the same counter, but each with its own comparator and PWM register. The on time of one is therefore independent of the other, but they share the same period.

The PIC 16C72 has a somewhat more complex, but also more flexible, PWM source. It forms part of the larger CCP (Capture/Compare/PWM) module of that microcontroller, and its block diagram is shown in Fig. 5.5(b). Central to the system remains a free-running counter driven via a prescaler from the clock oscillator. A Duty Cycle Register (named CCPR1 in

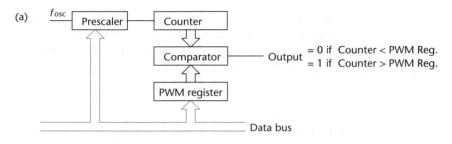

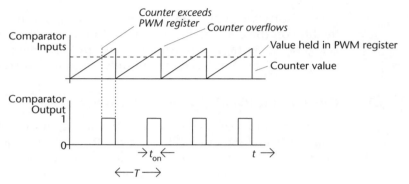

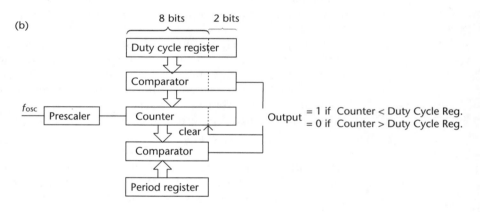

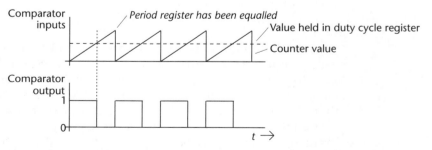

Figure 5.5 (a) Simple PWM structure; (b) an enhanced PWM structure.

PIC data) plays the role of the PWM Register of Fig. 5.5(a). Both the Counter and the Duty Cycle Register are 10-bit. The comparison action is such that when the Counter starts at 0, the PWM output is initially high, going low when the Counter reaches the value of the Duty Cycle Register. This means that the number fed to that register directly determines the output pulse width.

The main means for setting the PWM period is now a comparison between the higher 8 bits of the Counter and an 8-bit Period Register (named PR2 in PIC data). The Counter is reset when its value reaches that of the Period Register. This gives greater flexibility in setting the PWM period, as it need not be a binary multiple of the clock frequency. The full benefit of the 10-bit resolution is, however, only obtained if the Period Register is set to its maximum value. Full details on using this PWM source can be found in Ref. 5.2.

WORKED EXAMPLE ● **5.2**

The PWM output of a PIC 16C72 was set with a period of T = 96.5 μs, and connected to the circuit of Fig. 5.3(b), with R = 15k and C = 470 nF. The supply voltage was nominally 5 V. The filter turnover frequency for these values is 22.6 Hz. The output voltage values of Table 5.1 were noted.

Table 5.1 Some experimental PWM values.

M/S ratio	Average output	Ripple voltage (peak to peak)
1:9	0.5 V	5.0 mV
2:8	1.0 V	10.0 mV
3:7	1.4 V	13.5 mV
4:6	2.0 V	16.0 mV
5:5	2.4 V	17.0 mV
6:4	2.9 V	16.0 mV
9:1	4.4 V	6.5 mV
19:1	4.7 V	4.0 mV

These values agree reasonably well with Equations (5.3), (5.4) and (5.5), but show the limitations of this very simple circuit. First, the average output values are only approximately equal to those anticipated, and despite the limited bandwidth of 22.6 Hz, the worst-case ripple of 17 mV is still comparatively high. No special care was taken with power supply stability, and certainly more accuracy could be expected if this was done. A second- or higher-order filter could be applied to reduce the ripple significantly; this is the subject of Example 5.3.

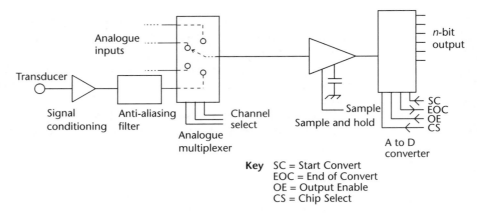

Key SC = Start Convert
EOC = End of Convert
OE = Output Enable
CS = Chip Select

Figure 5.6 A data acquisition system.

5.2 Data acquisition systems

We turn now to the question of digitising analogue information, done with an Analogue to Digital Converter (ADC). We generally need interfacing circuits to condition the incoming signal so that it is properly matched to the ADC input. The term 'Data Acquisition System' is used to describe the overall system. Such systems may be very sophisticated, and controlled from a desktop computer or workstation. Alternatively, they may be microcontroller-based, and contained on a single chip. The principles of both are, however, the same.

A typical system, controllable by a microprocessor, is shown in Fig. 5.6. Generally there is a need to digitise more than one channel. It would be expensive and space-consuming to commit an ADC to each input, so normally we switch between inputs in turn, using an *analogue multiplexer*, and digitise each when it is connected to the ADC.

We will shortly consider the principles of some of these elements, and then consider their implementation in a microcontroller system.

5.2.1 Acquiring AC signals: sampling rate and aliasing

Data acquisition systems are commonly used to make repeated samples of a changing analogue voltage. It is as if the ADC is taking a series of snapshots of the waveform. The result of the sampling process is a 'stepped' represen-tation of the original waveform (Fig. 5.7). If many of conversions are made per signal cycle, then the digitised representation is close to the original; the fewer the samples, the greater will be the relative size of the steps.

In Fig. 5.7 the sampling frequency is much greater than the signal frequency. Suppose it wasn't. Figure 5.8 shows an input signal which is sampled less than once a cycle. A 'reconstructed' signal would have a wave-form as shown in the dashed line, i.e. entirely different from, and *lower* than, the original signal. This effect is known as *aliasing*. It can be literally

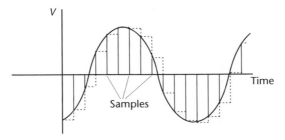

Figure 5.7 Digitising a sine wave.

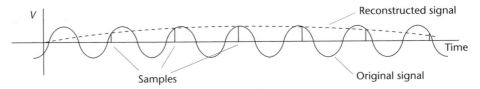

Figure 5.8 The effect of aliasing.

disastrous in a data acquisition system, as the original signal frequency, or a component of it, is replaced by its low-frequency 'alias'.

To avoid aliasing, Nyquist's sampling theorem must be applied. It states that *the sampling frequency must be equal to or greater than twice the maximum input frequency*. It is very important to ensure that this rule is *always* obeyed. It goes almost without saying that even unwanted input signals, like high-frequency noise or unwanted upper harmonics, will cause aliasing to occur. Many systems include an *anti-aliasing filter*, which filters out signal components that would cause aliasing to occur.

The maximum frequency at which an ADC can sample depends on the finite time that it takes to perform a single conversion. If this conversion time is T seconds, then its maximum *sampling rate* cannot be faster than $1/T$ samples per second.

5.2.2 Signal conditioning

Beyond the need to restrict the input frequency range to the limit set by Nyquist's sampling theorem, it is essential when matching a signal to an ADC input to ensure:

- that the input range is fully exploited, without risk of exceeding it
- that the resistance of the device or circuit driving the ADC input is not too high

It is possible therefore that any input signal conditioning will include both amplification and the addition of a DC offset. There are many techniques available to meet these needs, and they are well covered in Ref. 1.1. For simple applications a choice can often be made from the selection of

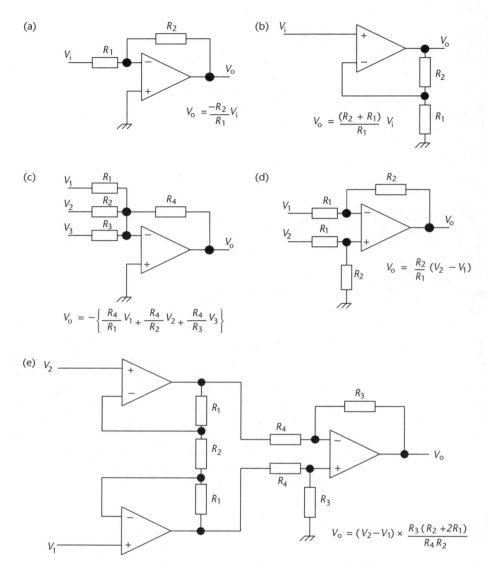

Figure 5.9 Standard op-amp configurations. (a) Inverting amplifier; (b) non-inverting amplifier; (c) summing amplifier; (d) difference amplifier; (e) instrumentation amplifier.

standard op-amp circuits, reproduced for convenience in Fig. 5.9. Circuits (a)–(d) can all be constructed from a suitable op-amp, chosen to match the needs of the application. Circuit (e) is a particularly useful enhancement of the basic difference amplifier of circuit (d). Its advantages lie in the symmetry of its inputs, its high input resistance and the fact that its gain can be controlled by only one resistor, R_2. While it can be constructed from standard op-amps, several manufacturers conveniently make it as one IC (e.g. Burr Brown INA114), with external pins for the R_2 connection. An INA114 is used in the case study of Appendix C.

5.2.3 The analogue multiplexer

The analogue multiplexer acts as a multiway switch, accepting many input channels, and selecting one to be connected to its output. The selection is made by channel select lines; these act as address lines, selecting one input channel. A disadvantage of the analogue multiplexer, as with all analogue switches, is the finite 'on resistance' of the internal switches. This tends to improve (i.e. fall) with increasing supply voltage, and ranges from tens to thousands of ohms. There can also be current leakage into the signal path, and digital interference from the channel select lines.

If a multiplexer is used to connect multiple input channels to a single ADC, then the sampling rate applied to each channel is reduced, as the one ADC is doing all the sampling. If the ADC sampling rate is f kS/s (kilo-samples per second), and there are n input channels, then each channel will be sampled at a rate of only (f/n) kS/s.

5.2.4 Filtering

Filtering is usually applied for two purposes. The first is to exclude signal components which are not wanted in the digitised record (for example high-frequency noise), the second is for anti-aliasing. The purpose of anti-aliasing filtering is to ensure that any signal component above the Nyquist limit is so small as to be insignificant (i.e. less than half of one lsb). Filter design is a huge topic, and well introduced in such books as Ref. 1.1. We restrict ourselves here to exploring the specification of suitable filters, and indicating certain types which are particularly suitable for the microcontroller environment.

When specifying filters the most important characteristics are the turn-over frequency f_o and the filter 'order'. In an op-amp-based filter implementation the order turns out to be the number of resistor–capacitor pairs used to define the filter characteristic. This is illustrated for a low-pass filter in Fig. 5.10. The order of the filter determines how rapidly the response falls off in the stop band; in this region the filter gain can approximately be represented as:

$$G = G_o(f_o/f)^n \tag{5.9}$$

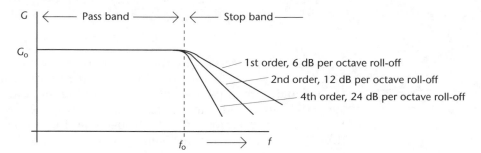

Figure 5.10 Low-pass filter characteristics.

where n is the filter order, f the frequency, and G_0 is the filter gain in the pass band. Alternatively, the stop band attenuation at frequency f is given by:

$$\text{Attenuation} = (f_0/f)^n \qquad\qquad (5.10)$$

WORKED EXAMPLE **5.3**

The ripple experienced in the PWM application of Example 5.2 is found to be excessive.

(a) If the turnover frequency is to remain the same, but a second-order filter applied, what will the worst-case ripple become?

(b) If the ripple is kept the same, but the bandwidth extended instead, what will the new bandwidth be?

Solution:

(a) The turnover frequency of the previous (first-order) filter was 22.6 Hz. Let us consider the ripple due to the PWM fundamental, which has a frequency of $(1/96.5\ \mu s)$, i.e. 10.36 kHz. Applying Equation (5.10), the attenuation of the second-order filter is given by $(10\ 360/22.6)^2$, i.e. 2.1×10^5. The 5 V_{p-p} PWM signal is attenuated by this factor, leading to a ripple due to the fundamental frequency of 24μV, which in this sort of application is negligible.

(b) To retain a ripple of 17 mV, but extend the bandwidth, we apply Equation (5.10) again:

$$\frac{0.017}{5} = \frac{f_0^2}{(10360)^2}$$

or

$$f_0 = 604\ \text{Hz}$$

Again, this calculation only takes into account the fundamental frequency, and does not therefore lead to a precise result. Clearly, however, there is great advantage to be gained from replacing the first-order filter. The second order filter can be a simple op-amp-based circuit (for example of Sallen Key type), with two R–C pairs.

Anti-aliasing filters are by their very nature low-pass, and generally must be of high order. As the filter order increases the component count of the standard op-amp-based filter becomes higher, and the dependence on accurately matched and stable components increases. This approach to filter design therefore tends to become inadequate for anti-aliasing filtering. To meet this problem part-way a number of ICs have been specially developed for linear filters, and contain the op-amps and appropriate interconnections already. The Maxim MAX274, for example, contains four second-order filter sections, each programmable by the addition of external resistors only. These can be cascaded to form filters up to 8th order, with high stability and low noise. Maxim provides design software to ease the design task.

An important development in filter design came with the 'switched capacitor' filter. Here a capacitor plus a pair of analogue switches are made to mimic the action of a resistor. If the switching speed of the switches changes, then the apparent resistance changes. Such 'pseudo-resistors' can replace chosen resistors in a conventional filter circuit, but their value is settable by a clock frequency. The great advantage of switched capacitor filters is that they can easily be integrated onto an IC (they take up less space than the equivalent resistor), and their cut-off frequency can be determined by an input clock frequency. Tracking and programmable filters can thus easily be made, and it is possible to make high-order filters on a single IC. The disadvantage, which is significant, is that there tends to be some breakthrough of the clock onto the signal line. Strange effects also take place if there is any component of the signal frequency at or near the clock frequency.

A good example of a switched capacitor filter is the Linear Technology LTC1062. This is an 8-pin IC which can be configured, with only three external components, as a 5th order low-pass filter. Alternatively, two can be cascaded to form a 10th order circuit. As the clock frequency is 100 times the cutoff frequency, it is not difficult to remove clock ripple by a simple extra R–C low-pass stage.

WORKED EXAMPLE   **5.4**

A voice signal is to be digitised by an 8-bit ADC which samples at a rate of 32 kHz. A signal bandwidth of approximately 4 kHz is required. What order of anti-aliasing filter should be used, and what is its cut-off frequency?

Solution: Let us assume that the frequency spectrum of the signal is flat, i.e. signal components at different frequencies all have the same maximum amplitude. Nyquist's sampling theorem requires that the maximum signal frequency should be less than half the sampling frequency, i.e. 16 kHz. The action of the anti-aliasing filter should remove all signal components above this frequency. In reality this is unachievable. A realistic requirement is for any signal component above the 16 kHz Nyquist limit to be less than half of one least significant bit, i.e. the filter attenuation introduced at 16 kHz should be 1/512.

This attenuation must take place in the two octaves between 4 kHz and 16 kHz. Applying Equation (5.10), with $f_0 = 4$ kHz,

$$\text{Attenuation} = (f_0/f)^n = (4/16)^n = 1/512$$

Solving this shows that n lies between 4 and 5. A fifth-order filter, with cut-off frequency of 4 kHz, should therefore be more than adequate.

5.2.5 The Sample and Hold

A simple form of the Sample and Hold (S&H) circuit is shown in Fig. 5.11(a). It is made up of an analogue switch, a capacitor, and a buffer amplifier, and acts like an analogue memory. When the switch is closed the capacitor charges up to the value of the input voltage, and this voltage appears on the circuit output. The output tracks the input as long as the switch remains closed. When the switch opens the capacitor holds its voltage until such time as the switch is closed again. If the single switch is replaced by an analogue multiplexer, then the functions of multiplexer and S&H can quite neatly be combined, as shown in Fig. 5.11(b).

The S&H operation is not, however, as ideal as the above description seems to suggest. Figure 5.12 shows input, output and control voltages. The

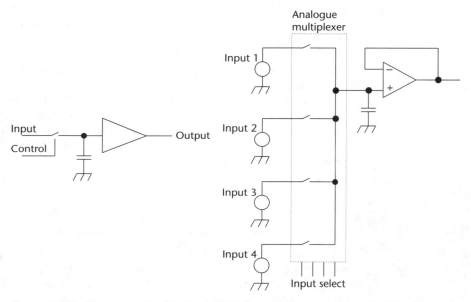

Figure 5.11 Sample and hold circuits. (a) Simple single channel; (b) four channel.

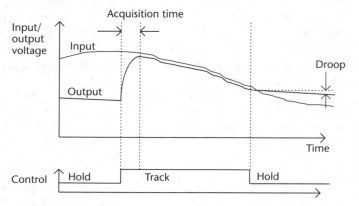

Figure 5.12 Sample and hold characteristics.

switch has finite resistance, so when the S&H goes into Track mode (i.e. the switch is closed) it takes a finite time for the capacitor to charge to the input voltage. This is known as the 'Acquisition Time'. While the S&H tracks, there is some gain and phase error, not least because the signal source is driving a capacitive load. Furthermore when the circuit is in Hold mode, there is leakage from the capacitor (into the amplifier, back through the switch, and through the capacitor itself), so its voltage does drift with time. This effect is known as 'droop'. As illustrated in Example 5.5, both the acquisition time and the droop rate depend upon the capacitor value. Droop rate *decreases* with increasing capacitance, but acquisition time *increases*.

WORKED EXAMPLE 5.5

A 4 channel S&H is made using the circuit of Fig. 5.11(b). The op-amp has an input bias current of 10 nA but is otherwise ideal. The switches have an 'on' resistance of 40 ohms, and the combined leakage from them is 1 nA. The capacitor is ideal. This is illustrated in Fig. 5.13. Estimate the acquisition time, to 0.1%, and the droop rate.

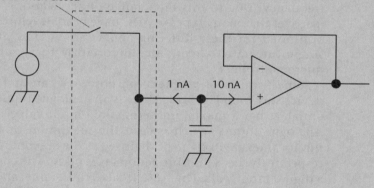

Figure 5.13

Solution: Let V_c be the voltage across the capacitor and V_{in} the input voltage. Then, when the switch is closed and the capacitor is charging up from zero

$$V_c = V_{in}\{1 - \exp(-t/RC)\}$$

where R is the switch resistance.
   For acquisition to within 0.1%, $V_c = 0.999 V_{in}$; hence

$$0.999 = 1 - \exp(-t/RC)$$

$$t/RC = +6.91$$

$$t = 6.91 \times 40 \times 10 \times 10^{-9}$$

$$= 2.76 \ \mu s$$

During hold mode, the total worst-case leakage current is 11 nA.

$$i = C.dV/dt$$

$$11 \times 10^{-9} = 10 \times 10^{-9} \ dV/dt$$

$$\text{Droop} = dV/dt = 1.1 \ V/s$$

5.3 Principles of analogue to digital conversion

An Analogue to Digital Converter (ADC) is an electronic circuit whose digital output is proportional to its analogue input. Effectively, it 'measures' the input voltage and gives a binary output number proportional to its size. The list of possible analogue input signals is endless, including such diverse sources as audio and video, medical or climatic variables, and a host of industrially generated signals. Of these some, like temperature, have a very low frequency content. Others, video for example, are very high frequency. Due to this huge range of applications, it is not surprising to discover that many types of ADC have been developed, with characteristics optimised for these differing applications. They are widely available as single integrated circuits or, importantly to us, are often embedded in microcontrollers.

Some features of a general-purpose ADC are indicated in the example device shown in Fig. 5.14. There is an analogue input, which may be differential. It has a maximum permissible input value, called the *input range*. Analogue inputs which exceed the maximum or minimum permissible input values are likely to be digitised as the maximum and minimum values respectively, i.e. a limiting action takes place. The conversion is initiated by a digital input, called here SC. It takes finite time, and the ADC signals with the EOC line when the conversion is complete. The resulting data can then

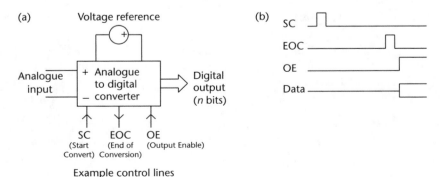

Figure 5.14 Essential features of the ADC. (a) The ADC; (b) timing diagram.

be enabled onto a data bus using the OE line. The input range of the ADC is usually determined by the value of a voltage reference. This may be included on-chip. If embedded onto a microcontroller, the control lines of the ADC are likely to be bits in SFRs.

There are many ways of designing ADCs, which optimise different features of the performance. Some, for example, are fast but low in accuracy, while others are slow with high accuracy. Let us first consider some principles of their application.

5.3.1 Interpreting specifications

5.3.1.1 Resolution and quantisation error

An ideal 3-bit converter has a characteristic shown in Fig. 5.15. The input voltage range (horizontal axis) is from 0 V to V_{max}, and is divided up into 8 (2^3) equal regions. If this voltage is initially set to zero, the digital output will be 000. If the input voltage is slowly increased, and the ADC is converting continuously, then at around $V_{max}/16$ the output changes to 001. If the input is further increased by $V_{max}/2^3$, then the output code will change to the next value, 010.

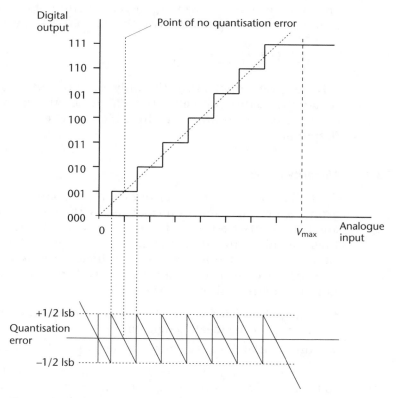

Figure 5.15 Ideal ADC conversion characteristic.

Table 5.2 Quantisation error as function of ADC bits.

n	No. of quantisation levels	Max. quantisation error as % of range	Quantisation error for range of 5 V
3	8	6.25	312.50 mV
4	16	3.13	156.25 mV
5	32	1.56	78.13 mV
6	64	0.781	39.06 mV
8	256	0.195	9.77 mV
10	1 024	0.0488	2.44 mV
12	4 096	0.0122	0.61 mV
16	65 536	0.00076	38.1 μV

Clearly, a *range* of analogue input values is represented by a *single* output code. The ADC *resolution*, which by definition is the amount by which the input can change without a change in output, is represented by this range. Ideally, the code should represent only one input value, the 'mid-step' value. All values elsewhere in the range are (mis)represented as the centre value, with an error equal to their distance from this centre value. This is known as *quantisation error*, e_q. It is a function of the input range and the number of bits output and, with these two fixed, it is irreducible.

$$\max e_q = \frac{V_{max}}{2^{n+1}} = \frac{\text{resolution}}{2} \tag{5.11}$$

For a given input range the only way to improve the resolution and reduce the quantisation error is to choose an ADC with a higher number of output bits. This is illustrated in Table 5.2, where n is the number of ADC output bits.

5.3.1.2 Linearity errors

On top of the unavoidable error due to the quantisation process come the errors due to inaccuracies introduced by the conversion. The simple ones are gain or offset errors. If these are recognised it is possible to eliminate them by adjusting either the analogue input or the digital result. What cannot be corrected for are the non-linear errors, when the regular 'staircase' of the conversion characteristic is disrupted. Two ways of defining non-linear errors, shown in Fig. 5.16, are recognised:

- integral linearity error – the maximum deviation of the transfer function from the ideal straight line
- differential linearity error – the deviation of any quantum from its ideal

If differential non-linearity becomes extreme, then it is possible for codes to be missed, as shown in Fig. 5.16. An even worse situation arises when a converter is *non-monotonic*. In this case an *increase* in analogue input leads

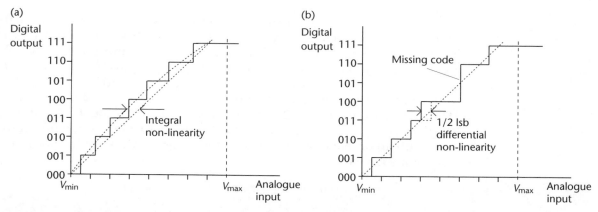

Figure 5.16   ADC non-linearity. (a) Integral non-linearity; (b) differential non-linearity.

for certain values to a *fall* in digital output. Most manufacturers are keen to proclaim for their converters 'no missing codes, monotonic output'.

Ultimately the overall accuracy of the ADC – that is, the difference between the *measured* value (as recorded in the digital output) and the *actual* input value – is due to the summation of quantisation error and any further linearity errors.

5.3.2   The voltage reference

In essence, an ADC compares an analogue input voltage with a reference voltage to determine its digital output. For a given ADC, the voltage reference is therefore one of the most important elements for ensuring operation within specification. The purpose of the voltage reference is to maintain a constant and known voltage value at a certain point in a circuit. Its weakness is that although its output should be entirely independent of load current and temperature, it never quite is.

In the past, references have mostly been based on the Zener diode. Such diodes do not give their best performance in the low-voltage supplies typical of the microcontroller environment (i.e. 5 V or less), and tend to be current-hungry for stable operation, requiring for example around 10 mA of bias current. An alternative device, of generally superior performance (excluding comparison with the temperature compensated or stabilised Zener diode, which is very good indeed) is the bandgap reference. This balances two voltages of equal and opposite temperature coefficient to produce a reference voltage of good stability, and has a characteristic voltage of 1.205 V. Bandgap devices have two big advantages over Zeners: they can operate at very low currents (down to 10μA), and their output voltage is lower than the Zener. They are thus particularly suitable for low-power, low-voltage circuits. In their unadorned form, bandgap devices are produced as two-terminal devices, and are connected in the same manner as Zener diodes.

Most ADC ICs come with their own internal voltage references. Their presence does not guarantee that the ADC is always best used with its own reference. Particularly when operating over a wide temperature range, it may become necessary to use an external, higher grade reference. For example the ICL7109 has an internal reference with a drift of 80 ppm/°C (where ppm = parts per million, 1000 ppm = 0.1%). The resolution of the ADC is one part in 4096, or 244 ppm. Thus a temperature change of only 3 °C will cause a 1 bit error. ADCs embedded in microcontrollers tend *not* to come with an on-board reference, and the designer will have to choose an appropriate device.

WORKED EXAMPLE   **5.6**

A signal from a microphone is to be digitised. Two ADCs are available: ADC1 has an input range of –2.5 V to +2.5 V and a conversion time of 200 μs, and ADC2 has an input range of 0 V to +4 V, with a conversion time of 12 μs. The microphone output has a bandwidth extending to 12 kHz, all of which should be digitised. Its maximum output signal is ±8 mV. Which is the most appropriate ADC, and what sort of signal conditioning is required?

Solution: To digitise a signal of 12 kHz, a sampling rate greater than 24 kHz is required. ADC1 has a maximum sampling rate of $1/(200 \times 10^{-6})$ samples per second, i.e. 5 kHz, and is therefore not suitable. ADC2 has a maximum sampling rate of $1/(12 \times 10^{-6})$ samples per second, i.e. 83 kHz, and can therefore be considered.

The microphone's output of 16 mV$_{p-p}$ must be amplified to match the ADC's input of 4 V$_{p-p}$, i.e. a gain of 250. It must also be offset by +2 V, as the microphone signal is bipolar, but the ADC input is unipolar.

A possible circuit to achieve this, in a single summing op-amp stage, is shown in Fig. 5.17. Preferred resistor values are used, so the gains are not precise. The microphone signal experiences a gain of 240 and is inverted. A bandgap voltage reference is used, whose output is –1.2 V. It experiences a gain of 1.6, which gives approximately the 2 V offset required.

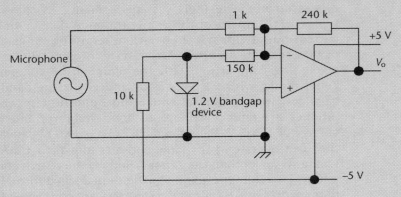

Figure 5.17

5.3.3 The successive approximation ADC

This type of ADC is very commonly found in microcontroller systems, and has conversion times typically down to a few microseconds. It is represented in the block diagram of Fig. 5.18.

The operation of the ADC can best be described in terms of the flow diagram of Fig. 5.19. A 'trial' digital word, held in the Successive Approximation Register (SAR), is converted to its analogue form by an internal DAC, and that analogue voltage is compared with the input (i.e. the voltage being converted). Depending on the outcome of the comparison the trial word may be changed. For each trial one bit of the SAR is set and the output then tested, starting with the most significant. The number of trials made is therefore equal to the number of bits in the output.

A 4-bit operation is illustrated in Fig. 5.20. Here we see the msb of the DAC set at the start of the conversion. The next bit is then set, but it is cleared when the DAC output is found to exceed the input value. The next bit is set, and remains set. The lsb is cleared after being set, as the value again exceeded the input. The final ADC digital output is 1010.

Clearly the successive approximation action illustrated in Fig. 5.20 relies on the input voltage remaining stable during the process of conversion. To avoid any error due to this, it can be shown that the input must not change by more than 1/2 lsb during the conversion period. This places severe restrictions on the input voltage rate of change, and a S&H is therefore commonly used.

The accuracy of the system depends on a number of its components, notably the DAC (including its voltage reference) and comparator. An error in either of these will be reflected as an error in the ADC output.

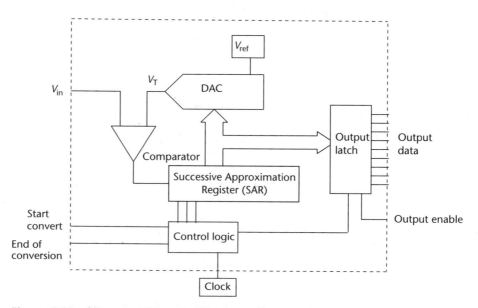

Figure 5.18 The successive approximation ADC.

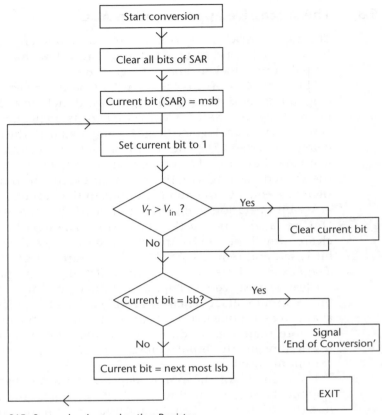

SAR: Successive Approximation Register

Figure 5.19 Flow diagram of the successive approximation ADC.

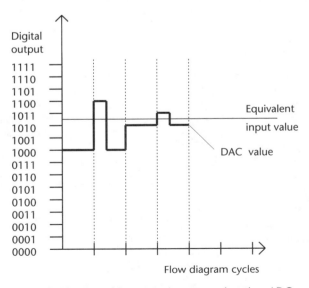

Figure 5.20 Four-bit successive approximation ADC.

WORKED EXAMPLE 5.7

A 12-bit successive approximation ADC has a conversion time of 10 μs and an input range of 0 to +5 V. The manufacturer specifies that the input must not change by more than 1/2 lsb during the conversion time. What is the maximum frequency, at maximum input amplitude, that the ADC can convert, without a S&H being used?

Solution: The equivalent of 1/2 lsb is $5/(2 \times 2^{12})$, i.e. 0.61 mV. The input must not change by more than this during the 10 μs conversion period, i.e. the acceptable voltage rate of change must be less than 61 V/s.

A sinusoidal input frequency of maximum input amplitude will be of the form $V = 2.5 + 2.5\sin 2\pi ft$. Differentiating this, we find that the maximum voltage rate of change is $2.5 \times 2\pi f$. In this case

$$2.5 \times 2\pi f = 61$$

i.e. $f = 3.88$Hz.

This is an extremely low frequency. In most situations of this type it is therefore necessary to include an S&H circuit before this type of ADC.

An implementation of this type of ADC, using a microcontroller as control element and SAR, together with a standalone DAC, is illustrated in the Case Study of Appendix C.

5.3.4 The switched capacitor successive approximation ADC

This very neat ADC is based on the Successive Approximation technique, but merges the functions of S&H, comparator and ADC into one circuit. Its operation is shown in simplified form in Fig. 5.21. The phases of operation are:

Sample
S2 is closed, and S1 is connected to V_{in}. V_x is maintained at 0 V due to op-amp action through S2. Charge Q_s accumulates on the capacitors, of value

$$Q_s = 16CV_{in} \tag{5.12}$$

Hold
All the capacitor switches are connected to ground, S2 is opened and voltage V_x takes the value $-V_{in}$. Charge Q_s will be retained on the capacitor plates (although it will be moved around) for the remainder of the conversion process; a Sample and Hold action has taken place. With S2 open the op-amp now acts as a comparator. S1 is switched to V_{ref}.

Bit cycle
The switch linked to the largest capacitor now changes state. Charge will redistribute itself, i.e.

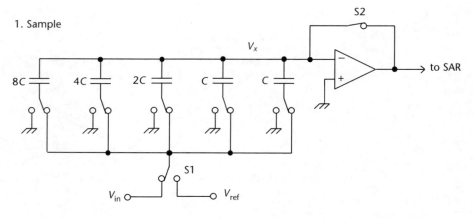

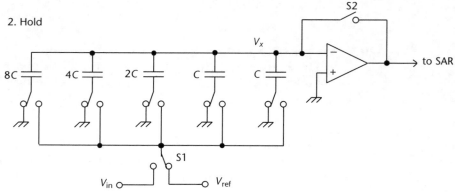

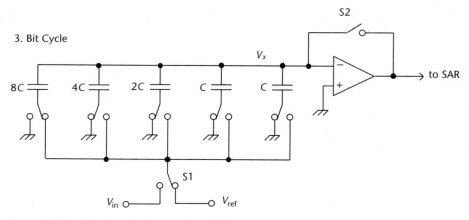

Figure 5.21 The switched capacitor ADC.

$$Q_S = 8C(V_{ref} - V_x) + (4C + 2C + C + C)(-V_x)$$

or

$$Q_S = 8CV_{ref} - 16CV_x$$

Substituting now for Q_s from (5.12)

$$16CV_{in} = 8CV_{ref} - 16CV_x$$

$$V_x = -V_{in} + (V_{ref})/2 \tag{5.13}$$

If at this stage $V_x > 0$, i.e. $V_{in} < (V_{ref})/2$, then the switch at $8C$ is reversed. In the next cycle the switch at the next smallest capacitor is reversed and a similar test is made. This successive approximation procedure is followed until the smallest capacitor has been switched.

This ADC is further detailed in Ref. 5.3. It is very attractive, because the S&H action is integral to it, and because it is easy to integrate the whole circuit onto an IC. It is commonly found both in microcontrollers, and in compact and low-power single-chip ADCs.

5.4   A to D conversion in the microcontroller environment

5.4.1   Where there is no ADC

It *is* possible to perform A to D conversion without a recognisable converter, and a number of ingenious methods have been described. Given a DAC and comparator, it is possible to implement a successive approximation ADC; Appendix C shows how it can be done. Other methods are based on a combination of the microcontroller's ability to measure time, together with the ability to convert voltage to time using a charging or integrating circuit. These can be cost-saving, but are often of limited applicability.

Figure 5.22 shows a circuit based around a comparator and voltage reference. An analogue switch allows selection of the measurand voltage V_{in}, or ground. When the switch moves from ground to V_{in}, capacitor C charges up with the characteristic curve shown. As it passes V_{ref} the comparator, whose output is initially low, switches high; this time is denoted t_1. When the switch is returned to ground, the capacitor discharges, and the comparator output switches after time t_2. These two times can be shown to be

$$t_1 = - RC \ln[(V_{in} - V_{ref})/V_{in}] \tag{5.14}$$

and

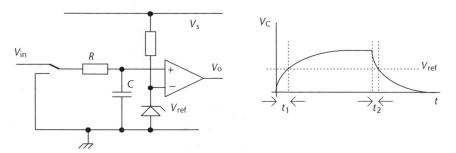

Figure 5.22   A to D conversion by measuring capacitor charging.

$$t_2 = - RC \ln(V_{\text{ref}}/V_{\text{in}}) \tag{5.15}$$

If t_1 and t_2 are measured by the microcontroller, and then one divided by the other, the RC term disappears, and the result can be used to determine V_{in}.

Ref. 5.4 details this approach further for an 80C51-based controller. Ref. 5.5 illustrates an alternative approach for the 68HC05.

5.4.2 ADCs in microcontrollers

ADCs are commonly found in the better-endowed microcontrollers, usually successive approximation, 8 or 10-bit, and in association with an input multiplexer. As with other controller peripherals, they are controlled through SFRs, and other SFRs permit the passing of data. Table 5.3 compares some ADC characteristics for three example controller families.

The table shows an interesting variety in ADC performance. The 80C552 offers potential for the most accurate conversions, with high resolution, differential input and externally applied voltage reference and power supply. It carries the disadvantage of slower conversion time. The highest speed is offered by the 16C74, with the interesting option of trading off reduced accuracy for even greater speed.

Both the 16C74 and 68HC11 permit a choice of clock sources, derived either from the internal oscillator, or an on-chip R–C oscillator. Both of these controllers offer clock frequencies down to DC (unlike the 80C552). Thus if the controller is running at a low clock frequency (perhaps for reasons of power saving) the ADC can operate from its own clock. This is necessary at slow clock frequencies, as the switched capacitor ADC suffers from voltage 'droop' in the same way as an S&H, and the conversion must be completed before too much charge has leaked from the capacitors.

For those ADCs with an integral S&H action, it is important to consider the time taken to charge the internal capacitance, applying calculations similar to those of Example 5.5. Both 68HC11 and 16C74 data gives analogue input models similar to that of Fig. 5.23, with values quoted for

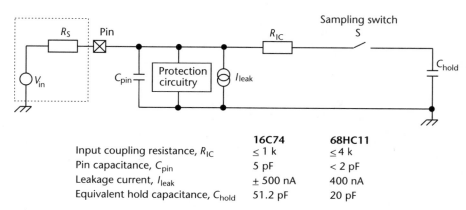

	16C74	68HC11
Input coupling resistance, R_{IC}	≤ 1 k	≤ 4 k
Pin capacitance, C_{pin}	5 pF	< 2 pF
Leakage current, I_{leak}	± 500 nA	400 nA
Equivalent hold capacitance, C_{hold}	51.2 pF	20 pF

Figure 5.23 ADC input model.

Table 5.3 A comparison of microcontroller ADCs.

	PIC 16C74	80C552	68HC11A8
Resolution	8 bit	10 bit	8 bit
Integral non-linearity[1]	<±1 lsb	±2 lsb	
Differential non-linearity[1]	<±1 lsb	±1 lsb	
Offset error	<±1 lsb	±2 lsb	
Type	Successive approximation, switched capacitor	Successive approximation	Successive approximation, switched capacitor
Input multiplexer	8 channels	8 channels	4/8 channels
Input S&H	Integral to ADC type	None	Integral to ADC type
Clocking	ADC Clock T_{AD} is derived either from oscillator or on-chip R–C oscillator; minimum period is 1.6 µs	Each bit cycle takes 4 machine cycles	Internal E clock or on-chip R–C oscillator
Conversion time	10 T_{AD}, e.g. 16 µs with 1.6 µs clock	50 machine cycles, e.g. 50 µs with 12 MHz clock	32 E cycles; e.g. 64 µs at oscillator frequency of 2 MHz
Voltage reference	V_{CC}, or externally applied, ground referred	Externally applied, differential, i.e. V_{ref-} to V_{ref+}	Externally applied, differential, V_{RL} to V_{RH}
Input range	0 to V_{ref}	V_{ref-} to V_{ref+}	V_{RL} to V_{RH}
Other features	Higher speed conversions permissible if reduced accuracy is tolerable	ADC has own supply pins	Four results registers allow conversions in sets of 4. Continuous scan mode possible

[1]See data sheets for conditions for which this data is valid.

each component. It is up to the user to ensure that adequate time is allowed for a sample to be made before the conversion process starts. The source resistance is part of the charging circuit, and a higher resistance will slow the charging process. In practice, sampling may be assumed to start when the input multiplexer selects the required input channel. The user then initiates the conversion process in software after a suitable delay.

In addition to the data given, graphical data is given for the 'on' switch resistance for the 16C74. The resistance is supply voltage dependent, and around 7 kΩ at 5 V supply. The 68HC11 has a charge pump, which raises the switch gate drive and accordingly reduces this resistance. The maximum source resistance recommended for the 16C74 is 10 kΩ.

The 80C552 also presents a capacitive load to the source. Although less in magnitude than those tabulated in Fig. 5.23, this must also be driven correctly. Its total capacitive load is quoted as ≤15pF, with a maximum recommended source resistance of 9.6 kΩ.

5.4.2.1 More on the 16C74 ADC

The general block diagram of this ADC is shown in Fig. 5.24. The eight lines available for analogue input may also be used for digital signals. Associated with the ADC are three SFRs: ADCON0, ADCON1 and ADRES. Most of the control is exercised through the former, and is shown in Fig. 5.25. It can be used to switch the ADC on or off, select the input channel and clock rate, start conversion and flag its end. Following a conversion, the ADC result appears in SFR ADRES. ADCON1 contains just three active bits, which are used to allocate input bits to analogue or digital, and also select the source of the voltage reference.

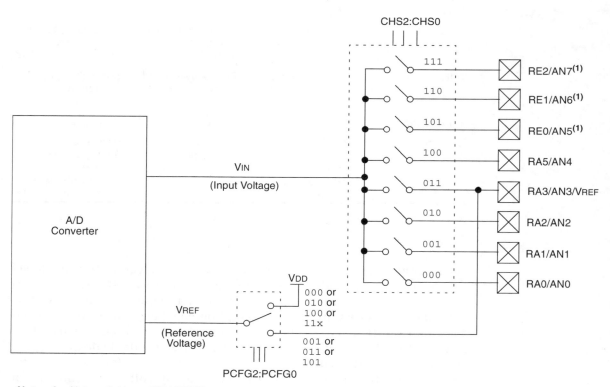

Note 1: Not available on PIC16C73B.

Figure 5.24 PIC 16C74 ADC block diagram. Reprinted with permission of the copyright owner, Microchip Technology Incorporated © 2001. All rights reserved.

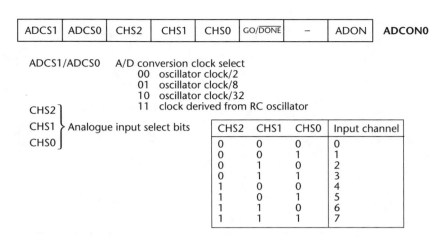

| ADCS1 | ADCS0 | CHS2 | CHS1 | CHS0 | GO/$\overline{\text{DONE}}$ | – | ADON | **ADCON0** |

ADCS1/ADCS0   A/D conversion clock select
    00   oscillator clock/2
    01   oscillator clock/8
    10   oscillator clock/32
    11   clock derived from RC oscillator

CHS2 ⎤
CHS1 ⎬ Analogue input select bits
CHS0 ⎦

CHS2	CHS1	CHS0	Input channel
0	0	0	0
0	0	1	1
0	1	0	2
0	1	1	3
1	0	0	4
1	0	1	5
1	1	0	6
1	1	1	7

GO/$\overline{\text{DONE}}$ Setting to 1 inititiates conversion; cleared to 0 on completion.

ADON   Setting to 1 switches ADC module on. When 0, ADC module is off, and consumes no power.

Figure 5.25 PIC 16C74 ADCON0 SFR.

5.4.2.2 *More on the 80C552 ADC*

The external connections to this ADC, and the SFR (called ADCON) which controls its operation, are shown in Fig. 5.26. The six lower bits of ADCON exercise control functions, while its two upper bits are the lowest two bits of the 10-bit conversion result. The upper eight bits of the result are placed in another SFR called ADCH. The result bits held in ADCON can be ignored if only 8-bit accuracy is required.

In keeping with the 10-bit ADC resolution, the general analogue environment is comparatively sophisticated. The ADC can have its own independent power supply, the input is differential (i.e. it need not be ground-referred), and an external differential voltage reference must be applied. This *can* be the power supply, although it is most unlikely that benefit could then be made of the full 10-bit accuracy. An interesting difference with the 16C74 is the possibility to initiate a conversion externally, via an IC pin connection.

The conversion result is given by the formula:

$$\text{Digital result} = \frac{V_{\text{in}} - V_{\text{ref}-}}{V_{\text{ref}+} - V_{\text{ref}-}} \times 1024 \qquad (5.16)$$

where V_{in} is the input voltage, and $V_{\text{ref}+}$ and $V_{\text{ref}-}$ are the reference voltage inputs. Clearly V_{in} is measured with respect to $V_{\text{ref}-}$, and need not therefore be ground-referred.

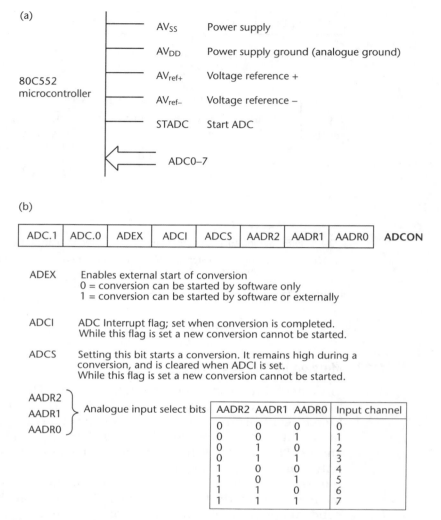

(a)

	AV$_{SS}$	Power supply
	AV$_{DD}$	Power supply ground (analogue ground)
	AV$_{ref+}$	Voltage reference +
	AV$_{ref-}$	Voltage reference −
	STADC	Start ADC

80C552
microcontroller

ADC0–7

(b)

| ADC.1 | ADC.0 | ADEX | ADCI | ADCS | AADR2 | AADR1 | AADR0 | **ADCON** |

ADEX   Enables external start of conversion
        0 = conversion can be started by software only
        1 = conversion can be started by software or externally

ADCI   ADC Interrupt flag; set when conversion is completed.
        While this flag is set a new conversion cannot be started.

ADCS   Setting this bit starts a conversion. It remains high during a
        conversion, and is cleared when ADCI is set.
        While this flag is set a new conversion cannot be started.

AADR2
AADR1   Analogue input select bits
AADR0

AADR2	AADR1	AADR0	Input channel
0	0	0	0
0	0	1	1
0	1	0	2
0	1	1	3
1	0	0	4
1	0	1	5
1	1	0	6
1	1	1	7

Figure 5.26   The 80C552 ADC. (a) IC pins associated with ADC; (b) SFR ADCON.

WORKED EXAMPLE   **5.8**

An ADC is required to measure the current I, of maximum value 10.00 mA, flowing in a circuit. The measured value will be used to drive a three and one half digit display (i.e. display range 0000 to 1999), and there should be no missing values. The current is passed through a precision resistor R, and the resistor terminals are available for connection to the ADC, as shown in Fig. 5.27.

Determine which processor ADC, of the 16C74 and the 80C552, is more appropriate for this application. Describe how the selected ADC should be

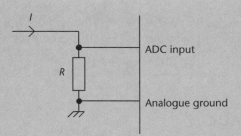

Figure 5.27

used, making reference to SFR settings. Suggest suitable values for the resistor R and the voltage reference input values.

Solution: As the measurement is only intended to drive a digital display, it may be assumed that there are no significant timing demands on the conversion. An 8-bit ADC, with output values from 0 to 255_{10}, will not have the resolution to supply data to the display without missing codes. The 80C552 is, however, a suitable choice for this measurement because of its 10-bit ADC. If the input is scaled so that a reading of 10 mA corresponds to a digital output of 1000_{10}, then the 10-bit range (i.e. 0000 to 1023) is almost fully exploited, there will be no missing values on the output, and no scaling in software will be required.

As the lower terminal of resistor R, across which the voltage measurement is being made, is connected to analogue ground, the negative reference input of the 80C552 (AV_{ref-}) should be connected to analogue ground as well. If R is set to 100.0 Ω, then the full-scale current value of 10 mA will develop 1 V across it. For this value of resistor the positive reference input (AV_{ref+}) should be connected to a stable voltage reference adjusted to 1.024 V.

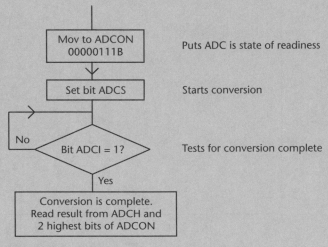

Figure 5.28

A flow diagram to achieve the measurement, which polls the Interrupt Flag ADCI to determine whether conversion is complete, is shown in Fig. 5.28.

5.5 Selecting an ADC

We turn (at last) to considering how the choice of ADC is made. We are concerned primarily with the following interrelated characteristics:

Accuracy (which includes resolution)   Number of input channels
Conversion time   Temperature stability
Cost   Input voltage range

The designer is likely to consider the three items in the left-hand column as being of great importance. Of these, it is in general true to say that excellence in any one of them tends to be at the expense of the other two (i.e. you can have very low cost *or* very high accuracy *or* very high speed, but not two or more together). The mutually exclusive nature of the first two is further illustrated in Fig. 5.29, which compares the performance of a number of popular low-cost ADCs, and illustrates the position in the arena occupied by the 16C74 and 80C552 ADCs. Clearly, the high-speed devices are of low resolution, while the high-resolution devices are slow.

An informal procedure for designing a microcontroller-based data acquisition system is given in Fig. 5.30. It is assumed that this lies within the broader context of the design process described in Chapter 12.

The fundamental system characteristics of sampling rate, number of channels and accuracy are first decided and cost constraints considered.

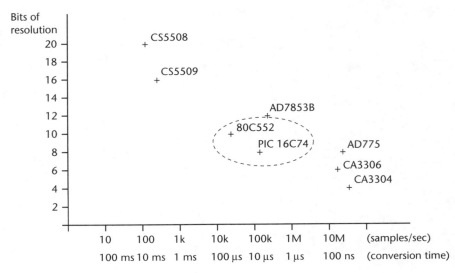

Figure 5.29 Comparison of example ADCs.

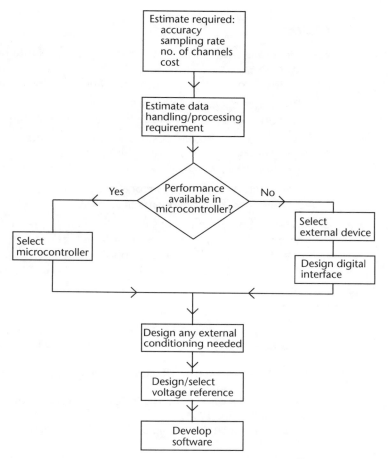

Figure 5.30 Informal procedure for ADC selection and design.

These will lead to knowledge of the required resolution and conversion time. The scale of data processing required should also be considered. If intensive computation is required, it may be that a microcontroller will be able to make the conversions fast enough, but not then be able to handle the subsequent processing. The major decision can then be made of whether such requirements can be met by an ADC embedded in a microcontroller. If so, choice can be made of a suitable microcontroller. If not, one must survey the field for possible external devices, and choose between serial and parallel data interconnection. Given successful selection of a device, the external elements can be added, noting carefully the issues already covered under the relevant headings. As the hardware system nears completion, attention should be given to the software development, again noting particular needs already covered (for example timing needs relating to acquisition and conversion times).

SUMMARY

1. There is a wide variety of commercially available digital to analogue converters suitable for the embedded system environment. These are compact, give good performance and offer low power consumption.

2. When the specification is undemanding Pulse Width Modulation offers a commonly used technique for digital to analogue conversion.

3. Even in a small-scale design, all the elements of a data acquisition system may be present. To apply such systems, it is necessary to have a good understanding of their principles.

4. Many microcontrollers incorporate analogue to digital converters, usually of the switched capacitor type.

5. External analogue to digital converters are also readily available; alternatively, techniques exist to achieve conversion, at low cost but to low specification, without a recognisable converter.

REFERENCES

5.1 Clayton, G. B. (1983) *Data Converters*. Basingstoke: Macmillan.
5.2 *Using the CCP Module(s)*. Application Note AN 594. Microchip Technology Inc. 1997.
5.3 Jones, D. A. and Martin, K. (1997) *Analog Integrated Circuit Design*. New York: John Wiley & Sons.
5.4 *A/D conversion with P83CL410 PCF1252-x*. Application Note AN91006. Philips Semiconductors. 1991.
5.5 *Simple A/D for MCUs without built-in A/D converters*. Application Note AN477. Motorola Inc. 1993.

EXERCISES

5.1 A PWM signal is to be generated from a CMOS microcontroller and converted by low-pass filtering into an analogue signal. The microcontroller is powered from a supply of 5 V ± 10%. It may be assumed that the PWM logic 0 output is 0 V, and its logic 1 output is equal to the power supply voltage. Beyond what resolution of the PWM source is it pointless for the designer to go? (Assume in this instance that any error introduced by supply voltage fluctuation should be less than the PWM resolution.)

5.2 The following analogue signals are to be digitised. Give the basic specification for an ADC which is suitable for the task, and indicate whether the application is appropriate for a small-scale embedded system:

    (a) a signal intended for a background music application, of bandwidth 50 Hz to 10 kHz, amplitude ±10 V

    (b) a temperature signal in a process control environment, range 0 V to +4 V representing 0 °C to 80 °C, with a resolution required of 0.2 °C

(c) a radar signal, bandwidth DC to 12 MHz, range 0 V to 5 V, with resolution of 5 mV.

5.3 For a certain 8-bit successive approximation ADC the manufacturer specifies that the input must not change by more than 1 lsb during the conversion time. It has a conversion time of 2.5 μs and an input range of −4 V to +4 V. It is hoped to use it to digitise an Amplitude Modulated signal of 950 Hz. What is the maximum signal amplitude at this frequency that it can accommodate without a S&H being used?

5.4 A voice signal of original bandwidth extending to 10 kHz is to be digitised with a sampling frequency of 28 kHz. Describe what precautions must be taken and for what reasons. If it is acceptable for the bandwidth to be limited to 4 kHz, and a signal to noise ratio of 60 dB is targeted, specify the characteristics of the anti-aliasing filter which must be used.

5.5 Fig. 5.31 shows the LM35 temperature sensor, which gives an output of 10 mV per °C between the terminals marked + and −. The diodes are silicon, and it may be assumed that the voltage across each is 0.6 V. It is to be connected to an ADC whose input range is 0 V to 4 V. Design a circuit which performs the necessary signal conditioning if the range of temperature measured is to be 0 °C to 40 °C.

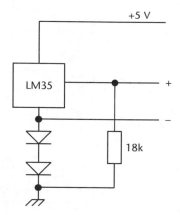

Figure 5.31

5.6 An 80C552 ADC has its voltage reference inputs set to +0.5 V and 2.5 V. Assuming no errors, what digital output will be received from the ADC if the input with respect to analogue ground is

(a) 1.0 V?
(b) 2.0 V?

5.7 Eight hot-wire anemometers are mounted on a wing section located in a wind tunnel, and it is required to digitise and record their output data. The anemometers are connected to suitable signal-conditioning circuitry, and the output of each of these is a varying voltage lying within the range −1 V to +1 V, with maximum frequency 20 kHz. The signals are to be recorded to an accuracy of 0.05%.

Describe a data acquisition system suitable for this purpose, giving:

(a) a well-labelled block diagram
(b) a description of the important features of each block, specifying each as far as you can from the information given
(c) system timing information
(d) a description of any features which may contribute to system error.

Cambridge University Engineering Department.
Module D8: Electronic System Design. 1996

5.8   A programmable gain amplifier is made using the circuit of Fig. 5.32.

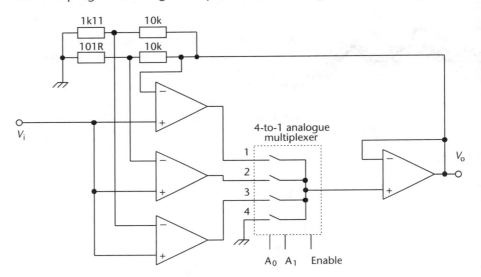

4-to-1 analogue multiplexer

A_0	A_1	Enable	Input channel selected
X	X	0	None
0	0	1	1
0	1	1	2
1	0	1	3
1	1	1	4

Figure 5.32

(a) For each address setting of the analogue multiplexer, determine the circuit gain.
(b) The circuit is to be adapted to incorporate a track-and-hold capability. Describe how this could be done, and calculate a hold capacitor such that the droop rate is less than 2.5 V per second.
(c) What is the acquisition time for this value of capacitor if the circuit output is to be fed to an 8-bit Analogue to Digital Converter?

Take note of the data given directly below. Assume to be ideal any device characteristic not specified.

All op-amps		4-to-1 analogue multiplexer	
Input bias current	±25 nA	'Off' Leakage Current (per switch)	18 nA
Capacitance at each input terminal	7 pF	Charge injection	20 pC
		'On' resistance	40 Ω
		'Off' capacitance, output to ground	40 pF

Cambridge University Engineering Department.
Module D8: Electronic System Design. 1997

5.9   The ADC of the PIC 16C74, operating with a 5 V supply, is to be used with the power supply as voltage reference. For true 8-bit resolution, how accurately should the power supply be controlled? If only 6-bit resolution is required, to what accuracy should the supply be controlled?

Strictly serial

IN THIS CHAPTER

There are fundamentally two ways to transmit digital data – parallel and serial. In parallel transfer, every bit of the data word being transmitted has its own connecting link, and all bits of each word are sent simultaneously. This leads to high-speed data transfer, at the cost of a large number of interconnections. In serial data transfer, on the other hand, each bit is sent in turn on a single data connection. There may be some other connections, for earth and control. This takes longer, but the amount of interconnection is dramatically reduced.

Most data transfer within conventional microprocessor systems is parallel. Microprocessors have multiple pins for data and address buses, and bus interconnections are made on printed circuit board and via multiway connectors. Data transfer between such systems has, however, often been serial, for example between computer and printer, using the well-known RS-232 standard. Over longer links the high-speed advantage of parallel transfer gives way to the simplicity and lower cost of serial transfer.

In the small-scale embedded environment the relative advantages are viewed differently. While microcontrollers are highly self-contained, there does remain a need to communicate with external ICs and other devices. To set aside eight pins (or 16, or 32) on a controller IC for the data bus, as well as a similar number for address lines, means that the microcontroller and peripheral ICs have to be physically bigger. Moreover, parallel buses take up space on printed circuit boards, not to mention needing ribbon cables between the boards. But the embedded system usually needs to be small, and speed, though important, is not necessarily so critical. Serial data links, with their small number of interconnect lines, are hence very important. They are space saving, and the longer time taken to complete a data transfer (as compared to parallel communication) is less of a problem in the typical embedded system.

This chapter describes how serial communication links can be implemented in the microcontroller environment. The specific chapter aims are:

- to review how serial data communication takes place, and distinguish between synchronous and asynchronous

- to describe some of the physical limitations within which electrical data communication must operate

- to introduce the serial standards and protocols which are important in the embedded system environment

- to describe the implementation of microcontroller serial ports

6.1 Serial communication overview

In serial data communication the data bits are presented in turn on a single data line. As an absolute minimum, a data communication link could be made of just two electrical connections, one for data and one for the earth return. With such a simple link two questions immediately arise:

- How do we know when one bit ends and the next one starts?

- How do we distinguish one group of bits (i.e. a byte or word) from the next?

To make the data transfer intelligible, any data link must therefore somehow provide this basic information, along with the actual data. Let us see how this is done by first considering the broad division of serial communication into synchronous and asynchronous transfer.

6.1.1 Synchronous data communication

In synchronous communication, both data transmitter and receiver are synchronised to the same clock signal. This is connected to both, and usually generated by the transmitter. Fig. 6.1 shows the timing diagram of a synchronous data stream transferring the 8-bit binary word 10101001. In this example a new bit of data is valid on every rising clock transition, the data can change at any time between these rising edges. As long as both receiver and transmitter are synchronised to the same clock, there is not

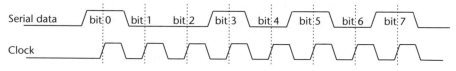

Figure 6.1 Synchronous serial data.

necessarily a requirement for the clock to run at a fixed or steady frequency, though other system requirements may introduce this need.

The rate of transfer of data in this type of link is sometimes called the *bit rate*, and expressed in bits per second.

6.1.2 Asynchronous data communication

Asynchronous communication is illustrated in Fig. 6.2. The transmitting device does not send a clock signal, but the rate at which data is sent has to be pre-arranged, within certain strict limits. The transmitted data is shown grouped into sets of bits, known as 'frames'. In many systems each frame has 10 or 11 bits. When data is not being sent the data link idles at a prede-termined level, either logic 1 or 0. So that a fault condition, for example a disconnected line, can be recognised, the idle state is generally *not* 0 V.

The sending of data is initiated by a 'start bit' of polarity opposite to the idle state. The receiver has its own clock generator, running at a fixed multiple of the data rate. For example, the receiver clock may be running at 16 cycles per data bit period. The receiver recognises the initial transient of the start bit, and then eight clock cycles later (i.e. in the middle of a bit period) it tests the value of the incoming bit. As long as it recognises the start bit as being of correct polarity at this instant, it will proceed to test all subsequent bits at intervals of one period until the frame is complete. The frame is terminated by a 'stop bit', which may be the start of another period of idling.

The receiver resynchronises its external clock repeatedly, at the start bit of every new frame, and any slight discrepancy in frequency between trans-mitter and receiver is tolerated due to this repeated resynchronisation. Garbled data, which may not immediately be detected as such, will of

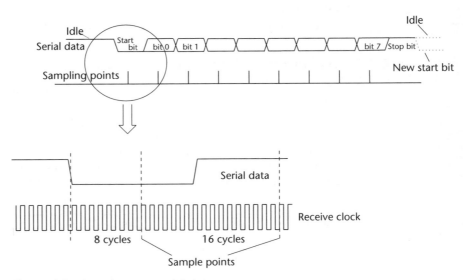

Figure 6.2 Asynchronous serial data.

course be received if the receiver baud rate has not been set to agree with that of the transmitter.

The rate of transmission of bits in this type of communication is traditionally called the *baud rate*. Implicit in this terminology is the fact that some bits are not data, but framing or parity. In some texts, baud rate and bit rate are used interchangeably.

6.1.3 So which do we choose?

From the above description, it can be seen that synchronous communication has the potential for higher speed and bandwidth efficiency, simply because it does not need the framing start and stop bits of the asynchronous alternative. The hardware design of its transmitter and receiver is also simpler. Its disadvantage is that a clock signal has to be transmitted along with the data.[1] Asynchronous communication has advantages for links when data is only sent sporadically, for example from a keyboard to computer, and for longer links, where limitations imposed by the data link make clock transmission difficult. Its bandwidth inefficiency becomes a disadvantage when large blocks of data have to be transferred. The outcome is that synchronous links are frequently used for short distances, for example within a piece of equipment. For medium to long distance, generally anything more than a few metres, asynchronous communication is more often found.

6.1.4 Some terminology

A simple serial link, for example from computer to printer, is called *point to point*. If data flows in only one direction it is called *simplex*. If there is a single communication link, but data can flow either in one direction or the other, it is called *half duplex*. If there are two links, so that data can flow simultaneously in both directions, it is called *full duplex*. A set of devices linked by the same serial network is called *multi-drop* or *multi-point*. A device communicating with a network is called a *node*. In a multi-drop system it is generally necessary to define one device which controls the system; this is then called the *master*, and the remainder are *slaves*. However, the master is not necessarily fixed, and devices can take it in turns to be the master. If there is competition for mastership, then a process of *arbitration* is required.

6.2 Physical limitations to serial transmission

This section is applicable to all categories of data transmission, not just serial. It is, however, appropriate to consider it here, as in the context of embedded systems the effects described are most likely to be experienced with serial data links.

---

1 Clock recovery from data is possible, but adds complexity at a level not generally acceptable in embedded systems.

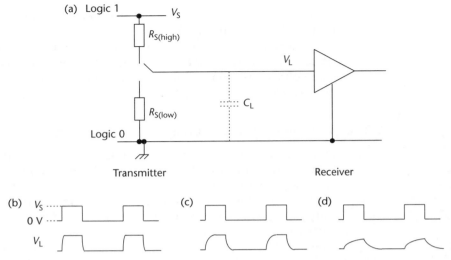

Figure 6.3 The 'short' serial link. (a) Simple model; (b) small time constant; (c) medium time constant; (d) large time constant.

6.2.1 The short data link

A simple model of a 'short' serial link is shown in Fig. 6.3(a). Here the transmitter output is modelled by a switch which connects either to Logic 1, or to Logic 0, in each case through a resistor. $R_{S(high)}$ is assumed to be the source resistance of the logic 1 state, while $R_{S(low)}$ is the source resistance of logic 0; the two may not be equal. It is assumed that the switch switches instantaneously, and that the two resistances shown are constant. The data link connects to a Receiver input, assumed to be of high input resistance. There is a certain amount of stray capacitance between data link and ground, and further capacitance present at the Receiver input. This is lumped together as C_L.

The predominant components are the source resistances and C_L. Simple electrical theory tells us that voltage changes on the data link rises and falls exponentially when the switch changes. When switching to logic 1, assuming here that logic 0 is represented by 0 V, the line voltage V_L is given by:

$$V_L = V_S\{1 - \exp(-t/R_{S(high)}C)\} \tag{6.1}$$

with rise-time $2.2R_{S(high)}C_L$. When switching to logic 0, V_L is given by:

$$V_L = V_S \exp\{-t/(R_{S(low)}C)\} \tag{6.2}$$

with fall time $2.2R_{S(low)}C_L$.

The product $R_{S(low)}C_L$ or $R_{S(high)}C_L$ is known as the time constant.

Figures 6.3(b), (c) and (d) show the impact of the time constant on a data pulse. If the time constant is small compared with the pulse width, then the only effect is a slight slowing of the rising or falling edge, which is generally negligible. As the time constant increases, the rise time can become a significant proportion of the overall pulse. Effectively there is now a delay before the pulse reaches its steady state value at the receiving end. As the time constant becomes proportionally greater still, the pulse is not able to reach its final value, and may not even be detected at the receiving end. This model is applied to the Inter-Integrated Circuit data link later in this chapter.

WORKED EXAMPLE 6.1

A certain logic family has output characteristics which can be modelled as Fig. 6.3, with $R_{S(high)} = R_{S(low)} = 800\ \Omega$. It is to drive a synchronous link in a coaxial cable of 56 pF/m. The clock speed is to be 200 kHz, and it is specified that the clock rise time must be less than or equal to 2 µs. Considering only these parameters, what length of cable can the logic gate drive?

Solution: From above, the rise and fall times will be equal to $(2.2 \times 800 \times 56 \times l)$ ps, where l is the cable length. Setting this equal to the maximum permissible rise/fall time, we can find the maximum length:

$$2.2 \times 800 \times 56 \times l \times 10^{-12} = 2 \times 10^{-6}$$

Solving this leads to a maximum cable length of 20 m.

6.2.2 Transmission line effects

The simple model described above assumes that the data connection is so short that it has no significant series inductance or resistance. In longer data links, and/or when working at high frequency, we can no longer make this assumption. The line is often then modelled as in Fig. 6.4. This figure represents an elemental length of the line, and recognises the fact that resistance, inductance and capacitance are distributed along its length. The actual line we imagine as being made up of a great number of these elements connected together. This model can be applied to applications of widely varying scale: from a short length of PCB track to a cable linking two pieces of equipment, or a telephone cable linking house and exchange, up to a trans-national power line. The analysis of transmission lines is a

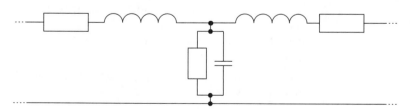

Figure 6.4 Elemental length of transmission line.

complex art, and can be found in texts such as Ref. 6.1. Here we just take note of some of their characteristics.

A data link acting as a transmission line has a number of features, which depend on the values of the components in the line model of Fig. 6.4. The line has a *characteristic impedance*. If there is a change of impedance at any point along the line, a reflection will occur. This can occur at the receiving end itself, or at connectors or other discontinuities. If a reflection back to the transmitter does take place, then the transmitting end may itself reflect the already reflected signal. To avoid any reflection at the receiving end, the receiver impedance must be made equal to the characteristic impedance. This is frequently done by including a terminating resistance of appropriate value across the receiver terminals. Coaxial cable, for example, is often made with characteristic impedance of 50 Ω, and in proper use must be sourced and terminated with this impedance. Other forms of electrical data links have characteristic impedances up to several hundred ohms.

There is also an *attenuation coefficient*, which quantifies by how much a signal is attenuated, usually in dBm^{-1}, and a *phase velocity*, which is the velocity with which a sinusoidal voltage travels along a line.

6.2.3 Electromagnetic interference (EMI)

While this can occur with data links of almost any length, its effect is more pronounced with longer links, and of course with links in a hostile electromagnetic environment. Unfortunately, the latter includes many places where embedded systems are to be found, including most industrial environments and the motor vehicle. It must also be noted that embedded systems are themselves a source of EMI.

EMI is generated due to high voltage rates of change, which may be due to high-frequency systems, low-frequency but high-voltage systems, or any logic system.

There are a number of ways of minimising this problem, including:

- at source of interference:
  - reducing the voltage rate of change, for example by limiting rise and fall speeds of logic or data signals
- within communication link:
  - physical separation from the source of interference
  - increasing the data voltage; the higher the voltage transmitted, the less is the proportional effect of the interference
  - screening
  - use of optical, rather than electrical, links
- at receiver:
  - analogue or digital filtering of incoming signal
  - error checking of received data

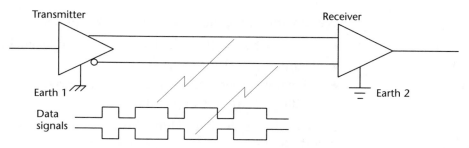

Figure 6.5 Differential data transmission.

6.2.4 Ground differentials

With long links, there can no longer be a certainty that the ground potential at one location is the same as at another, even if they are connected together. The concept of the signal being a voltage referred to a single earth is then no longer valid.

A possible solution to this is to transmit the data as a differential signal (Fig. 6.5). Two identical signals are sent, except that each is the inverse of the other. The received voltage is dependent only on the voltage difference between the two data connections, and is independent of the local ground voltage. Electrical isolation by pulse transformer or opto-isolator is also a possibility. The problem can of course also be solved, usually at the expense of greater complexity, by applying optical communication for the complete data link.

6.3 Serial standards and protocols

6.3.1 Exploring protocols

If electronic devices, whether integrated circuits, or pieces of equipment, are to communicate successfully with each other, then there must be agreement about how the data is to be sent and received. In simple terms, they must speak the same language. A communication protocol is essentially a set of agreed rules which govern how a data link can be established.

As a starting point, let us consider what the capabilities of a conventional parallel bus are, operating in a microprocessor environment. Working within specified voltage levels, and within defined timing constraints, we expect it to be able to:

● transmit or receive data

● identify the intended receiver, using address data

● send control information and synchronisation data, for example Read/ Write

We can anticipate that a serial protocol will need to add the following:

- define whether the link is synchronous or asynchronous
- define permissible data rates
- signal whether a device is ready to receive data
- indicate start and stop of data stream

Beyond this we could develop a wish-list of desirable characteristics, which might include:

- detect error conditions
- specify physical aspects, for example connector and cable type
- accommodate devices of different speeds

This list appears to be a promising start, but the actual situation is somewhat more puzzling. Far from having a standard list of characteristics which every protocol should define, different protocols seem to stress different aspects. Some (like RS-323) are specific about voltage levels. Others, like the CAN (Controller Area Network) bus, do not actually define voltage levels at all. Some protocols are very simple, while others are extremely complex. Some are very well defined, while others (RS-232 again!) have been through numerous revisions, and are open to widely varying implementations.

In the late 1970s, when the manufacture of computer systems was mushrooming, the ISO (International Organization for Standardization) devised a protocol model. This was to encourage manufacturers to produce equipment which would communicate with products made by other manufacturers – even their rivals! Their 'Open System Interconnect' (OSI) model, standard ISO 7498, defines seven protocol layers, as shown in Fig. 6.6.

Each layer provides a defined set of services to the layer above, and each accordingly depends on the services of the layer below. The lowest three layers are network dependent. The physical layer defines the physical and electrical link. The link layer is meant to provide reliable data flow, and includes activities such as error checking and correcting. The network layer places the data within the context of the network, and includes activities

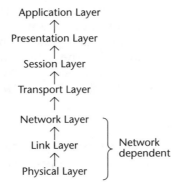

Figure 6.6 The ISO 'Open System Interconnect' model.

such as node addressing. This model forms a framework against which new protocols can be defined, and a useful point of reference when studying the various protocols already available. A new protocol tends to identify which layers of the ISO model it addresses, and those with which it has no concern. Further information on the ISO OSI model may be found in Ref. 6.2.

We will now take a tour of a few of the standards and protocols important in the embedded system environment.

6.3.2 Microwire™ and SPI (Serial Peripheral Interface)

These are simple synchronous serial interface standards, developed by National Semiconductors (Ref. 6.3) and Motorola (Ref. 6.4) respectively. Essentially they derive from the structure of the serial ports built into the microcontrollers of each of these manufacturers: 68HC05 and 68HC11 for Motorola and COP800 for National Semiconductors. Both are intended for short distance communication between a microcontroller and its peripherals, or between several microcontrollers. Both have seen many years of service, coming into existence in the mid-1980s, and both have seen development and evolution (for example MicroWire/Plus™; Ref. 6.5).

Table 6.1 summarises the capability of each of SPI and MicroWire/Plus. Interconnect is achieved in each case with two data lines and a clock. For unidirectional data flow, only one serial line is necessary. To these lines can be added one or more chip select lines derived from parallel port bits. Master and Slave mode is permitted. The Master always controls the system clock, which is divided down from the host microcontroller internal frequency, with division ratios selectable as shown. SPI has an extra

Table 6.1 Characteristic features of SPI and Microwire/Plus™.

	SPI	MicroWire/Plus™
Clock	SCK	SK
Serial data	MISO (Master In, Slave Out) MOSI (Master Out, Slave In)	SI (Serial In) SO (Serial Out)
Chip select lines (from master)	Use port pins	Use port pins
Clock source, internal oscillator divided by:	2, 4, 16 or 32	2, 4 or 8
Clock relative to data	Polarity and phase controllable	SI clocked in on rising edge of SK, SO clocked out on falling edge of SK Polarity reversible in COP888
Data sequence	msb first	msb first
Number of control registers	3	1
Fault detection flags	WCOL (Write Collision) MODF (Mode Fault)	

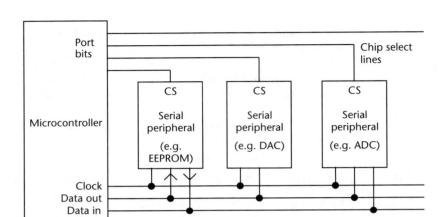

Figure 6.7 Serial devices individually selected.

dedicated input line $\overline{SS}$, which allows an external device to switch an SPI port between slave and master. It can be deduced that SPI is possibly more flexible, but complex, than Microwire/Plus.

A serial DAC, the Maxim MAX538, was introduced as Fig. 5.2 in Chapter 5. This is both SPI- and Microwire-compatible, and is representative of a wide range of ICs, made by many manufacturers, which are compatible with one or both of these standards.

Using either of these serial ports, several serial slave devices with inter-face capability similar to the MAX538 can be driven by a single master, as shown in Fig. 6.7. Some devices (for example the DAC) will just receive data, while others (like the EEPROM memory) will both receive and transmit. This configuration allows independent access to each peripheral IC, as each has its own unique Chip Select line. As each requires its own select line, however, the space-saving advantage of the serial link disappears as the number of peripherals grows.

Devices can also be daisy-chained, as shown in Fig. 6.8. This configuration minimises the number of Chip Select lines used. Serial data must be formatted into a long sequence, which is then sent until the last eight bits

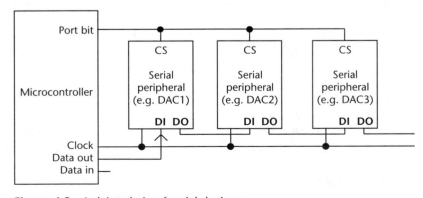

Figure 6.8 A daisy chain of serial devices.

received (assuming 8-bit communication) by each device is the word intended for it. The CS line can then be raised to latch it in. This approach is most suited to requirements where there are a group of identical serial peripherals, for example DACs, as shown. The Data Out from the last peripheral can be looped back to the Microcontroller Data In, for use either if the peripherals are actually returning data to the controller, or as a diagnostic check to ensure the link is complete.

Figures 6.7 and 6.8 hint at the weakness of these simple interfaces. While networks at this level of complexity are simple to implement, it becomes increasingly difficult as either the number of slaves, or the number of masters, increases. Selecting or addressing slave devices is cumbersome, and determining which device should be master at any moment, when there is more than one, is not readily catered for. There is also no built-in handshaking, so that a master can be certain that its data is being received and not just transmitted out to a broken link.

6.3.3   The Inter-Integrated Circuit (I²C) bus

This is a bidirectional, half duplex, two-wire, synchronous bus, originally intended for interconnection over short distances within a piece of equipment. The two lines are called SCL (serial clock) and SDA (serial data). The bus is defined by Philips, and has been through several revisions which have dramatically increased the possible speeds and reflected technological changes, for example in reduced minimum operating voltages. The original version, standard mode, allowed data rates up to 100 kbit/s. Version 1.0, in 1992, increased the maximum data rate to 400 kbit/s. This latter is very well established, and still probably accounts for most I²C implementation. It is defined in Ref. 6.6, and forms the basis of the description which follows. Version 2.0 in 1998 increased the possible bit rate to 3.4 Mbit/s. The most recent version, 2.1, introduced minor revisions; it is defined in Ref. 6.7.

Nodes on an I²C bus can act as Master or Slave. The Master initiates and terminates transfers and generates the clock. The Slave is any device addressed by the Master. A system may have more than one Master, although only one may be active at any time. An arbitration process is defined if more than one Master attempts to control the bus.

A data transfer is made up of the Master signalling a *Start Condition*, followed by one or two bytes containing address and control information. The Start condition (Fig. 6.9(a)) is defined by a high to low transition of SDA when SCL is high. All subsequent data transmission follows the pattern of Fig. 6.9(b). One clock pulse is generated for each data bit, and data may only change when the clock is low. The byte following the Start condition is made up of seven address bits and one data direction bit, as shown in Fig. 6.9(c). A 10-bit addressing mode is also available. A Slave device which recognises its address will then be readied either to receive data or to transmit it onto the bus.

All data transferred is in units of one byte, with no limit on the number of bytes transferred in one message. Each byte must be followed by a 1-bit acknowledge from the receiver, during which time the transmitter

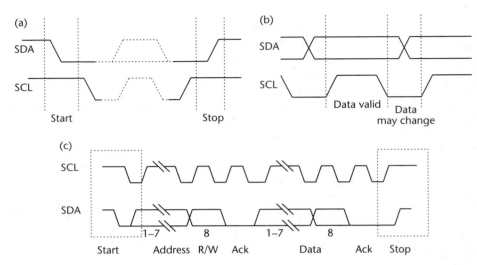

Figure 6.9 I²C data transfer. (a) Start and Stop conditions; (b) Clock and Data timing; (c) a complete transfer of one byte.

relinquishes SDA control. To pause data flow, a receiver may hold the clock line low. A low to high transition of SDA while SCL is high defines a *Stop* condition. Figure 6.9(c) illustrates the complete transfer of a single byte.

Figure 6.10 illustrates many of the electrical features of the Version 1.0 I²C bus. The bus is designed to be tolerant of different device technologies and power supply rails. The two bus lines are driven by open drain or open collector outputs, and are therefore active low. Externally connected pull-up resistors, shown here as R_{PU}, provide the idle high condition. On each line there is a capacitive loading, lumped together as C_L on the diagram, made up of stray interconnection capacitance, and device input

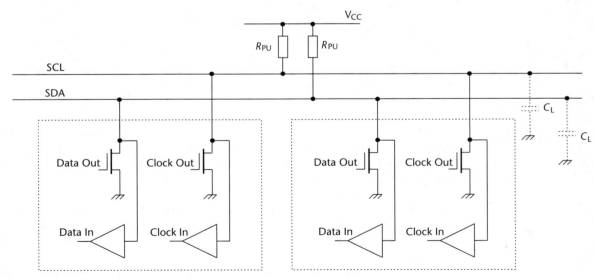

Figure 6.10 Electrical interconnection on the I²C bus.

capacitances. Each line acts essentially as the simple data link of Fig. 6.3. In Version 2.0, the pull-up to logic 1 is no longer by resistor, but is active. This allows the much faster bit rate.

The maximum time permitted in the standard-mode specification for both SDA and SCL lines to rise to logic high input is stated as 1000 ns. In a 5 V system, the minimum logic high input is defined as 3.5 V. As shown in Equation (6.1) and following, this time is dependent on C_L and R_{PU}. Although the maximum bus load capacitance is defined as 400 pF, in practice it will usually be lower than this. The pull-up resistors may therefore be selected according to the estimated line capacitance.

Applying the standard equation for voltage response to a step function input (i.e. Equation (6.1)):

$$\text{line voltage } V_L = 5\{1 - \exp(-t/R_{PU}C_L)\}$$

Setting V_L to 3.5 V gives:

$$\exp(-t/R_{PU}C_L) = 0.30$$

$$(t/R_{PU}C_L) = 1.20$$

And substituting for t = 1000 ns leads to

$$R_{PU}C_L = 830 \text{ ns} \tag{6.3}$$

This time constant must not be exceeded for any combination of C_L and R_{PU}. It follows that for a small system, with low load capacitance, the value for R_{PU} can be high. It must, however, be reduced for high load capacitances. For example, if the line capacitance is estimated to be 100 pF, R_{PU} can be 8k; If the line capacitance is the maximum of 400 pF, R_{PU} must be no more than 2k.

The I^2C bus is a natural choice for many microcontroller-based systems. Many recent microcontrollers have I^2C ports, and many I^2C peripherals are available, including static RAM, EEPROM, display drivers and real-time clocks. Nevertheless, its full implementation is not entirely simple to apply.

Like all good standards, applications of the I^2C bus have gone way beyond what was originally intended for it. While it was perceived as a standard for data distribution within a system, once its value was recognised, applications were developed to use it between equipment. This led to a range of 'bus extender' ICs (e.g. the Philips 82B715), which allowed transmission over greater distances. A good example of a standard being stretched outrageously beyond its original concept can be found in Ref. 6.8, which describes I^2C communication over the distance of one mile!

6.3.3.1 An I^2C device example

Figure 6.11 shows the PCF8574, a 'remote 8-bit I/O expander for the I^2C bus'. This can be used as a remote parallel port. The port bits, P0 to P7, have the 'quasi-bidirectional' characteristic illustrated in Fig. 2.7, and therefore require minimal setting up. An interrupt is generated on a change of state of any of the port bits. The overall interface to the host microcontroller can

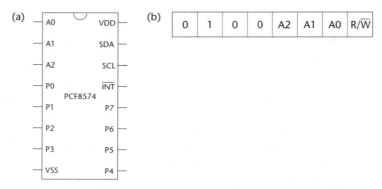

Figure 6.11 The PCF8574, a remote 8-bit I/O expander for the I2C bus. (a) IC pinout; (b) address byte.

therefore be Interrupt, SCL and SDA, which makes a very powerful combination.

The user determines the device address by setting the logic values on pins A0, A1 and A2, generally by hardwiring them to ground or supply. Figure 6.11(b) shows how these bits are integrated into the address byte transmitted from the host, i.e. the first byte of Fig. 6.9(c). The first four bits of the address are preset internally to 0100. Therefore eight different units of the PCF8574 can be included on a single bus, which is usually more than adequate. Other devices are addressed in a similar way, but are given different higher order bits. Therefore an overall system can be made up of numerous slaves, generally each with three address bits user defined, with the higher bits defined by IC manufacturer.

6.3.3.2 An I²C system example

Figure 6.12 illustrates an I²C system used to provide data transfer within an intelligent voltage source, itself controllable by user, or from an IEEE[2] bus. Microcontroller 1 communicates with the user interface, provided by a keypad and display. The keypad and display are interfaced using PCF8574 ICs, with an interrupt from the keypad device to flag when the pad has been pressed. The voltage output is controlled by the DAC, and output current is monitored by the ADC. The IEEE interface is handled by a dedicated controller which has the capability to generate an interrupt when IEEE data is received.

The User Interface section of this system is described in Chapter 9 and illustrated in Fig. 9.14. The pull-up resistors, which do not appear in Fig. 6.12, appear here.

---

2 The IEEE bus (so-called because it is defined by the US Institute of Electrical and Electronics Engineers) is a parallel control interface, widely used to allow automatic computer control of laboratory instruments, for example signal generators, voltmeters and frequency counters. It is also known as the GPIB (General Purpose Interface Bus) bus.

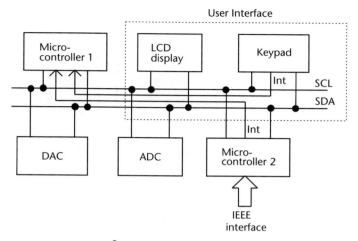

Figure 6.12 The I^{2}C bus applied within an 'intelligent instrument'.

6.3.4 RS-232/EIA-232

This asynchronous standard dates back to the early days of computer communications. Arising from needs recognised in the 1960s for a standardised means of data communication, it was first defined by the Electronics Industries Association (EIA) in 1971. It is still very widely used. It has been through numerous revisions and, being robust, has been subject to many informal adaptations, not all of which are compatible with each other. Along with changes of content in the standard there have been changes in name. Updates were identified by alphabetic suffix, with RS-232C being a particularly well-recognised stage of development. Most importantly, RS-232 became EIA-232 in 1991. The most recent implementation is EIA-232E. Despite these name changes, RS-232 is still widely recognised as a generic title, and continues to be used informally.

The basic EIA-232 data pattern looks like Fig. 6.13(a). As expected, it is similar to Fig. 6.2 except that it is inverted. The nominal signal levels recognised by the receiver are +3 V to +25 V for logic 0 and –3 V to –25 V for logic 1. The transmitter output voltage is specified as being between +5 V and +15 V. Higher voltages give higher noise immunity, and 12 V is commonly found. Baud rates are not actually defined by EIA-232. The common ones in use are, however, 300, 1200, 2400, 9600 and 19 200. Less commonly applied are 110, 600 and 4800.

The standard defines 25 different signals, which include data, timing, control, diagnostic and earth interconnection. These signals are listed in Refs. 6.2 and 6.9. All 25 are rarely used in any modern system, and almost certainly never in an embedded system. Connectors are also specified, either D-type 25-way or 9-way.

EIA-232 carries with it a certain amount of terminology, drawn both from the wider world of data communications and from the labels given to the

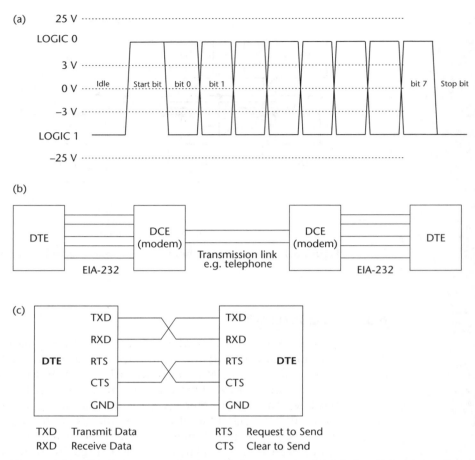

Figure 6.13 EIA-232 communication. (a) Data format; (b) full data link; (c) a null modem link.

data links. Use is made of the terminology Data Terminal Equipment (DTE) and Data Communications Equipment (DCE). These relate to the full data link for which RS-232 was originally intended, depicted in Fig. 6.13(b). Here a computer interfaces to a modem using EIA-232, which communicates with another modem, perhaps over a telephone connection. That modem then communicates with another computer with EIA-232.

Applying EIA-232 is not always simple, as numerous implementations are possible. These include different combinations of signals used, different interpretations put on these signals, and the different ways connectors are applied. In many applications now it is not clear whether a device is acting as a DCE or DTE, and many links are viewed as DTE to DTE. In this case a *null modem* connection is used. A simple implementations is shown in Fig. 6.13(c). RTS and CTS lines are used to signal readiness to send and receive. A minimalist approach is to dispense with RTS and CTS altogether and use three only, i.e. TXD, RXD and ground. This approach is very commonly used in simple links.

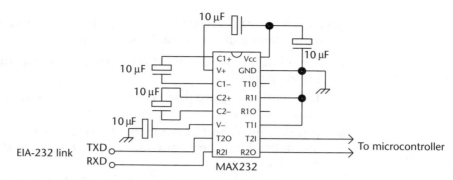

Figure 6.14 EIA-232 voltage levels achieved using a MAX232.

As EIA-232 signal levels are greater than standard logic, it is necessary to boost them to the required voltages. Many ICs, powered by 5 V only, are available for this application. Figure 6.14 shows the popular MAX232, which is one of a family of many similar devices. It is powered only from 5 V, but generates +12 V with internal charge pumps (this useful circuit is explained in Chapter 10). It supports two full-duplex EIA-232 links, with two transmit inputs (T1I, T2I), two transmit outputs (T1O, T2O), and a similar arrangement for receive inputs and outputs. Only one link is implemented in the figure. The lines labelled TXD and RXD form the EIA-232 link, with the associated lines, T2I and R2O, going to the microcontroller serial port.

EIA-232 has been developed into many other similar standards. EIA-422 is not ground referred, but sends the data stream as a differential pair (i.e. similar to Fig. 6.5). The minimum valid signal at the receiver is +200 mV. It supports bit rates up to 10 Mbit/s over a distance of 12 m, or 100 kbit/s over a distance of 1200 m. EIA-423 is ground referred (like EIA-232), but allows longer links and higher data rates due to improved transmitter capability. EIA-485 transmits differential data, and allows for multiple drivers and receivers.

6.3.5 Controller Area Network (CAN) bus

With increasing electronic control in cars, the need for a totally new, flexible, high-reliability serial link was recognised, able to transfer data between a large number of nodes while operating in an electromagnetically hostile environment. No serial standard described so far offers any real data security. I^2C has good addressing capability, but is not proof against EMI, and lacks the line driving capability to distribute signals throughout a vehicle. While EIA-232 or a derivative could be considered, it too is prone to EMI-induced error, and also lacks the ability to address and prioritise nodes in a robust way.

The CAN bus was developed by Bosch to meet the needs of in-vehicle data transfer. It emerged as the front runner among a number of standards developed for the same purpose, and has now become internationally recognised and applied. Its main features include:

- asynchronous communication, with fixed bit rate (up to 1 Mbit/s) for a given system
- half duplex, peer-to peer configuration (i.e. master and slave designation is not used)
- physical interconnect *not* defined, except that logic values on the bus are defined as 'dominant' or 'recessive', where dominant overrides recessive
- flexible bus access: arbitration is applied in the case of simultaneous access, *without* loss of time or data; the arbitration process recognises prioritisation
- unlimited number of nodes
- bus nodes do not have addresses, but apply 'message filtering' to determine whether data on the bus is relevant to them
- high level of security, with exhaustive error checking

6.3.5.1 The CAN data frame

The CAN data frame is shown in Fig. 6.15. The data link idles in the recessive state, and when any node wishes to initiate a data frame it outputs a dominant bit. The arbitration field which follows is made up mainly of an identifier, which both identifies the message and encodes its priority. The control field includes a data length code, which indicates the number of bytes of data, up to a maximum of eight, contained within the following data field. The CRC field contains data used in the cyclic redundancy check.[3] The Acknowledge field is just two bits long. A receiver which successfully receives a message asserts a dominant bit here. The End of Frame is made up of seven recessive bits.

The arbitration process is illustrated in Fig. 6.16. Here, nodes A, B and C are all simultaneously attempting to gain access to the bus, and are all hence putting out an identifier. The connection of nodes A, B and C to the bus *could* be done with the wire-or arrangement of Fig. 6.16(a). In practice,

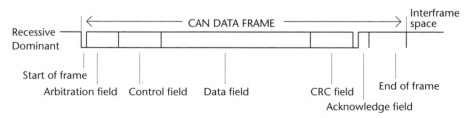

Figure 6.15 The CAN data frame.

---

3 Cyclic redundancy check is a powerful error detection mechanism. It is detailed in Ref. 6.2.

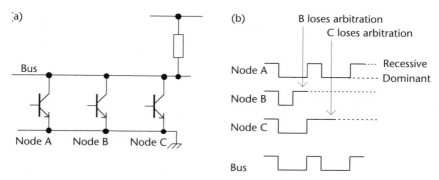

Figure 6.16 (a) Possible electronic implementation; (b) the arbitration process.

an active pull-up arrangement will probably be applied. Whenever there is a difference of value in the identifier, the dominant value (logic 0 in this case) is asserted. First Node B will sense that it has lost arbitration and will cease outputting data. This is followed by Node C. Node A then has control of the bus, and there has been no loss of time or data. It will be seen that the node putting out the lowest value code during arbitration will win. Thus low-value identifiers are given to high-priority messages.

6.3.5.2 CAN error correction

The ability of CAN systems to identify and contain errors is one of the strengths of the bus. In summary these are:

- *Bit Error*: A transmitter monitors the value of the bus during transmission and detects any difference between the value transmitted and the value actually on the bus. Such a difference is acceptable only at certain instances, for example during arbitration.

- *Stuff Error*: Sequences of bits of the same value are interrupted with 'bit stuffing'. An error is detected if a sequence of bits exceeds the maximum stipulated, for example in the case of the bus going open or short circuit.

- *CRC Error*: A cyclic redundancy check is conducted on the data value. An error is detected if the calculated value is not the same as the transmitted one.

- *Form Error*: This is detected when a fixed form bit field contains illegal bits.

- *Acknowledgement Error*: detected when a transmitter does not receive an acknowledge during the ACK slot.

Nodes which find themselves to be faulty can automatically disconnect from the system.

Because of its high reliability, CAN has become widely adopted, not only in the motor vehicle environment but also in other applications, for example factory automation, marine, medical and scientific. It is moderately complex to implement, and the mass of control data which accompanies the actual data in any one data frame forms a high overhead when estimating actual data rates.

6.4 Serial ports

Standalone ICs have been used for many years to implement serial (especially asynchronous) transmission. These have commonly been given names such as:

- **UART**: Universal Asynchronous Receiver Transmitter
- **ACIA**: Asynchronous Communications Interface Adaptor

Such devices are highly programmable and designed for easy interface with a conventional microprocessor system. Modern microcontrollers usually do not need these, but have one or more serial ports on-chip. One may be intended for synchronous communication, another for asynchronous, and possibly another for a more specialist protocol like I^2C or CAN. Alternatively, more general-purpose microcontroller ports are available, which can be configured under user control for either synchronous *or* asynchronous communication. Terminology used in the serial port context includes:

- **SPI**: Serial Peripheral Interface (Motorola, synchronous port, used in 68HC05/08 and 68HC11)
- **SCI**: Serial Communications Interface (asynchronous port, also used in 68HC05/08, 68HC11)
- **SSP**: Synchronous Serial Port (PIC, used for example in the 16C74). This port can operate both in SPI and I^2C modes
- **USART:** Universal Synchronous Asynchronous Receiver Transmitter (PIC, used for example in the 16C74). This port can operate in both full duplex asynchronous mode or half duplex synchronous

All of these are based around the action of a shift register used to convert the external serial I/O to the parallel world inside the controller.

6.4.1 Simple synchronous serial ports

The block diagram of a simple synchronous serial port is shown in Fig. 6.17. The principles it illustrates may be found in the PIC SSP module and the 68HC05/08 SPI. Each, however, has its own significant embellishments. It uses three external pins of the microcontroller, one each for Serial Input, Serial Output and Serial Clock.

Data can be clocked into or out of a shift register, which is loadable/readable from the processor data bus, usually via a buffer. The shift register clock may be derived from an external or internal source. If the external clock is chosen, then it is acting in Slave mode, and can only receive or transmit data when it receives a clock. If the internal clock is chosen, the port is acting in Master mode, as it will initiate and control data transfer. The clock is then also connected to the external clock pin, so that it can be

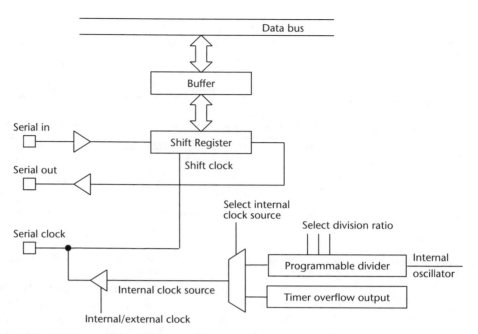

Figure 6.17 A synchronous serial port.

connected to the slave device. If the internal clock source is chosen it may be directly divided down from the processor clock, or it may be derived from a Counter/Timer unit, typically in Auto Reload mode. The former method is used for the faster baud rates. Note that the port can receive and transmit simultaneously, whether in Master or Slave mode.

As with other microcontroller peripherals, one or more Special Function Registers (SFRs) will be associated with the serial port to set up its operating characteristics.

Figure 6.18 shows a common application of the synchronous port. Two identical microcontrollers, each having a serial port like the one of Fig. 6.17, are connected to communicate serially. One is Master and the other Slave. The Slave processor disables its own serial clock, and responds to clock signals from the Master. When the Master clock is active, it clocks data out to the Slave and at the same time clocks data in at the lsb end of the Shift Register. Whether that data is valid will depend on the programming of the controllers. For example, the Master can clock out valid data, but clock in dummy data, or it can clock out dummy data and clock in valid data, or it can clock out valid data and clock in valid data.

6.4.2 A complex synchronous port: the I²C port

An I²C port, though still synchronous, is considerably more difficult to design and operate than the simple ports just described. It must be able to recognise the start and stop conditions, distinguish an incoming address,

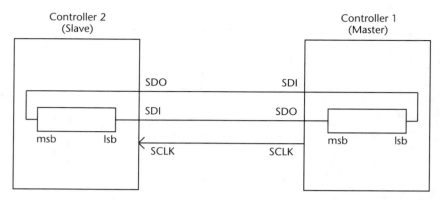

Figure 6.18 Simple synchronous serial link between two microcontroller ports.

recognise and interpret acknowledge bits, and implement the arbitration process.

Most major microcontroller families, especially those of Philips, its inventor (but also PIC and 68HC05/08), now offer I^{2}C ports. The I^{2}C port of the 80C552 can operate in four modes: Master Transmit, Master Receive, Slave Transmit and Slave Receive. To do this it has four SFRs which control the port logic. These are:

- A Data Register, S1DAT. This is the central shift register. Incoming data is clocked in here and outgoing data is clocked from here.

- A Control Register, S1CON. This enables the port and controls start and termination of transfer, bit rate, address recognition and acknowledgement.

- A Status Register, S1STA. This read-only SFR holds a 5-bit code, indicating which of the 26(!) possible states the port is in.

- An Address Register, S1ADR. This is used only in Slave mode, and holds the slave address of the port, determined by the programmer.

6.4.3 Asynchronous serial ports

Ports designed for asynchronous serial communication must be able to generate and interpret data of the form represented in Fig. 6.2. Like the I^{2}C port, they are considerably more sophisticated than the simple synchronous port. Asynchronous ports are usually used to implement standards like EIA-232.

On the transmit side, the port needs to be able to generate a clock at recognised baud rates and insert the start and stop framing bits. Preferably they will also include the choice of inserting a parity bit, and allow the option of transmitting seven or eight bits of data. The transmit side of an asynchronous link is similar to Fig. 6.17, except that the clock signal is not transmitted and circuitry is included to insert the start and stop bits.

On the receive side, it is necessary to have a clock running at a suitable multiple of the baud rate (for example ×16), to detect and synchronise to a

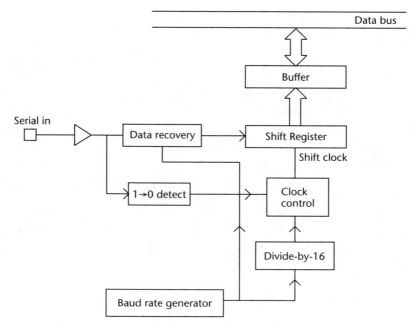

Figure 6.19 Asynchronous receive port.

start bit and, ideally, to have a simple filtering mechanism to reduce the effect of noise pick-up. The receive side of a possible asynchronous link is shown in Fig. 6.19. The Baud Rate Generator is, as with the synchronous port, divided directly down from the internal clock or a Counter/Timer running in Auto-Reload mode. In the case of the latter, the frequency may be derived from Equation (2.5). The resulting baud rate, if there are 16 cycles per bit, is given by:

$$\text{Baud rate} = \frac{f}{16 \times (2^n - R)} \tag{6.4}$$

where f is the Counter/Timer clock frequency, n its number of bits and R the reload value.

The port works like this. The input line idles at logic 1. This is also the value of the stop bit which signals the end of one character. A start bit is detected on the input by a $1 \rightarrow 0$ transition, occurring either after a stop bit to start bit transition or after a period of idling. The clock control is then triggered to commence clocking in a bit, according to the diagram of Fig. 6.2. The data recovery circuit detects the bit value at the centre of the bit time, and this value is then clocked in to the Shift Register. After seven or eight bits (the number depending on the receiver configuration) have been detected the receiver anticipates a stop bit, i.e. a logic 1. If one is detected, then the received word is transferred to the buffer, and the port signals to the processor (usually by interrupt) that a word has been received. A framing error is returned if a stop bit is not detected.

WORKED EXAMPLE   **6.2** _____

An asynchronous port is to be used for EIA-232 communication, with a baud rate of 4.8k. The baud rate is set by an 8-bit Timer/Counter, with Auto Reload capability. The Timer/Counter is clocked from a 1.8432 MHz source. What reload value should be set?

Solution: The baud rate of 4.8 kHz implies that every data bit has a period of $[1/(4.8 \times 10^3)]$ s, i.e. 208.33 µs. The sampling clock must run at 16× this speed, i.e. with a period of 13.02 µs or frequency 76.8 kHz. This is exactly 1/24 of the Counter/Timer clock frequency. Therefore it must be preloaded with a value $(256 - 24) = 232$, or $E8_{16}$.

Note: the factor of 24 discovered above was not just a happy accident! Clock crystals are specifically supplied with frequencies which are direct multiples of EIA-232 baud rates.

6.4.4 'Bit banging'

Having described the availability of various types of hardware serial ports, it is tempting to imagine that without such a port it is impossible to implement a serial link. This is far from the truth. As long as the programmer has a few bit-programmable I/O pins available, a range of serial connections can be implemented, but the software has to do the work. This process is sometimes known colloquially as 'bit banging'. It is a technique that was commonly used in the early days of microcontrollers, and is still sometimes necessary. It is easy to implement simple synchronous links in this way. It is more difficult, but not impossible, to implement asynchronous links like RS232. Even complex protocols like I^2C can be implemented with bit banging. Both Microchip and Motorola publish Application Notes proving that it can be done (Refs. 6.11–6.13).

Figure 6.20 shows the basis of a synchronous output port. Just about all microcontrollers have a 'Rotate through Carry' instruction, which acts on the accumulator, and possibly on a register or memory location. The byte to be transferred is rotated out according to the flow diagram of Fig. 6.20(b). The actual mechanism of transferring the carry bit to the port bit will be microcontroller dependent. The frequency of transmission can be controlled by inserting delay instructions or making use of the Counter/Timer, as discussed in Chapter 3. A 16-bit example of software-generated synchronous serial output can be found in the subroutine dacsend of the Case Study of Appendix C.

Figure 6.21 shows the flow diagram of an asynchronous receive port implemented in software. The program waits to detect a falling edge on the receive line. It then sets a Bit Counter to the number of bits expected to be received (generally seven, eight or nine). It waits half a bit period, and tests whether the start bit is still valid. If not, a framing error is flagged. If yes, the data bits can be clocked in at intervals of one bit period. When the last bit

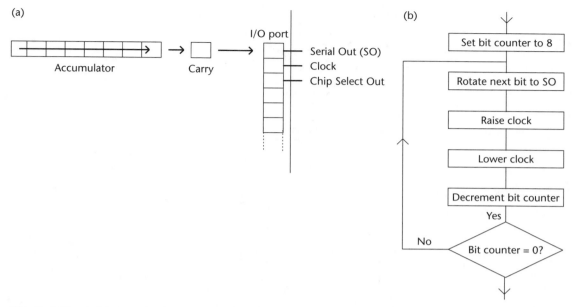

Figure 6.20 A 'bit-banging' synchronous serial output port. (a) Allocating parallel port bits; (b) the flow diagram.

has been received, the routine tests for a stop bit and flags a framing error if this is not found. Again, standard timing techniques will need to be applied to implement this flow diagram.

Bit banging remains a valid technique, and in an undemanding environment it can be cost-saving. It should, however, be noted that it does not readily allow fast baud rates, its flexibility is limited, it takes up processor time, and of course it increases the use of program memory.

SUMMARY

1. Serial data communication is of great importance in the embedded system environment, principally because it places limited demand on interconnection resources.

2. Serial links are broadly categorised into synchronous and asynchronous; the former is transmitted with a clock signal, the latter without.

3. The physical limitations of electronic data links are governed by their underlying electronic principles; these set limits on operation of any standard.

4. There are a very large number of serial data standards and protocols; those important in embedded systems include Microwire, I^2C and CAN bus.

5. Many microcontrollers have serial ports as peripherals. It is also possible to transmit and receive serial data without such ports.

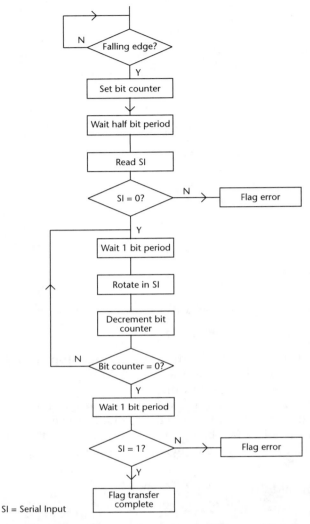

SI = Serial Input

Figure 6.21 Asynchronous receiver implemented in software.

REFERENCES

6.1 Hayt, W. H. (1981) *Engineering Electromagnetics*, 4th edn. New York: McGraw-Hill.

6.2 Green, D. C. (1995) *Data Communication*, 2nd edn. Harlow: Longman.

6.3 *MICROWIRE™ Serial Interface*. Application Note 452. National Semiconductor. January 1992.

6.4 *Using the Serial Peripheral Interface to Communicate Between Multiple Microcomputers*. Application Note AN991. Motorola Inc. 1987.

6.5 *MICROWIRE/PLUS™ Serial Interface for COP800 Family*. Application Note 579. National Semiconductor. May 1989.

6.6 *The I²C Bus and How to Use it*. Application Note. Philips Semiconductors. January 1992.

6.7 *The I²C Bus Specification. Version 2.1.* Philips Semiconductors. January 2000.

6.8 *One Mile Long I²C Communication Using the P82B715.* Application Note. Philips Semiconductors. October 1994.

6.9 Seyer, M. D. (1991) *RS-232 Made Easy*, 2nd edn. Englewood Cliffs, NJ: Prentice Hall.

6.10 *CAN Specification. Version 2.0.* Robert Bosch GmbH. 1991.

6.11 *Using a PIC16C5X as a smart I²C Peripheral.* Application Note 541. Microchip Technology Inc. 1997.

6.12 *HC05 MCU Software-Driven Asynchronous Serial Communication Techniques Using the MC68HC705J2A.* Application Note AN1240. Motorola Inc.

6.13 *Software I²C Communication.* Application Note AN1820. Motorola Inc. 1999.

EXERCISES

('+' indicates a simple review question)

+**6.1** What are the relative advantages of serial and parallel data transfer:

    (a) between items of equipment?

    (b) within an item of equipment?

+**6.2** Describe the difference between synchronous and asynchronous serial communication. What are their relative advantages? Under which category do Microwire, RS232, I²C and CAN bus fall?

+**6.3** State the typical applications of Microwire, RS232, I²C and CAN bus, listing two possible environments in which each would commonly be found.

6.4 Serial data is being transmitted in each of Microwire, RS232, and I²C, each at a bit/baud rate of 100 kHz.

    (a) Estimate the total time taken to transmit one byte of information in each protocol, including all framing/addressing/error correcting data that the protocol requires. State any assumptions made.

    (b) Estimate the total time required by each if 80 bytes are to be sent, optimising each transmission in any way possible.

6.5 Design the hardware for a simple signal synthesiser, based around the 16F84, which has four analogue output channels, each made of a MAX538. The serial interface to each is made by 'bit banging' port bits. All DACs share the same clock input, which is also a port bit.

6.6 Four PCF8574 I/O expander ICs (Fig. 6.11), and two X24165 memory ICs (Fig. 4.13), are connected to a common I²C bus. Two of the PCF8574 ICs are also connected to a single interrupt line.

    (a) Sketch a complete circuit which shows all IC connections which relate to the bus or to the interrupt connection.

    (b) Indicate for each IC its address, ensuring that it is unique.

    (c) For one PCF8574 IC of your choice and one X24165, sketch the timing diagram for the first byte that it would receive if addressed on the bus, if data is to be written to the device.

6.7 Draw a flow diagram for a routine that *transmits* asynchronous serial data and which complements the routine of Fig. 6.21. Discuss the possibility of a full duplex terminal, applying this pair of routines. Assume all timing is done in software loops and that the microcontroller is undertaking no other tasks.

6.8 The block diagram of a conventional successive approximation Analogue to Digital Converter (ADC) is shown in Fig. 6.22(a). An experimental ADC, based on this block diagram, is to be built, with the logic

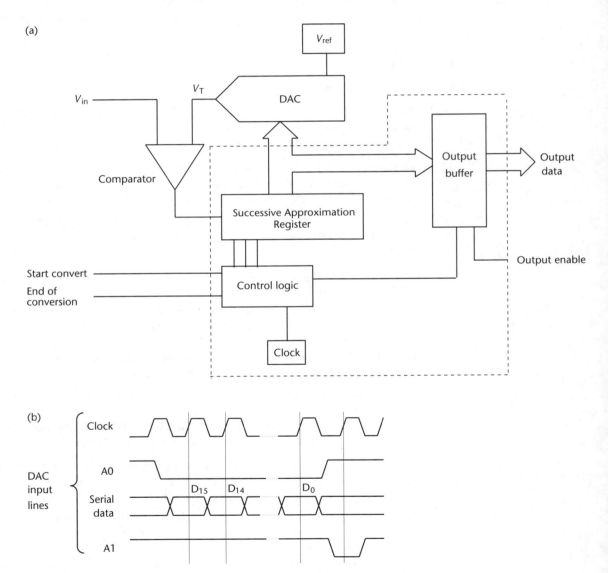

Figure 6.22 In part (b), if A0 is at a logic low, then serial data is clocked in to the input shift register of the DAC on the rising edge of the clock, MSB first. With A0 high and A1 low, data is transferred from the shift register to the DAC buffer, and a digital to analogue conversion commences.

elements of the circuit (i.e. all those parts enclosed by the dashed line) replaced by a PIC 16C84 microcontroller. The DAC is a 16-bit serial input device whose input timing diagram is given in Fig. 6.22(b). A comparator is available whose characteristics can be viewed as ideal. An analogue to digital conversion will be initiated by the 'Start Convert' control line, and should take the form of an interrupt routine.

Show how you would use the 16C84 for this application, giving:

(a) diagrams indicating how you would connect and configure the microcontroller input/output to the DAC, comparator and control lines
(b) one or more flow diagrams showing the actions of the microcontroller during an analogue to digital conversion; indicate clearly how the serial link to the DAC is implemented

To illustrate your answer to part (b), you may wish to make *occasional* reference to suitable 16C84 instructions.

Cambridge University Engineering Department.
Module D8: Electronic System Design. 1997[4]

6.9   How long does the microcontroller in the Case Study of Appendix C take to transmit a 16-bit word to the DAC714 ADC? (In other words, what is the duration of the subroutine dacsend?) If this arrangement was to be used to synthesise a sine wave with 32 samples per cycle, drawn from a look-up table, what would be the approximate maximum frequency?

---

4 The 'answer' to this question forms the basis of Appendix C.

Systematic software

IN THIS CHAPTER

Programming in Assembler was introduced in Chapter 3. There we selected instructions from the microprocessor instruction set, and put them together in sequence to achieve the processor actions required. We introduced simple means of structuring the program, using the flow diagram to perform the initial design and subroutines and interrupt routines to provide some structure.

Once we start developing programs for more advanced systems, we find they have a tendency to get very long very quickly! Even simple applications may require a thousand lines or more of code, while complex ones need considerably more. The simple programming methods of Chapter 3 then begin to fail us. The flow diagram no longer gives adequate insight into how the program should be structured, and the line of code is too small a building block from which to build the overall structure.

In all but the smallest programs, therefore, it becomes necessary to apply systematic principles to the development of programs, similar to those applied in conventional engineering design and development. This has led to a discipline called 'software engineering'. While the aim of software engineering is still to produce a program (usually called 'the software') to run on the target hardware, the process is not simply one of writing lines of code. Software engineering implies systematic design procedures, applying recognised techniques to structure the program, to use and reuse standard building blocks, and to test and commission the program in a complete way.

The aims of this chapter are therefore:

- to identify the aims of software development

- to explore systematic methods of software development and possible program structures

- to consider the relative advantages of some programming language options

- to introduce the main features of the C programming language and how it is applied in the embedded environment

- to consider means of refining the software development process

7.1 Overviewing software development

Before considering the detail of software development, we need an overall picture. What really are our aims in developing software, and what happens to the software once it is developed? These are the questions we try to answer in this section. When they are answered, we will be in a better position to explore the actual process of software development.

7.1.1 The software life cycle

The software of an embedded system project is likely to follow the life cycle pattern shown in Fig. 7.1. If we make certain assumptions about the development process we will adopt, it starts with the specification, a formal statement of the intended program function. It then moves on to defining the program structure and developing the actual program code. After an exhaustive cycle of test and further development the program is released.

Given a successful project there is then likely to be a substantial product life. While on the market there *may* be further software errors to be removed. During this time the available technology will develop and competitors will move in with their own alternative products. Significant upgrades will therefore be required. In all successful projects the time spent in this final phase of the cycle is much longer than the time spent in development. For example, a year may be spent in development, while the product may be on the market for five years and in use for a further five. The people working on the product are likely to change, and those who undertake the later upgrades are unlikely to be those who did the original development.

The product will have a good chance of success if it is early to market and is effective and reliable in the market. In other words, if the development time is short but effective, the total life is likely to be long! Conversely, if too much time is spent on development, there will be no life beyond that phase; the competitors will have got there first.

This chapter focuses on the second and third stages of the software life cycle of Fig. 7.1, i.e. the development of structure and of program code. It is put into the larger context of system development in Chapter 12, which covers the overall design process, as well as test tools and procedures.

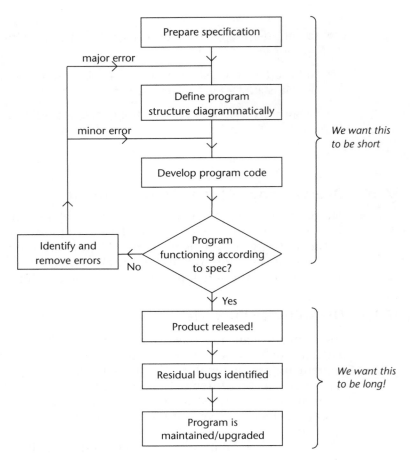

Figure 7.1 The software life cycle.

7.1.2 The aims of software development

Before embarking on the process of software development, let us consider its aims. The apparent overall aim is to produce a program which meets the target specification. It is, however, now simplistic to adopt this as our only aim. Instead we broaden the single aim into five:

- *Aim 1: To develop a program that fully meets the target specification*
  This remains the primary aim. It implies that we must start with a good program specification and we must know how to test that it is met.

- *Aim 2: To develop the program as quickly as possible*
  The development process adopted should be refined to lead rapidly and without deviation to a successful conclusion. Time should not be spent in writing program sections which are already available elsewhere. As the process of finding errors is itself time-consuming, the code should be error-free in the first place, as far as this is possible.

- *Aim 3: To enable subsequent maintenance and upgrading to take place*
  The software must be comprehensible to the human reader, both to its own originator and to a co-worker, with a simple and logical structure and excellent accompanying documentation. The program structure itself must be tolerant of upgrade and maintenance.

- *Aim 4: To make the most efficient use of the system hardware resources*
  The program must optimise the use of hardware, with a view to minimising resource demands and hence possibly minimising cost. Compact programs lead to reduced demands on memory size and hence reduced hardware costs. Replacing a hardware function by its equivalent in software leads to similar saving.

- *Aim 5: To make the most efficient use of time during program execution*
  The program, in its execution, must optimise the use of time. This may enable increased functionality or, given a fixed function, a relaxation in CPU specification or reduced clock speed.[1]

Clearly there is some conflict between these aims. It is, for example, usually found that strategies which lead to fast and efficient program execution tend to be more time-consuming to develop. Conversely, strategies which give quick program development tend to produce code that is less efficient and which executes slower.

Stated overall, in developing software, we need to start with a clear idea of the end result and work quickly and efficiently towards a fully functioning and testable program that is robust and flexible enough to sustain further development. We achieve this essentially by developing a good specification, by choosing the most appropriate programming language for the need in hand, and by *designing* a clear structure for the program.

7.1.3 The process of software development

A number of different software development procedures have been identified. Four possible alternatives, described in Ref. 7.1, are:

- *Linear*: We start at the beginning and go on writing until we reach the end. We could call this the beginner's mode of operation.

- *Evolutionary*: Development starts from some small but complete initial program, and grows by the *ad hoc* addition of new features. We could call this the hacker's mode of operation.

- *Bottom Up*: Program modules designed to fulfil certain identified tasks are written, and later added together to form the overall program.

- *Top Down*: the overall function of the program is first defined, and then successively broken down into smaller and smaller units, until each of these is converted into a module of code. By this process the purpose of each module is understood and defined, as is their interrelationship.

---

1 Reduced clock speed leads to reduced power consumption (as we shall see in Chapter 10), as well as reduced electromagnetic radiation.

Jumping ahead a little, we could call this the professional's mode of development.

Of these development modes, *all* may possibly lead to successful programs, and all therefore have the chance of satisfying Aim 1. The first two will, however, almost certainly lack any recognisable structure, and may fail Aim 3. Without the benefit of any design process, they will probably fail Aims 2, 4 and 5 as well. The third process, of bottom-up development, has some validity, particularly in situations where the program components have to be demonstrated before the overall program can be developed. The program is bolted together from known-good sub-sections, but lacks any real thought about overall structure, so again is likely to fail several of the later aims. The fourth development mode, used well, gives us good hope of achieving all our software development aims. We will therefore adopt it as the vehicle for further development.[2]

7.2 Developing program structure

Any computer program is essentially a series of actions which act in some way or other on the data in the system. These actions follow each other in direct sequence, or in a sequence determined by program jump or branch instructions. The manner in which the sequence is determined is sometimes called the program flow. In order to achieve the aims of software development, we need to explore both how we define the program actions themselves and how they follow each other.

7.2.1 Program flow

The microcontroller instructions which allow program branches and jumps were identified in Section 3.4. These are essential instructions, but the danger with using them indiscriminately is that the resulting program is a tangle of jumps and branches leading anywhere. Such a program is very hard to follow, understand or update, and is likely to contain errors which are hard to trace. It would therefore not meet either of Aims 2 or 3.

To bring order to the inherent dangers of developing program flow, it has been recognised that all programs can be made up using only three forms of flow: *sequence*, *selection* and *iteration*. These are illustrated in Fig. 7.2. Of these, sequence is entirely straightforward. An example of selection is the flow diagram of Fig. 3.10, in which the choice is made between switching a light on or switching it off. Selection is not, however, limited only to a choice of two actions. It may be a choice between taking an action or omitting it. This is represented in Fig. 7.2(b), and is effectively the construct implicit in the PIC test and skip instructions, such as btfsc or

---

2 In practice, we should be honest enough to admit that we all use all of these development modes at one time or other, at least in writing short and maybe undemanding programs. We adopt mode 4 when we recognise that the situation is such that each of the software development aims must rigorously be met.

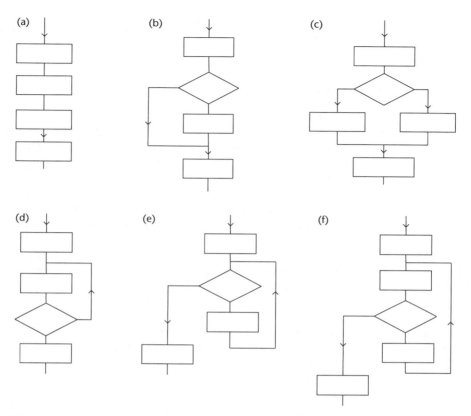

Figure 7.2 Options in program flow. (a) Sequence; (b) selection (action or skip); (c) selection (choice of actions); (d) iteration (post-check); (e) iteration (pre-check); (f) iteration (mid-check).

decfsz. Alternatively, it may be a choice of two or more actions, as in Fig. 7.2(c). However many the number, they are always mutually exclusive.

Three versions of iteration are shown, each of which is conditional. In the *post-check*, the conditional looping branch is made at the end of the loop. This loop will always iterate at least once, and is the most common. An example is the delay loop flow diagram of Fig. 3.14(a), in which a counter is *first* decremented within a loop and its value *then* tested. The loop is repeated if the counter has not reached zero. *Pre-check* and *mid-check* are also possible, as shown. A pre-check loop is used if the possibility must be made of the loop not even being executed once. Iteration can also be endless, with no check whatsoever. An example of this is the Wait loop in the Case Study of Appendix C.

These three primary constructs deliberately do not include the unconditional jump or branch (for example the 16F84 goto), where program execution is transferred from one program point to another, unrelated point. The goto interrupts sequential program flow and makes a program difficult to follow, and hence to develop or maintain. The exception to this is if it is used as part of an unconditional iteration loop, where it continuously leads back to itself. Application of these three constructs leads to controlled program flow, and hence contributes to a well-structured and robust program.

7.2.2 Modules

In the simple assembler programming of Chapter 3, the building block was the assembler instruction, and the program was constructed by adding lines of assembler code to each other. An exception to this was the application of subroutines, which were recommended as potential program building blocks.

Our aims of software development require rapid program development, easy testing and a visible and robust structure. These cannot be achieved if the building block is too small. An improvement over assembler is offered by any High-Level Language (HLL), as a single program line of the HLL will generate at least several lines of machine code. Even with an HLL however, the single program line is a small unit from which to build.

Just as we build an engineering system from a number of sub-systems, each of which has a defined identity and function, so we need to build a program from identifiable sub-systems. In general terms we call these program sub-systems modules. This helps to give the program a clear structure, and thereby aids testing and further development. Importantly, it also allows programs to be built up from pre-tested modules which are already in existence. This in turn greatly assists us in meeting development Aims 2 and 3.

In order to use modules effectively, it is important to define their characteristics as follows:

- they should have a clearly identifiable function
- they should be testable as standalone entities
- they should have defined single entry and exit points, and defined data input and output
- they should be moderately independent of other modules; for example not depending on them excessively for data interchange, or causing them unexpected side-effects
- the programmer should be able to use them as a 'black box', without necessarily needing detailed knowledge of their internal workings.

The sections which follow will show how modules can be identified and developed.

7.2.3 Taming the flow diagram – the module flow diagram

The disadvantage of the traditional Flow Diagram is that it does not lead to the production of well-structured programs, nor does it systematically encourage the production of good code. This is partly because there is no restriction on how the flow diagram symbols are connected together – branches can take place from one point to almost anywhere else.

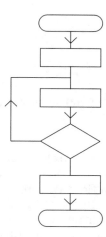

Figure 7.3 A modular flow diagram.

So do we now reject the flow diagram as being of no further use as a programming tool? The answer is a qualified 'No'. It would be a shame to discard a tool which is so easy to learn and use. Instead we can 'tame' the flow diagram by constraining its use to the design of modules, and apply other methods to develop the overall structure.

A *module flow diagram* applies the module constraints listed above to the module design. An example is shown in Fig. 7.3. The diagram must start and end with a terminator symbol, and there must be no entry or exit apart from via these symbols. Such simple diagrams can be used to design the code that goes within a module.

Flow diagrams are not recommended for designing overall program structure, except in simple cases. The section which follows gives a preferred method for this and for determining the role of each module.

7.2.4 Top-down decomposition: the structure diagram

Let us now explore how we can apply the principles of constrained program flow and the characteristics of the software module to developing the structure of an actual program. Top-down decomposition, described as the preferred means of program development earlier in the chapter, is a process of starting 'at the top', with the simplest possible overall statement of the program function. This overall function is then systematically broken down into smaller component functions, which themselves can be broken down into their component functions. At the same time the way that the program moves between these sub-units is determined.

Techniques to do this have been proposed by a number of different individuals and groups. The one outlined here is an informal application of Jackson's Structured Design (JSD); Ref. 7.2. JSD was first developed for data processing programs, and builds on the premise that the structure of a program is closely linked to the structure of the data that it processes.

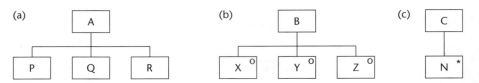

Figure 7.4 JSD program flow representation. (a) Sequence; (b) selection; (c) iteration.

Hence the same notation is used to represent both data and program structure. Here, however, we apply the method only to program structure.

In JSD, the three modes of program flow are represented as in Fig. 7.4. In each case, the upper module is called the parent, and the lower the child or children. It is understood that the parent module is *made up of* the children ones. In the sequence, program module A is made up of modules P, Q and R, which follow each other in turn. When R is complete, then A is complete. There must be at least two modules in the second layer of a sequence, and all are executed when the parent module is executed. In the selection, module B is made up of module X *or* module Y *or* module Z. Like the sequence, there must be at least two modules in the second layer of a selection, but only one is executed as the parent module is executed. Selection is symbolised by a small circle in each child module. In the iteration, module C is made up of a certain number of iterations of module N. There can be *only* one child per parent for iteration. Iteration is symbolised by an asterisk in the iteration block.

Two important rules relating to these diagrams are:

- Any module can have only one parent.
- All child modules of a parent must be of the same type: selection and iteration types for example cannot be mixed. If this appears to present a problem, then dummy modules can be introduced.

JSD leads to a form of block diagram. The most generalised statement of program function is represented in a block at the top of the diagram. This statement can then be broken up into the main program sub-sections, which themselves can be further subdivided, so that a number of layers are created. Generally around three to five layers are recommended for programs of moderate complexity.

The diagrams of Fig. 7.4 seem to raise as many questions as they answer, however. How are the number of iterations determined, for example, and how are conditions for selection expressed? The answer is that conditions, both for selection and iteration, are written into the module which forms the parent of the modules that are executed if that condition is met. This is illustrated for three sportsmen of varying sporting prowess, who choose their sports according to the conditions shown in Fig. 7.5. The sportsman of (a) will swim if it is sunny, and otherwise do nothing. This is an example of the 'action or skip' construct shown in Fig. 7.2(b). The sportsman of Fig. 7.5(b) will swim if it is sunny, and otherwise play squash. This corresponds to Fig. 7.2(c). The sportsman of (c) is similar, except his choice is of three. If

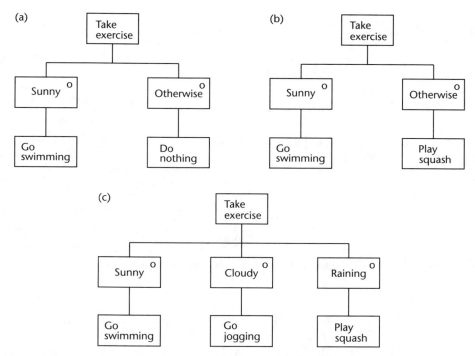

Figure 7.5 Forms of selection: (a) action or skip; (b) choice of two; (c) choice of more than two.

sunny he swims, if raining he plays squash, but if just cloudy he goes jogging.

To develop an example already introduced, the simple flow diagram of Fig. 3.10 is now redrawn as a Structure Diagram in Fig. 7.6. The action is one of unconditional iteration, and for this a 'For Ever' module box has been introduced. Within every iteration there is a sequence of two events, and within the first of these, Switch Light, there is a selection between switching the light on or off.

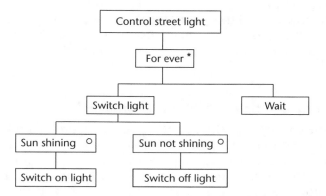

Figure 7.6 Street light controller represented as JSD.

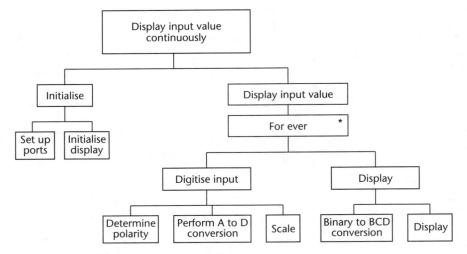

Figure 7.7 Digital panel meter represented with JSD.

A slightly more complex example, the digital voltmeter case study of Appendix C, is represented in Fig. 7.7. The overall task is to display (in digital form) the value of the input voltage, which is represented as *Display Input Value Continuously*. This is then decomposed into a sequence of two events. The first of these, initialisation, occurs only once, while the second, *Display Input Value*, becomes an endless iteration. To obey the rule that children of a parent module must be of the same type, an extra layer, the *For Ever* block, is inserted into the decomposition. This is made up of two sequence stages, which are repeated endlessly, the *Digitise Input*, and *Display*. The children modules of these are the lowest level that this decomposition goes.

By the time we have reached this level we have come across something very useful, for each of these lowest modules has the potential to become a software module, showing the characteristics of software modules listed in Section 7.2.2. Each has a clearly identifiable function, and each should be a testable entity.

Having reached this point, the flow diagram of each module can be developed with a module flow diagram and then converted to code, using whichever programming language has been chosen.

7.3 Assembler or high-level language?

We turn now to considering the question of programming language. It goes almost without saying that its choice is one of the most fundamental decisions of the software development process. There is no 'right' answer to this question; the need must be matched by the most appropriate language. It is, moreover, possible to some extent to mix languages to good effect within one program.

7.3.1 Assemblers, compilers and interpreters

Essentially in the world of the small-scale embedded system we are choosing between one low-level language (Assembler), and several HLLs (C, C++ or conceivably Java). The fundamental difference between Assembler and an HLL is that in Assembler there is a one-to-one correspondence between program instruction and machine code instruction. In an HLL, each program 'instruction' generally leads to several lines of machine code. The programmer develops the program in the language of choice and then a *compiler* or *interpreter* converts that program into machine code. A compiler does this *before* the program is run, so that the program is stored in machine code in program memory, just as with a program developed with Assembler. C and C++ are both compiled languages.

An interpreter actually converts the program to machine code *as the program runs*. The best-known example is the language Basic. This allows very flexible program development, as the program does not need to be compiled or assembled every time a change is made. Its disadvantage is that the interpretation process, taking place as the program runs, is time-consuming. An interpreted program is therefore never likely to achieve fast execution, and is not usually considered seriously for embedded system applications.[3]

7.3.2 Assembler evaluated

Assembler programming allows the programmer to work in a very direct way with the microprocessor instruction set, and the assembler itself handles a number of routine tasks, such as determining program memory addresses. Its advantage is that it gives the most direct control of the microcontroller programming resources possible, and leads therefore to the fastest and most compact code possible. Importantly for us, it is a suitable programming medium to use when learning about microprocessors and controllers. But it is not without disadvantage! Assembler programming requires the programmer to develop every small detail of the program; for example, even a simple multiplication has to be developed as a subroutine. It becomes all too easy to introduce small and almost undetectable errors, which are difficult to track down and remove, and may not even be revealed until months after the program was written. Program development is comparatively slow. It is also easy to write programs which have no structure, and which are incomprehensible to all but the programmer, and incomprehensible even to him or her after a few months. It is particularly difficult to write very large programs in Assembler. Moreover Assembler, by its very nature, is not portable. An Assembler program written for an Intel processor, for example, simply cannot run on a Motorola device.

Despite its weaknesses, Assembler programming is not going to go away, particularly for small programs or program sections which have to be fast

---

3 As always, there are exceptions. The well known 'BASIC Stamp' controller module, based on a PIC, uses a simple Basic interpreter.

and/or compact. For the larger program, however, it is barely a viable option, and there has been, over the past decade or more, a migration towards applying HLLs to microcontrollers.

7.3.3 A high-level language survey

It is generally recognised that programmers write lines of program at approximately the same speed, whatever the language. As any one line of an HLL achieves much more than any line of assembler, this leads directly to the first advantage of the HLL: faster programming. Most HLLs furthermore have flow control features which constrain program structure into 'legal and reliable' constructs. They also have features which control data use and storage. This leads to their second advantage: the structure of the language itself does not permit the spaghetti coding, or programming error based on data misuse, which can cause such heartache in Assembler. Because a good HLL encourages good structure and regulates data use, it follows that software developed in an HLL should also be much easier to maintain.

The disadvantage of a program written in any HLL, when compared with a well-written Assembler program, is that the final machine code for any given task will almost certainly be less compact and efficient. This means that the HLL program will execute somewhat slower and require more memory.[4] The extent to which this happens depends on the HLL itself and the efficiency of the compiler used. While still significant, these disadvantages are becoming of less importance because of the increasing availability of low-cost and high-density memory and the higher speed of processors. Also, compilers are becoming increasingly efficient in their coding.

In the embedded environment we require an HLL which compiles efficiently, i.e. with a low 'bloat factor', and which allows us to remain close to the underlying hardware. The most popular one for this purpose, at the time of writing, is undoubtedly C. It was conceived as a simple and general-purpose language, and because of this it is flexible and can operate close to the hardware. Indeed, unlike most HLLs, it can directly manipulate hardware and memory addresses. Unlike Assembler, it is portable.[5] This means that a program developed for one machine can, perhaps with a little adjustment, be transferred to run on another. Despite its simplicity, however, it provides the flow control features necessary to develop well-structured programs. C allows the fairly rapid development of large working programs. Because it is so important in our environment, we will consider it in some further detail in the following section.

---

4 The term 'bloat factor' is sometimes informally used to indicate by how much an HLL has 'bloated' the memory demands of the program compared with the Assembler equivalent.
5 Because of its ability to access system hardware, C has sometimes been called a 'portable assembler'.

At a certain magnitude of program, C becomes unwieldy. There are therefore a small number of other languages that are becoming increasingly important for the larger scale embedded system and are working their way down into smaller systems.

C++ is an example of *object-oriented* programming. As such it represents a radically different approach to the whole question of program design and visualisation. This approach is well described, and contrasted with the structured program design we have so far considered, in Ref. 7.3. C++ is an evolutionary step on from C, and effectively contains all the features of C as a subset. It requires a larger hardware base on which to run, particularly in terms of memory, and usually leads to slower overall execution. It can, however, be used as the basis for programs very much larger than possible for a C-based product. C++ can with some success be applied in the small-scale embedded environment, particularly when there is an understanding of how to minimise its perceived disadvantages, as described for example in Ref. 7.4.

Embedded C++ is a subset of C++, targeted towards the embedded product. It aims to find a balance between the efficiency of C and the resource demands of C++. Few compilers are available, but it may yet emerge as an important HLL alternative.

Java is a product of Sun Microsystems, and because of this single ownership it is a comparatively stable language. It was originally intended for the embedded world, with a view to being as far as possible machine-independent. To maintain this independence, it introduces the concept of the *Java Virtual Machine*, for which a significant hardware overhead must be paid. With the advent of the World Wide Web, with many machines of different types communicating with each other, this concept becomes of great importance, and Java has become widely used in Web applications. It is a 'semi-interpreted' language, and therefore does not lead to particularly fast code execution. Like C++, Java is object-oriented, but unlike C++ it was designed from scratch, so it was able to adopt completely novel solutions in its design.

A fully featured Java-based system is most likely to be at home on a machine with the computing power of a personal computer. Nevertheless, Java is penetrating further into the small-scale embedded world. Certain subsets of the language have been defined, including *Personal Java* and *Embedded Java* (Ref. 7.5). As an example of the resource demands of these subsets, Ref. 7.5 estimates that Embedded Java can run on a system having less than 512 Kbyte of ROM, less than 512 Kbyte of RAM, and using a 16 or 32-bit processor. This shows that its hardware demands are still some way beyond the scale of designs considered in this book.

7.4 The C programming language – an overview

We now turn to look at C in further detail. What follows is *not* intended to teach the C language; for that a full course or textbook is needed. The purpose is to introduce the reader to C in a general way, allowing you to recognise its advantages over Assembler, to read a simple C program with some understanding, and to evaluate a microcontroller C compiler. The application of C in the embedded environment is then considered.

C was developed in the early 1970s at the Bell Telephone labs in order to write the Unix operating system. Its first 'formalised' version was actually in the book *The C Programming Language*, by Kernighan and Ritchie (Ref. 7.6). This book was so influential that one continues to find reference to the 'K&R' version of C. In 1989 C was defined as an ANSI (American National Standards Institute) standard, and most commonly one now finds reference to ANSI C. A little later this same version was adopted by the ISO (International Organization for Standardization) as an international standard. The ISO subsequently issued an amendment which developed the support for handling large character sets. A full description of this internationally adopted version of C can be found in many standard texts, for example Ref. 7.7. Because of this long period of development, the C language has the advantage of being stable.

7.4.1 Data types

Starting with the mundane but essential, all variables in a C program must be declared before use and the data type defined. The type of data determines the size of memory, and hence the possible number range. Once the data type of a variable has been defined, it is up to the compiler to allocate it a memory location. The basic data types are shown in Table 7.1, along with *typical* storage size. These types appear in different forms; for example, integers may be *short* or *long*, *signed* or *unsigned*. These latter qualifiers determine the size of memory storage (which is ultimately set by the compiler), and hence the possible number range. It is important to check storage sizes before using a compiler. They are *not* all standardised, and errors can occur due to range overflow, which the compiler may fail to detect.

Data names must start with a letter. Usually the name is prefixed with a letter identifying their type. This gives no further information to the C compiler, but is an *aide-mémoire* to the programmer, and limits errors due to incorrect data typing.

7.4.2 C functions

The building block of the C program is the *function*, which may be custom written or drawn from a standard library. Every C program must have at least one function, which is always called main. Program execution starts at the beginning of main.

Table 7.1 C data types.

char	character	1 byte
int	integer	4 bytes
float	floating point	4 bytes
double	floating point, double precision	8 bytes

Functions are made up of declarations and statements. Declarations can be used to define variables used within that function, and hence usually appear at the beginning of the function. A statement defines an operation, and is terminated by a semicolon. Curly brackets { and } are used to define a group of declarations and/or statements as a block, for example as a function.

Functions have a close similarity to assembler subroutines. Unlike their assembler cousins, however, data transfer to C functions is much more strictly defined. Data is passed to functions by *arguments*, which are enclosed in brackets immediately following the function name. At the top of the function is a header, which takes the form:

```
returntype   name   (list of arguments)
```

Here `returntype` is the data type of the data returned from the function. This format forces the programmer to be specific as to the data which is transferred to the function and that which is returned from it. Functions are called in the program simply by naming them and attaching to the name the appropriate list of arguments. A function ends with a `return` statement, which passes control back to the calling environment.

The passing of arguments is done on the Stack, and it is the actual value that is passed, rather than the address of where it is held. This means that the function is not able to alter the original version of that variable, only its own temporary version.

7.4.2.1 *Function libraries*

Because of the limited functionality of the language itself, C has a heavy dependence on libraries of standard functions, which are supplied with every compiler. Functions in libraries are saved as pre-compiled object files. The main program interfaces to these through *header files* which are included in the program. Header files contain declarations of functions and constants used by that particular library. For example, a value for π is included in the maths header file. Commonly used (in practice ubiquitous) function libraries are input/output (header file `stdio.h`), general utilities (header file `stdlib.h`) and maths (header file `math.h`).

7.4.3 Keywords

C has 32 reserved 'keywords'. These are used for program flow control and declaration of data types. They cannot be used by the programmer as names. A selection of keywords are shown in Table 7.2. It can immediately be recognised that a number of them have already been mentioned in connection with data types; others will be introduced below.

7.4.4 The C pre-preprocessor and pre-preprocessor directives

When a C program is compiled, it is initially scanned by a *pre-processor*. This responds to a set of directives, similar to assembler directives, which give

Table 7.2 Example C keywords.

break	double	goto	return	switch
char	else	if	short	unsigned
const	float	int	signed	void
continue	for	long	static	volatile
do				while

Table 7.3 Example preprocessor directives.

#define	#error	#include
#else	#if	#line
#elif	#ifdef	#pragma
#endif	#ifndef	#undef

instructions to the compiler, rather than ending up as part of the final program. Example directives, each of which starts with a # symbol, are shown in Table 7.3.

The #define directive can be used to define constants. The #include directive is used to insert another named file at its place in the program. There is also a set of directives, for example #if, #endif and #ifdef which allow conditional compiling of selected sections of code.

7.4.5 Program layout

Unlike Assembler, the page layout of C is not defined by any particular format; the programmer can to some extent use the layout to improve program clarity. For example, code can be indented and blank lines included to improve readability. The exact positioning of the curly brackets enclosing a function is also not defined.

A simple C program[6] is shown in Fig. 7.8. Comment lines are contained within the symbols /* and */, and are used initially to define the program title block. There follows two #include directives, which causes the insertion of header files stdio.h and stdlib.h at that point.

The function main starts with a header, which follows the form shown above. As there are no arguments being sent, the keyword void is entered within the arguments bracket. It continues by defining the data type of the two variables it will use. In this example both are integers.

The program then proceeds to print a message to the screen using the function printf, which is in the stdio library. Using scanf, it waits for a response, which it places into memory location iweeks. Formats for printf and scanf are indicated in the library documentation. The program now tests whether the number of weeks entered is excessive. If so, it prints an error message, and exits the program using function exit. Otherwise, it

---

6 This program was compiled on a personal computer using the Pacific C Compiler, currently available as freeware (Ref. 7.8).

```
/*
==========================================
Program to compute memory requirement
for a data logger recording 2 bytes of
data every hour. Logger is not required
to record for more than one year.
                            TJW Dec'98
==========================================
*/

#include <stdio.h>
#include <stdlib.h>

int main (void)  {
int iweeks;
int ibytes;

/*Input number of weeks*/
printf("Input number of weeks:");
scanf("%d",&iweeks);

/*check for excess weeks*/
if(iweeks>52){
        printf("Too many Weeks\n");
        exit (1);
        }

/*compute memory requirement*/
ibytes = iweeks*7*24*2;
printf("Memory Requirement is %6i \n", ibytes);
return (0);
}
```

Figure 7.8   Example C program.

goes on to compute and display the memory size required. The function is terminated with statement return(0). This indicates successful program completion.

7.4.6 C operators

A reflection of C's simplicity is its limited range of operators. There are, for example, no trigonometric ones, and where these are needed they must be done by a function call to a library. It does, however, have logical operators, and operations on single bits. Operators are allocated an order of precedence, so that when several are used in a single expression the high-precedence ones are evaluated first. Example operators are shown in Table 7.4.

7.4.7 Flow control

C statements follow each other sequentially, except when certain flow control keywords are applied. Using some of the keywords mentioned above, C has the constructs shown in Table 7.5. These are grouped together, to reflect the main branching options shown in Fig. 7.2.

Table 7.4 Example C operators.

add	X+Y	bitwise AND	X&Y
subtract	X-Y	bitwise OR	X|Y
multiply	X*Y	left shift	X<Y
divide	X/Y	right shift	X>Y
unary plus	+X	postincrement	X++
unary minus	-X	postdecrement	X--
greater than	X>Y	preincrement	++X
less than	X<Y	predecrement	--X

Table 7.5 C keywords applied in control flow.

Selection	Iteration	Other
if	while	
if...else	do...while	
else...if	for	
switch	break	goto

A simple example is included in the program of Fig. 7.8. If the number of weeks entered is greater than 52, then the statements within the following curly braces are executed, i.e. the error message 'Too many Weeks' is printed and the program exits. This construct allows implementation of the 'action or skip' selection construct of Fig. 7.5(a). To implement a choice of two, the if else construct is applied.

7.4.8 Files and file structure

A typical C file structure, which also illustrates the program development process, is shown in Fig. 7.9. It is similar to that of Assembler (i.e. the similarity with Fig. 3.9 is not accidental!), except that a compilation process replaces assembling. A C source file, with extension .c, is developed as a text file. This almost certainly will include one or more library header files. It is compiled to an object file. Other object files may have been created by other C compilations or from Assembler sources. Alternatively they may be drawn from a library. A linker links these together before creating the final executable file. File extensions for source, header and object files are normally standard as shown. Those for the executable file vary among compilers.

7.5 Using C in the embedded environment

When C leaves the personal computer or workstation and moves over to the embedded system, it has to adapt to the culture shock of limited processing

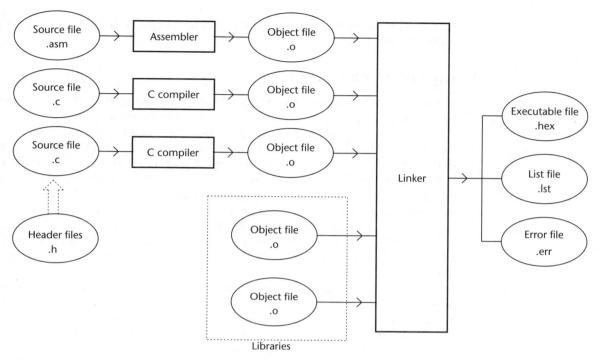

Figure 7.9 Example C file structure.

power, limited memory resources and extensive input/output demands. In outline, the compiler needs to be able to:

- accommodate the limited processing capability
- allow easy interaction with the peripheral devices
- as part of this, enable ready use of interrupts
- optimise use of limited memory
- permit the embedding of Assembler program sections

C Compilers are available for most microprocessors and micro-controllers. Generally they are part of, or link into, a larger Integrated Development Environment, which will also include an Assembler, Simulator and Linker. These compilers may be produced by the processor manufacturer or by a third-party supplier.

In response to the microcontroller environment, the typical C compiler may add keywords, restrict data types, offer pre-compiled object files for peripheral initialisation and handling, and encourage focused memory modelling.

Two example C compilers are cited in Refs. 7.9 and 7.10. The MPLAB™–CXX Compiler, supplied by Microchip itself, comes in two versions: the C17 for 17- series PIC microcontrollers, and the C18 for the 18- series. The IAR compiler is one of many offered by this company, who supply compilers for many popular microcontroller families, including both PIC and 8051.

7.5.1 The MPLAB-CXX compilers

The MPLAB-CXX Compilers add six keywords to the 32 of C. These are _asm and _endasm, to permit the easy inclusion of assembler blocks; far and near, to aid in the handling of banked memory; and ram and rom, to respond to the Harvard memory structure of the PIC. It restricts data types, particularly in the MPLAB-C17, which does not support double, float or long data types. It does, however, make use of the volatile storage class for variables, like input port values, which are changed outside the control of the program. Variables are passed to functions on a software stack set up in data memory.

MPLAB-C17 offers four memory models, as shown in Table 7.6. This leads to a wide range of libraries, as for many functions there are different libraries for each memory model, as well as for each microcontroller for which they are offered.

To specify the target processor and define the SFRs, MPLAB-CXX uses processor-specific header files and associated register definition files. System initialisation takes place immediately after reset (i.e. before main). This includes setting up the software stack, and can also include setting up data in data memory and peripheral initialisation.

Assembler sections may be included into a MPLAB-CXX program in two ways. Large sections of Assembler may be written in MPLAB in the usual way and assembled into an object file for later linking into the final executable program. Small sections of assembler can be included directly in the program, contained between the _asm and _endasm keywords.

There are certain microcontroller instructions which it is more or less impossible to replicate in C. It would be tedious, but possible, to repeatedly drop into assembler to implement these. MPLAB-CXX therefore includes a set of macros which allow these to be accessed directly within the C program. They are shown in Table 7.7. All of these are drawn directly from the PIC 18XXX instruction set, and many are familiar from the 16F84 instruction set of Appendix B.

Table 7.6 MPLAB-C17 memory models

Memory model	Memory size
Small	Program memory $\leq$ 8K Data memory $\leq$ 256
Medium	Program memory > 8K Data memory $\leq$ 256
Compact	Program memory $\leq$ 8K Data memory > 256
Large	Program memory > 8K Data memory > 256

Table 7.7 C macros for assembler instructions.

Macro	Assembler equivalent, or action
Nop ( )	no operation
Clrwdt ( )	clear watch dog timer
Sleep ( )	enter sleep mode
Reset ( )	device reset
Rlcf (var)	rotate *var* left through carry
Rlncf (var)	rotate *var* left, not through carry
Rrcf (var)	rotate *var* right through carry
Rrncf (var)	rotate *var* right, not through carry
Swapf (var)	swap upper and lower nibbles of *var*

MPLAB-C17 has an extensive selection of pre-compiled object files. These include libraries to handle on-chip peripherals. For example, there are six functions to deal with the ADC, with self-explanatory names: BusyADC, CloseADC, ConvertADC, OpenADC, ReadADC and SetChanADC. The arguments of these, as appropriate, set the ADC to the desired operation. Beyond this, there are further functions for dealing with common external devices. There are, for example, a set of functions just to deal with the popular Hitachi HD44780 LCD (liquid crystal display) controller described in Chapter 9.

The IAR PIC C compiler has many characteristics which reflect those of MPLAB-C17, including its own language extensions. Unlike the MPLAB compilers, a single compiler is able to compile for different PIC microcontroller families, and the full range of data types is implemented for each family. In connection with its language extensions, an interesting feature is its ability to switch these off, using the strict_ansi option.

7.5.2 Portability issues

Reading the section directly above will cause anyone to question the belief that C is a portable language, at least as far as microcontrollers are concerned. With header files, initialisation, and assembler infixes all being processor-specific, and data types to some extent compiler-dependent, there appears to be little chance of just moving a C program written for one processor over to another. In many applications this will simply not matter: the programmer or company has no wish to change processor. In others, however, it is important. A processor change may be forced due to considerations of availability or cost-saving, or because a product is being upgraded and the current processor is no longer adequate.

If a C program is to be transferred between processor families, there will need to be a painstaking process of rewriting. Header files, initialisation procedures and all Assembler sections will have to be replaced. All applications of processor-specific macros, such as are shown in Table 7.7, will also need to be changed. As these directly reflect the processor instruction set, it will be found that there is unlikely to be a one-to-one correspondence between them.

If a change is anticipated before a program is written, then some care can be taken to restrict the use of these processor-specific aspects, and to clearly identify them by commenting all aspects which remain dependent on the microcontroller.

7.5.3 Safety-conscious C – MISRA C

C has been widely adopted in the motor vehicle industry, in many cases as the only practical language to choose. Assembler is discarded for reasons already discussed, and no other HLL offers the code and time efficiency of C, as well as its ability to access the underlying hardware. Because of its extreme cost competitiveness, a high level of portability is required in this environment, with manufacturers ready to switch microcontroller supplier if the opportunity to make any savings presents itself.

It is, however, recognised that C carries a number of features which do not lead to the best reliability. Just as with Assembler, it is easy to introduce typographical errors which are not readily detected, but still lead to valid code (for example, entering == instead of =, or vice versa). Programmers can make mistakes or write code that is difficult to follow. Compilers themselves can interpret various 'grey areas' of the language in different ways.

These problems are exacerbated when an attempt is made to transfer a program from one controller family to another. All those special customised extras of the compiler developed for one family simply aren't portable! Alternatively, they may appear portable but carry hidden errors. With compilers using different interpretations of data types, this does not take long to happen.

Nevertheless, as a mature language the shortcomings of C have at least been studied and understood. The Motor Industry Software Reliability Association (MISRA) has defined a set of rules, leading to a subset of C which aims to facilitate the production of C in safety-conscious applications. These are intended to lead to reliable portability, good program structure and explicitly correct coding. The document, Ref. 7.11, is short and readable, and even if not formally applied, leads to a valuable insight into good C program development. There are 127 rules, which are all categorised as either *advisory* or *required*. A few examples of the simpler and more self-explanatory rules, drawn from the MISRA-C rule summary, are quoted in Table 7.8. In every case further explanation is given in the document.

It is recognised by MISRA that a company adopting these rules may phase them in over a period of time, or may adopt only a subset of them.

7.6 Optimising the development process

The second of our aims for software development requires the program to be developed as fast as possible. While we do this partly by working with

Table 7.8   Example MISRA C rules.

Required

1. All code shall conform to ISO 9899 standard C, with no extensions permitted.

14. The type char shall always be declared as unsigned char or signed char.

50. Floating point variables shall not be tested for exact equality or inequality.

56. The goto statement shall not be used.

57. The continue statement shall not be used.

Advisory

10. Sections of code should not be 'commented out'.

15. Floating point implementations should comply with a defined floating point standard.

47. No dependence should be placed on C's operator precedence rules in expressions.

82. A function should have a single point of exit.

the most appropriate programming language, and by building a simple and effective structure, we also do it by evolving development procedures which minimise time.

7.6.1   Version control

While developing the program, the developer will spend considerable time in the loop of Fig. 7.1. Many versions of the program are likely to be developed in quick succession. In the process, superseded versions of a program may be lost, yet they can give valuable information about the development process. For example, sometimes an error will be detected but a new one introduced as the first is fixed. This new one may not be detected until much later in the development process. Knowledge of when it may have been introduced can be very valuable. When the product is on the market there will be further program iterations, each with its own performance characteristic. An old model may come in for maintenance, with information needed on its software version. During this time the design tools themselves, for example the compiler, are likely to be superseded.

It is important therefore to be able to track software (and hardware) versions, and maintain records of those that are superseded and of their lifespan. A useful start is of course to date each software issue rigorously and give a version number. If these are simply overwriting each other, however, that may be of limited help. An individual or company can devise a means of archiving all versions, and even the software development tools, as these become superseded. Version control systems, which maintain a database of all program versions, are also commercially available. These enforce discipline in the archiving of records, and if used properly ensure that the project history is held in its entirety.

7.6.2 Style and layout

Some consideration was given in Chapter 3 to the question of style and layout in Assembler programs. With the freeform nature of C, there is considerable scope to vary the layout. If used to good advantage, it can greatly improve the readability of the program. Systematic use of indentations and placing of braces are commonly used means of improving C program layout. Formal style and standards guides, for example Ref. 7.12, are applied in many companies.

7.6.3 Learning from experience

Every time a program error is discovered, it is worth pausing to consider how it got there in the first place. Is it typographical, or a misunderstanding of the language, or something more complex? It is worthwhile keeping a log of all errors and seeing whether a pattern of common ones emerges. If so, consideration should be given to how that type of error can be eliminated.

SUMMARY

1. It is essential to apply systematic design methods to the development of embedded system software. These methods will lead to programs structured from modules, with restricted program flow options.

2. Assembler is a suitable programming language for short to medium length programs, or programs which must execute fast or occupy minimal memory. It is also a suitable language for learning about microprocessors and controllers. In many applications, however, Assembler is unlikely to lead to the fastest development time or the most reliable program.

3. An efficient high-level language, such as C, is the preferred choice for much embedded software. It is readily able to access the system hardware, and leads to programs of good code and time efficiency.

4. C has been adapted to the embedded environment in a large number of compilers. These apply considerable ingenuity towards optimising the use of the target microcontroller in the C environment.

REFERENCES

7.1   Heaven, S. (1995) *An Introduction to Software Design*. London: Edward Arnold.
7.2   King, D. (1988) *Creating Effective Software: Computer Program Design Using the Jackson Methodology*. Englewood Cliffs, NJ: Prentice Hall.
7.3   Cooling, J. E. (1997) *Real-Time Software Systems; An Introduction to Structured and Object-Oriented Design*. London: International Thomson Computer Press.
7.4   Kreidl, H. (2000) Using C++ in embedded applications on Motorola's 8- and 16-bit microcontrollers. *Embedded System Engineering*, 8(5), 42–60.

7.5   Perrier, V. (2000) Adapting Java for embedded development. *IEE Review*, **46**(3), 29–35.

7.6   Kernighan, B. W. and Ritchie, D. M. (1978) *The C Programming Language.* Englewood Cliffs, NJ: Prentice Hall.

7.7   Plauger, P. J. and Brodie, J. (1995) *Standard C, a Reference.* Englewood Cliffs, NJ: Prentice-Hall.

7.8   Pacific C. http://www.htsoft.com/products/pacific.html.

7.9   *MPLAB™–CXX Compiler User's Guide.* Microchip Technology Inc. 2000. Document number DS51217B.

7.10 *PICmicro™ C Compiler, Programming Guide.* IAR Systems. October 1998. Part no ICCPIC-1.

7.11 *Guidelines for the Use of the C Language in Vehicle Based Software.* Motor Industry Software Reliability Association. 1998.

7.12 *A Firmware Standards Manual.* http://www.ganssle.com/.

EXERCISES

('+' indicates a simple review question)

7.1   Draw up a list of the five aims of software development. Identify against each aim the other aim(s) which conflict with it. Explain the reason for the conflict.

7.2   A machine for sorting waste containers consists of a conveyor belt along which the waste items are presented in turn. The items may be plastic, glass or metal. The initial sorting is done acoustically: the waste item falls a short way onto a metal bar, and the resulting sound is analysed to identify the material. The machine then directs the item into the appropriate channel. Plastic items immediately go into a plastics bin. For glass there is a further identification stage, done optically, to distinguish between clear, brown and green glass, for each of which there is one bin. For metal there is a further stage, to distinguish between ferrous and non-ferrous material, each again being allocated one bin. All sorting is done sequentially, i.e. one item is completely categorised before the next one is dealt with. After every 500 items, the machine is stopped for automatic cleaning and calibration of the sensors. At this time it sends, to a central controller, logged data on how many of each item has been identified. The machine is controlled by a single microcontroller. Design a structure for its software.

+**7.3**   List the main advantages and disadvantages of Assembler programming.

7.4   Using only the information given in this chapter, give five reasons why C is preferred as a programming language to Assembler and two reasons why it may not be. Explain your reasons.

7.5   In the embedded environment the programmer must work close to the system hardware, but any high-level language aims to be completely portable, i.e. hardware-independent. Explain what steps the MPLAB-CXX C compiler takes to solve this dilemma.

7.6 Using only the information given in this chapter, comment on the potential ability of the MPLAB-CXX and IAR C compilers to comply with the MISRA-C guidelines.

Dealing with time

IN THIS CHAPTER

Almost every embedded system has a strong dependence on time. This may mean that a timely response must be made to an external event, that certain time-related measurements must be made, or that certain time-dependent outputs need to be generated.

For all time-related activities, the hardware tools of interrupts and Counter/Timer are important, and need to be applied with good understanding. From a software point of view, if the requirements are not too demanding it is possible to satisfy them with the programming techniques that have already been described. In general this is the case for a system dedicated to only one activity.

Most systems, however, undertake more than one activity at the same time, a situation sometimes called multi-tasking. This means that there may be multiple interrupts enabled or that multiple outputs have to be generated, all of which are time-dependent. Great care and understanding then have to be exercised with both hardware and software if all the demands placed on the processor are to be successfully met.

This chapter aims to explore the principles of working with and designing time-dependent systems. The first half is devoted to developing a greater understanding of interrupt and Counter/Timer structures, first considered in Chapter 2. The second half introduces the concepts of multi-tasking and of the real-time operating system.

Specifically the chapter aims to:

- introduce the concept of real-time operation

- describe advanced interrupt structures and their applications

- describe advanced Counter/Timer structures and their applications

- introduce the concept of multi-tasking and its requirements

- introduce in simple terms the real-time operating system and its applications

8.1 What is 'real-time'?

Systems which have a strong dependence on time are often said to operate in 'real' time. Let us try to clarify what this terminology means. Many people entering the world of embedded systems think they have a good notion of what a real-time system is, but then find it difficult to offer a good definition. Suggestions of rapid response, or responsiveness to time requirements imposed from outside, are readily made. Failing to find a definition, one could try the alternative, and attempt to define what is *not* a 'real-time' system. Here people suggest 'a system for which time is not relevant', or 'a system which has no time dependence on any other system'. This becomes dangerous ground itself, for where is the computer or embedded system which has no time dependence whatsoever? Yet despite this apparent uncertainty, it is not too difficult to list examples of systems which are most definitely 'real-time'. The motor vehicle environment gives many examples, including the engine management or braking systems. An example sometimes quoted of a 'non real-time' system is the company payroll accounting program.

The above paragraph hints at the difficulty of defining real-time systems. This difficulty need not surprise us, because there is not necessarily a hard-and-fast distinction between those systems which are real-time and those which are not. The variation is really the extent of time dependence. A number of different definitions for 'real-time' have been proposed. A simple and workable one (Ref. 8.1), which we adopt, is as follows:

A system operating in real time must be able to provide the correct results at the required time deadlines.

Notice that the only requirement in this definition is that time deadlines are met. There is no absolute requirement for high speed, although it does usually give an advantage. If the time deadline is short, then the system must be fast. If the deadline is long, the system can be slow. Because *all* systems have some time dependence, a further distinction is frequently made between *hard* real-time and *soft*. In hard systems, the above definition is a non-negotiable requirement; in soft systems there is some flexibility in how well the deadlines are met. After all, most of the world's working population have a strong view that their payroll accounting *is* a real-time system!

Programming for real-time applications is characterised by:

- the ability of the program to respond at appropriate speed (usually rapidly) to external events occurring at unpredictable times

- the ability to measure time intervals accurately, and hence to initiate time-based activity

- in complex systems, the ability to distinguish between needs which are time-critical and those which are not, and to respond to them accordingly

We spend much of the rest of the chapter exploring how these characteristics can be achieved.

8.2 Advanced interrupt structures

The interrupt structure of a simple microcontroller, the 16F84, was introduced in Chapter 2. Its capabilities are now contrasted in Table 8.1 with two much more complex controllers. The difference is first of all in scale. Four interrupt sources in the 16F84 are replaced by 15 and 23 in the 80C552 and 68HC11 respectively. Instead of the single interrupt vector of the 16F84, the '552 and 'HC11 have 15 and 18. Each vector can have its own entirely independent Interrupt Service Routine (ISR). New possibilities, such as

Table 8.1 Comparison of interrupt capabilities.

	16F84	80C552	68HC11
Interrupt sources	4 in total, all maskable	15 in total, all maskable	23 in total, 3 non-maskable, 20 maskable
Interrupt vectors	One; source must be determined by polling within interrupt routine. Vector *is* start address of ISR[1]	15 Vectors *are* start address of ISRs	18 (5 serial port sources share one vector) Vectors *hold* start address of ISRs
Prioritisation	None in hardware; can be determined in software by sequence of polling of flags	Interrupts can be assigned to one of two priority levels; within each level hardware determines the priority sequence	Fixed in hardware, except that any one maskable source can be raised to highest (maskable) priority.
Context saving on Stack	Program Counter only	Program Counter only	All CPU registers saved
Nested interrupt strategy	On response to an interrupt all other interrupts are automatically masked	A low-priority interrupt may be interrupted by one of high priority	By default not possible, but can be enabled by clearing mask bits while within ISR

[1]ISR = Interrupt Service Routine

interrupt prioritisation in hardware and nested interrupts[1] – one interrupt interrupting an already executing ISR – are also introduced. We spend some time now exploring the opportunities and challenges offered by these more advanced structures.

8.2.1 Prioritisation

When interrupts occur simultaneously, it is necessary to decide which one to service to first. The same applies if several interrupts have not occurred simultaneously, but have accumulated during a time when the processor has been unable to respond to them, for example if all interrupts have been disabled through the Global Interrupt Enable. Again the processor must have a mechanism to decide which one to service first.

In the 16F84 there is a single interrupt vector, and it is up to the programmer to determine the interrupt source by polling the interrupt flags within the interrupt routine. The order that the flags are polled can be used to determine the priority. This is illustrated in Example 8.1 below. In larger processors, however, where there are multiple interrupt vectors, the processor scans interrupt sources in turn in the hardware. In so doing it sets a fixed priority. Some processors, including both the 68HC11 and the 80C552, allow the user to modify this order of priority, but not to set it with complete freedom.

As an example of an advanced structure, the 80C552 interrupt system is illustrated in Fig. 8.1. All the switches shown in the diagram are controlled by SFR bits. The left-hand column shows all possible interrupt sources. These are the anticipated mix of external and internal sources, with most peripheral devices being represented. A staggering *nine* interrupt sources come from *one* peripheral module, the Timer 2. These are described later in this chapter.

The first column of switches shows that all interrupts can be individually enabled (i.e. there are no 'non-maskable' interrupts). The second column of switches represents the 'global enable' capability, by which all interrupts can be simultaneously disabled.

The prioritisation mechanism of Fig. 8.1 takes some understanding. There are two priority levels, represented by the two rightmost columns labelled 'polling hardware'. The user can switch each interrupt source into one fixed position in either of these through the Interrupt Priority Registers. Sources are scanned in the direction shown by the large downward arrows. Response is made to the first activated interrupt encountered in this scanning sequence. Timer 2 Overflow can, for example, be switched to be the lowest priority source of all, or it can be switched to be the lowest of the high-priority grouping.

---

1 While nested interrupts are allowed in some processors, they are not considered anywhere in this chapter.

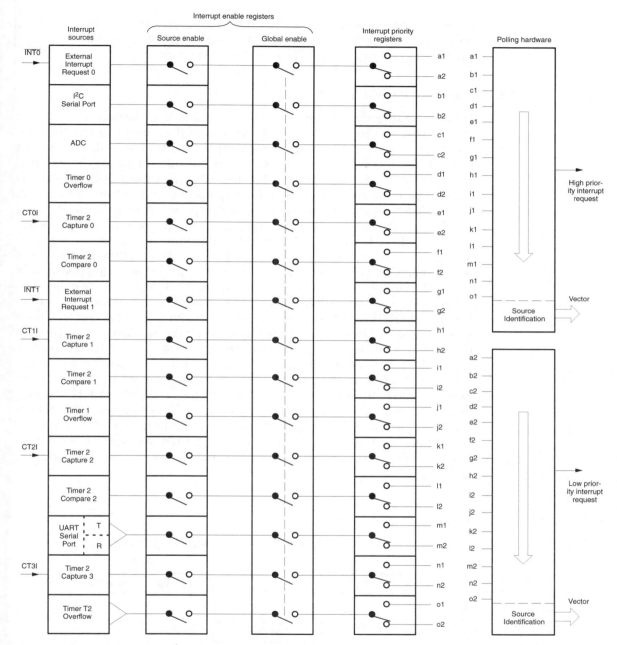

Figure 8.1 Interrupt structure of the 80C552. Reprinted with permission of the copyright owner, Philips Semiconductors. All rights reserved.

8.2.2 Interrupt latency

As the nature of an interrupt is to disrupt the normal activity of the controller with an urgent requirement, it follows that the interrupt requires a rapid response, and that it should be serviced as soon as possible. The delay between the instant at which the interrupt is asserted and the instant the ISR starts being actioned is called the *latency*.[2]

8.2.2.1 Latency of single interrupt

For a single interrupt that is permanently enabled, the latency arises due to the requirement to complete the execution of the current instruction and any housekeeping tasks, for example context saving. This is illustrated in a general way in Fig. 8.2. It is illustrated specifically for the 16F84 external interrupt in Fig. 8.3. We notice that the incoming interrupt (INT pin) sets the interrupt flag, which is polled by hardware at a certain instant in every instruction cycle. In this example the interrupt occurs midway through the first instruction cycle shown and is recognised in the second. The processor then completes the current instruction and, following a dummy cycle, it fetches the instruction at the Interrupt Vector (address 0004_{16}). This is the first instruction of the ISR, which is executed in the following instruction cycle. The final latency (taking the worst case) is four instruction cycles.

Processors like the 16F84, for which there is not a wide variety of instruction times (all instructions are executed in either one or two cycles), have an easily defined latency, as there is not a high dependence on the instruction being executed. It is, however, extended by any software polling required, as shown in Example 8.1. For others, whose instruction execution time varies greatly (the instruction execution time of the 68HC11 for example varies from two to 41 machine cycles), a highly variable latency results. The worst case value (the one which catches you out when you least want it), must take this longest instruction execution time into account.

If the interrupt input is sometimes masked, then the maximum effective latency will be increased by the maximum duration of the interrupt masking (Fig. 8.4).

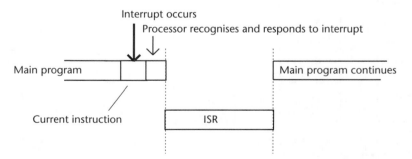

Figure 8.2 Generalised processor response to (unmasked interrupt).

---

2 Use of the word *latency* is not always clearly defined. In this book we will take it to mean the *worst-case* time, for a particular interrupt configuration, between an interrupt being asserted and the *start* of its effective ISR.

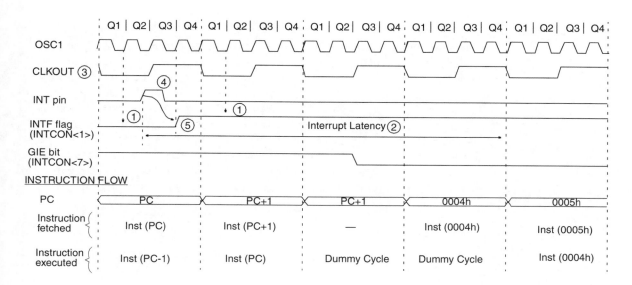

Note 1: INTF flag is sampled here (every Q1).
   2: Interrupt latency = 3-4Tcy where Tcy = instruction cycle time.
      Latency is the same whether Inst (PC) is a single cycle or a 2-cycle instruction.
   3: CLKOUT is available only in RC oscillator mode.
   4: For minimum width of INT pulse, refer to AC specs.
   5: INTF is enabled to be set anytime during the Q4-Q1 cycles.

Figure 8.3   Timing of 16F84 external interrupt. Reprinted with permission of the copyright owner, Microchip Technology Incorporated © 2001. All rights reserved.

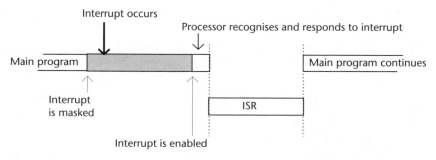

Figure 8.4   Interrupt response with occasionally masked single interrupt.

WORKED EXAMPLE   **8.1**

All interrupts on a PIC16F84 are enabled. Write the opening section of the interrupt routine, which is placed at the end of the program. Poll the interrupt flags to determine the source, with priority External Interrupt (highest), Interrupt on Port Change, Timer Overflow Interrupt and EEPROM Write Complete (lowest). What is the effective increase in latency of each source if the clock frequency is 8 MHz?

Solution: A possible routine is shown in Fig. 8.5. It starts with a goto instruction to satisfy the requirement that the ISR is placed at the program's end. Prioritisation lies in sequence of the flag tests. If both External Interrupt and Timer Overflow Interrupt have been asserted simultaneously, for example, it will be the external interrupt that is serviced. When the external interrupt has been serviced, and if it does not reoccur, then the timer overflow interrupt can be serviced. When the active flag is found, the ISR branches to the program section that will service it. Some time during that section the calling flag must be cleared, as shown. The latencies, including the initial goto, are shown in Table 8.2.

```
        org 0004
        goto int
...
int     btfsc intcon,1      ;test for external interrupt
        goto ext_int
        btfsc intcon,0      ;test for port change
        goto pch_int
        btfsc intcon,2      ;test for timer overflow
        goto tof_int
eeprom_int ...              ;must be EEPROM write complete
...                         ;effective EEPROM ISR starts here
...
        bsf status,5
        bcf eecon1,4        ;clear flag
        return
...
ext_int                     ;effective ext int ISR starts here
...
        bcf intcon,1        ;clear flag
        return
pch_int                     ;effective port change ISR starts here
...
        bcf intcon,0        ;clear flag
        return
tof_int                     ;effective timer overflow ISR starts here
...
        bcf intcon,2        ;clear flag
        return
```

Figure 8.5 Polling of interrupt flags in a 16F84 ISR.

Table 8.2 Latency calculations.

Source	Initial goto	Flag test(s)	Second goto	Increase in latency
External	2	1	2	5 cycles, i.e. 2.5 µs
Port change	2	2 + 1	2	7 cycles, i.e. 3.5 µs
Timer overflow	2	2 + 2 + 1	2	9 cycles, i.e. 4.5 µs
EEPROM write complete	2	2 + 2 + 2	–	8 cycles, i.e. 4.0 µs

Processor recognises and responds to Interrupt 1

Interrupt 1   Int. 3   Int. 2

Main program

Current instruction

ISR 1

ISR 2

ISR 3

Main program continues

Figure 8.6   Interrupt response with multiple interrupts.

8.2.2.2 Latencies with multiple active interrupts

If more than one interrupt is used, then the latency of any one interrupt is further increased by the time required to service any interrupts which may need to be serviced before it. This will depend on the interrupt priority. For example, in Fig. 8.6 there are three interrupt sources; Interrupt 1 is highest priority and Interrupt 3 is lowest. Although Interrupt 3 occurs before Interrupt 2, it must wait until Interrupt 2 has been serviced before it can be attended to.

The action taken by a processor as it completes one ISR and then detects that another interrupt is pending requires careful interpretation of manufacturers' data sheets. It is not uncommon to find (for example the 80C552 structure) that the processor needs to return to the main program and execute one instruction before it moves to the next ISR, even if that interrupt has been pending throughout the previous ISR. With the 16F84, the retfie instruction, which is of two cycles duration, sets the GIE bit within the first cycle, thereby enabling response to further interrupts to commence during its second cycle.

The situation may be even more complex than the above diagram suggests. If a low-priority interrupt occurs, and within it a higher priority interrupt takes place, then even the highest priority interrupt will have to wait for its completion (Fig. 8.7).

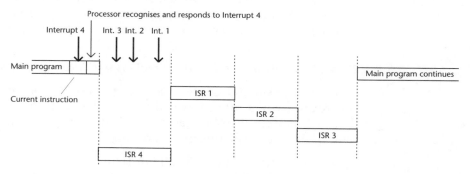

Processor recognises and responds to Interrupt 4

Interrupt 4   Int. 3 Int. 2   Int. 1

Main program

Current instruction

ISR 1

ISR 2

ISR 3

ISR 4

Main program continues

Figure 8.7   Interrupt response with multiple interrupts; lowest priority interrupt occurs first.

From this discussion we can see that the general worst-case latency for any one interrupt is the sum of:

the execution times of all higher priority interrupts,
    plus the execution time of the longest lower priority interrupt
        plus the longest masked region
            plus the processor interrupt response time(s)

8.2.3 Interrupts without tears

As far as the main program is concerned, the ISR must be more or less 'invisible'. It should occur where necessary, but the only trace left behind should be whatever was intended. Yet the nature of the interrupt is that it can occur at any time, and the danger is that it will sometimes occur at an instant in the program which causes maximum disruption. Two important precautions must be applied to avoid the disruption which misuse of interrupts always brings.

8.2.3.1 *Avoiding disruption 1: context saving*

The risks of untoward effects of interrupts can be greatly reduced by ensuring correct context saving, i.e. the saving of registers which define the current state of the processor.

When a response is made to an interrupt, certain CPU registers are saved on stack. The minimalist approach (16F84 and 80C552) is to save the Program Counter only. This is essential, as the processor then has a record of where to return to when the ISR is over. Yet in many cases this is not enough. If the interrupt occurs within a multi-instruction calculation, and the ensuing ISR then causes changes in the Status Register and/or Accumulator, a meaningless calculation will result. Therefore it is often necessary to save the Status Register and Accumulator as well, and possibly other registers. The 68HC11 does this automatically. Does this make it a 'better' processor from this point of view? The answer is no. It removes a responsibility from the programmer, and makes the processor potentially more reliable. It also makes it slower, because this context saving takes time. The 16F84 and 80C552 approach is that of living 'fast but dangerously'. The interrupt response time should be faster, but the responsibility is placed on the programmer to arrange as much context saving as is necessary.

8.2.3.2 *Avoiding disruption 2: critical regions and interrupt masking*

Even with suitable context saving arranged, there remain moments in the program when interrupts cannot be tolerated. The situations when interrupts may need to be masked (and ultimately it depends on the relative importance of the interrupt) are when:

- the system is not yet fully initialised
- a critical time-sensitive task is being performed, for example a delay loop or output pulse timed in software

- certain port outputs are being changed near simultaneously in a sequence of instructions

- a variable, which is accessed by the interrupt routine, is being calculated by the main program as a sequence of instructions

As an example of the fourth case, suppose two 2-byte numbers were being added, and the result was to be used by the ISR. If the interrupt occurred in the middle of the addition (say the lower byte of the result had been updated but not the upper), then the value used by the interrupt routine would be meaningless.

Those parts of the main program where interrupts are disabled are called *critical regions*. In some cases single interrupts will be turned off; in others it will be more appropriate to disable all interrupts globally.

This approach does not apply to interrupts which reflect life and death situations (at least from the embedded system's point of view). In these situations, for example loss of power or major system malfunction, the niceties of maintaining program order must anyway be set aside. Such interrupts should always be kept enabled. It is for these that the Non-Maskable Interrupts of the 68HC11 are intended.

8.2.4 Conclusions on interrupts

It should be clear that considerable care must be taken when working with interrupts, particularly if several are used. To summarise:

- Complications increase when multiple interrupts are used; use interrupts only where essential, and minimise the number of sources.

- When executing an ISR the processor is committed only to that activity; other interrupts and other activities are blocked. ISRs should normally be of minimal length.

- Understand the interdependence between interrupt routines and the main program; hence identify all critical areas, mask interrupts as appropriate, and ensure that correct context saving takes place.

- Estimate latencies for each interrupt source and ensure that timing requirements remain satisfied.

8.3 Advanced Counter/Timer (C/T) structures and applications

8.3.1 Time measurement and Counter/Timer specification

In Chapter 2 we considered in simple terms the application of a counter to the measurement of time. Yet time measurements come in all shapes and sizes, with durations from microseconds to hours or more, i.e. a factor in excess of 10^9. There are also a variety of demands in terms of measurement precision.

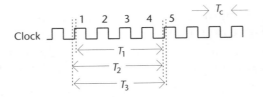

Figure 8.8 Timing resolution.

Let us first consider the essential characteristics of the time measurements we will be attempting to make, i.e. their accuracy, resolution and range. This is similar to the analogue to digital conversion process already encountered in Chapter 5, for when making a time measurement with a counter we are also undertaking an analogue to digital conversion.

A time measurement based on a C/T is simply a count of the number of clock cycles during a certain period of time. If the counter counts n clock pulses, and the clock period is T_c, then the time duration is understood to be nT_c, as in Fig. 2.12(b). The accuracy of the measurement therefore depends initially on the accuracy of the clock itself. The requirements for clock accuracy and stability are analogous to the requirements for voltage reference accuracy and stability in the analogue to digital converter.

In both cases a further error arises due to the quantisation process, which is dependent on the system resolution. For time measurement this is illustrated in Fig. 8.8. Here three very slightly different time periods, of nominal duration four clock cycles, are measured with a counter which is clocked on the rising edge. Depending on how the measurement period is synchronised, counts of 3, 4 or 5 may result. It follows that error due to the measurement resolution is ±1 least significant bit. Ideally the resolution should be small compared with the total count. The clock period should therefore be very small compared with the total time duration measured.

If a counter is n-bit, then it can count to $(2^n - 1)$. The time *range* it can therefore represent, if it is not to overflow, is from 0 to $(2^n - 1)T_c$.

The above discussion accounts for aspects of measurement due to the hardware. Further errors may be introduced in software, for example due to the interrupt latency of a timer overflow response.

WORKED EXAMPLE   **8.2**

An ultrasonic ranging system measures the time between an ultrasound pulse and its echo received from a reflecting surface. To do this its signal conditioning circuit produces a logic pulse whose duration is equal to the time between the transmitted and received pulse. This is represented in Fig. 8.9. The measurement range is to be at least 2 m and the resolution better than 0.25 mm. Both 8-bit and 16-bit C/Ts are available. Each can be clocked at the controller clock frequency of 2 MHz, or this value can be prescaled by

2, 4 or 8. Determine which C/T is appropriate for the task of measuring the pulse width T, and how it should be applied. Deduce the actual range and resolution of the resulting system. Take the speed of sound in air to be 1 mm per 3 µs.

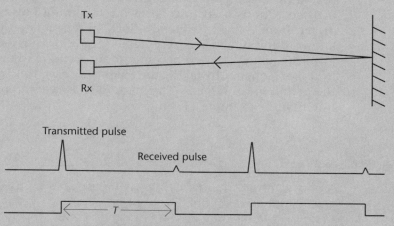

Figure 8.9 A pulse–echo ranging system.

Solution: With a measurement range of 2 m, the sound pulse will travel twice this (as the pulse must travel to the target and return). The maximum measured time is therefore $4 \times 10^3 \times 3 \times 10^{-6}$ = 12 ms. The required resolution expressed in time is (0.25×3) µs = 0.75 µs. To achieve this resolution the clock period must be less than this, i.e. its frequency should be greater than 1.33 MHz. Applying this minimum clock frequency, the C/T numerical range will need to exceed $[(12 \times 10^{-3})/(0.75 \times 10^{-6})]$ = 16000. This is beyond the 8-bit range (256) but within the 16-bit range (65536).

Let us therefore evaluate the possibility of the 16-bit C/T being clocked at 2 MHz. The time between increments at this frequency will be 500 ns, which will lead to a system resolution of $(0.5/3)$ = 0.167 mm. This is adequate for the resolution part of the specification. The maximum count represents $(2^{16} \times 0.165)$ mm, i.e. 10.813 m. The measured range is half this, i.e. 5.4 m approximately.

If we had tried to use the 8-bit C/T, then to achieve the range the clock period would have to be no more than (12 ms/256) = 46.875 µs, or a frequency of 21.3 kHz. This would give a resolution of (46.875/3) = 15.6 mm, which is way outside the specified value.

The solution above evades the question of how the C/T is enabled so that it counts only for the duration of time T. Most C/Ts do not allow gating by an external signal of the internal clock signal, and it may therefore be necessary to use an external clock source, gated externally by the measurand pulse. The section below on Capture Registers gives an alternative approach to this sort of measurement.

8.3.2 Frequency measurement

The standard means of frequency measurement is to count incoming cycles of the measurand frequency over a precisely defined period of time, typically 1 or 10 seconds. The frequency is then the total count divided by the period of measurement. A controller with two C/Ts can be used for this method. One is used to totalise the incoming pulses, while the other is driven from a fixed frequency clock, and generates a counting period of known duration. This is acceptable for all measurements except those of lower frequency. Here it becomes more accurate to measure a period, and find the frequency by taking the reciprocal. Now only one C/T is needed for the period measurement. The method is, however, computationally more intensive, as Example 11.3 illustrates.

WORKED EXAMPLE   **8.3** _____

A microcontroller has two 16-bit C/Ts, each having Interrupt on Overflow capability, and each with the option of prescaling by a factor of 2, 4 or 8. It is to be used to measure the rpm of a motor whose operating speed range is from 100 rpm to 1000 rpm. Either C/T can be clocked with a stable clock frequency, whose value can be selected within the range 500 kHz to 8 MHz. A sensor on the motor produces 60 pulses per cycle. Show how the C/Ts can be used to implement a rpm counter.

Solution: Noting that each revolution of the motor produces 60 pulses, the range 100 rpm to 1000 rpm translates to 100 Hz to 1000 Hz. We will apply one C/T to provide a stable timing period, within which pulses from the motor can be counted. If this is one second, then applying the resolution deduced above, a maximum error of ±1 Hz (to be read as 1 rpm) is possible at the low-frequency end. If one C/T is to overflow in precisely 1 second, then the clock period must be $1/2^{16}$, i.e. 15.259 μs, or a clock frequency of 65.536 kHz. This is too low for the permissible clock range, but if a divide-by-8 prescaler is used, then the clock frequency can be set at (8×65.536) kHz, i.e. 524.288 kHz, which is acceptable. The Interrupt on Overflow output of this C/T is enabled, and an Interrupt routine written which reads the value of the other C/T, and then clears it. This value can be displayed directly as rpm.

8.3.3 Counter/Timer enhancements

One hardware enhancement of the basic C/T structure, the Auto Reload, was described in Chapter 2. Two other enhancements are now described. For situations such as those outlined in Examples 8.2 and 8.3, these enhancements take some of the burden from software to hardware. By eliminating software-induced delays, they can improve accuracy of operation.

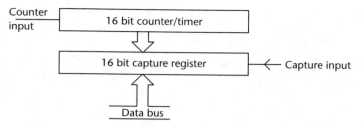

Figure 8.10 A Capture Register structure

8.3.3.1 *Capture Register(s)*

These are registers linked to a C/T which 'capture' its value every time an external input is asserted; see Fig. 8.10. Thus if the C/T is running in Timer mode, the moment at which an event occurs can be recorded with no immediate action required from the program. With more than one Capture Register, time intervals between two different events can easily be determined. Capture Registers are normally accessible as SFRs.

8.3.3.2 *Compare Register(s)*

This technique has already been mentioned in the context of PWM sources, for example Fig. 5.5. The Compare Register, appearing as a SFR, can be preset to a particular value. The C/T runs in Timer mode, and the Compare Register value is continuously compared with it, as shown in Fig. 8.11. When they become equal an output is generated. This output may be an interrupt, but in many cases the purpose of the time delay is actually to initiate an external event, so the processor need not even 'know' that the delay period is up. Therefore in some controllers this output is connected directly to an external pin of the processor.

With several compare registers a complex timebase can be set up for repetitive events, in contrast to the simple Auto Reload.

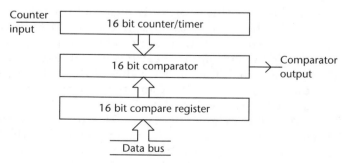

Figure 8.11 A Compare Register structure.

WORKED EXAMPLE **8.4**

A microcontroller is required to poll a digital input every 100 ms precisely, and then update the value of a parallel port output 1.8 ms later. It has a 16-bit C/T with three Compare Registers and an Interrupt on Overflow capability. Its input clock frequency can be selected within the range 1 MHz to 8 MHz, and this can be prescaled by 2, 4 or 8 if required. How should the C/T be used to meet this requirement, assuming there is no Auto Reload capability?

Solution: Let us arrange the input polling as an interrupt routine called by the C/T Interrupt on Overflow. For interrupts every 100 ms, the period of the Counter clocking frequency must be $(100\,000/2^{16})$ μs, i.e. 1.526 μs. This corresponds to a frequency of 655.36 kHz, i.e. below the lower clock limit. The clock could be set to twice this value, i.e. 1.31 MHz, with a divide-by-2 prescaler. One Compare Register can be used to generate a further interrupt 1.8 ms after overflow. With the Counter counting with a period of $(1/655.36)$ ms, the number of cycles before 1.8 ms is reached will be (1.8×655.36), i.e. 1179.6. The closest possible will be to load the Compare Register with 1180_{10}.

8.3.4 Device example: the 80C552 Timer 2

The 80C552 Timer 2, shown as a simplified block diagram in Fig. 8.12(a), is a major subsystem in itself. It is a 16-bit Timer, which carries most of the enhancements that have just been described. It contains nine different interrupt sources and no fewer than 20 SFRs. The interrupt sources can be identified both here and in Fig. 8.1. All have their own interrupt vector, except for the 8-bit and the 16-bit overflows, which share one vector. The main control SFR for this C/T, TM2CON, is shown in Fig. 8.12(b). Other SFRs control interrupt features, and the capture and compare registers.

Despite its extra sophistication, the central features of this C/T are very similar to much simpler structures. The two possible clocking sources are an external input or the internal oscillator input divided by 12 (i.e. the machine cycle rate). An external reset is available, which is enabled by an internal SFR bit. This is the only means of reset, apart from the controller reset itself. There is a prescaler, with division rates 1 to 8. The Counter itself appears as two SFRs, which are readable and resettable, but *not* writeable.

Each of the four Capture Registers has an external input associated with it. When asserted, the current value of the C/T is saved and an interrupt request is generated. If the Capture capability is not used, these pins can just be used as extra external interrupts. Each Capture Register occupies two SFRs, readable by the processor. The three Compare Registers are all writeable, and generate an interrupt request when the C/T value is equal to their value. The C/T also includes a facility, not pictured, to set, reset or toggle output pins when a Compare register match is found. This gives the

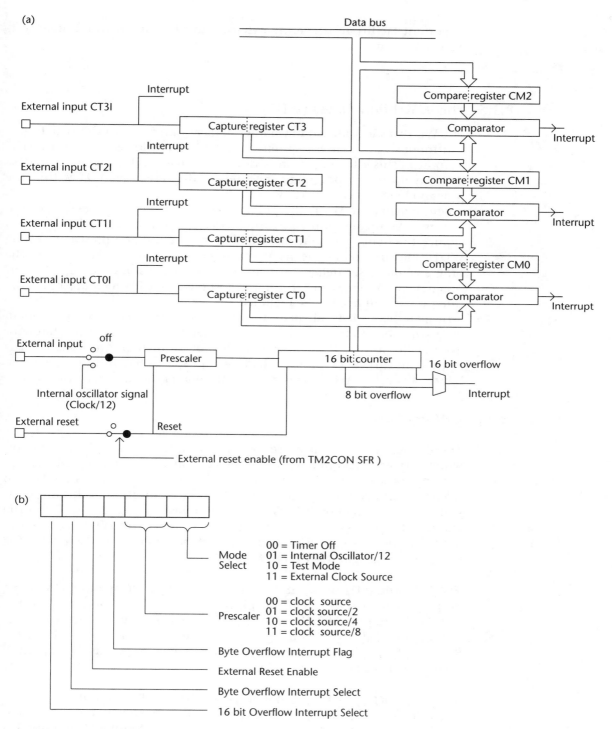

Figure 8.12 The 80C552 Timer 2. (a) Block diagram; (b) Control SFR, TM2CON.

important facility of triggering external timed events with minimal software overhead.

8.3.5 The Watchdog Timer (WDT)

The WDT was introduced in Chapter 2. In summary, it is a C/T which runs continuously and causes a system reset if it is ever allowed to overflow. The purpose of this is to allow a processor to reset itself if its program ever locks up. It is up to the programmer to repeatedly reset the WDT within the program before it can overflow.

Before using the WDT, the programmer must know its timeout period. This may be predetermined by the processor hardware. Usually there is a chance to change its value, for example by the use of a prescaler. A long WDT timeout period means fewer resets need to be included in the program, but reset delay is longer if the program does lock up. A short timeout period leads to more rapid response to CPU lock-up, but more timer resets will need to be included.

In many cases it is adequate to include a WDT reset once in the main program loop. Care must be taken of delay loops, and wait states for external responses. WDT resets should not be including in ISRs, as it is possible in this case for a periodic interrupt source to continue operating, hence causing the WDT to reset even if the main program itself has failed.

8.3.5.1 WDT structures

16F84 Here the WDT is based on a free-running internal *R–C* oscillator, which drives a counter with a nominal timeout period of 18 ms. An optional post-scaler, which is also the TIMER0 prescaler, can be used to stretch the timeout period to around 2.3 s. The allocation of this pre/post-scaler is determined by bit 3 of the Option register (Fig. 2.13(b)), and the scaling rate is set as shown. If the scaler is allocated to the WDT, it cannot be used for the TIMER0. The WDT is enabled by the WDTE bit in the Controller Configuration Word (Fig. 2.19).

68HC11 The WDT of the 68HC11 is provided by its 'Computer Operating Properly' (COP) sub-system. Enabling of this WDT is controlled by the NOCOP bit in the EEPROM-based CONFIG register. The internal E clock is initially divided by 2^{15}, with a further optional scaling of 1, 4, 16 or 64. Resetting takes place in a two-stage sequence. First 55_{16} must be written to the COPRST register, followed by AA_{16}.

80C552 (Timer 3) This sub-system is enabled by the state of an external controller pin, $\overline{EW}$. It is driven from the oscillator signal divided by 12, and consists of an 11-bit prescaler followed by an 8-bit counter. The counter is mapped as a writeable SFR. The WDT period is determined by the number reloaded into the SFR.

WORKED EXAMPLE 8.5

A microcontroller has a 16-bit WDT which is clocked from the 1 MHz system clock. A system reset occurs if the WDT overflows. With what (minimum) frequency must the WDT be reset?

Solution: The WDT will fill in 65 536 cycles (16 bit range), i.e. 65.536 ms. Therefore the WDT must be reloaded approximately every 65 ms or less.

8.4 Multi-tasking

Many programs so far considered in this book are developed to perform essentially one activity, or a very closely related group of activities. The reality of most programs, however, is that they deal with more than one activity, which may be independent, or partly independent, from other activities which need to be run. Consider the situation of Fig. 8.13, which shows a power source

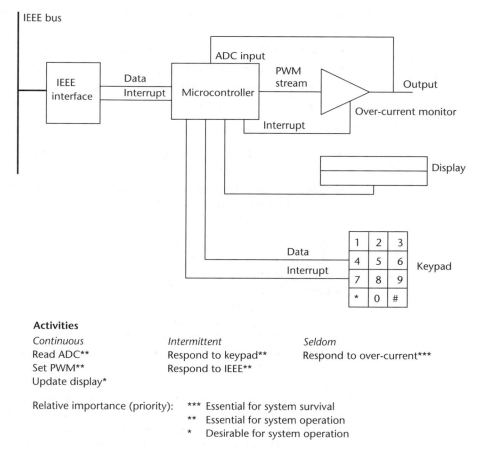

Activities

Continuous	*Intermittent*	*Seldom*
Read ADC**	Respond to keypad**	Respond to over-current***
Set PWM**	Respond to IEEE**	
Update display*		

Relative importance (priority): *** Essential for system survival
  ** Essential for system operation
  * Desirable for system operation

Figure 8.13 A simple multi-tasking system.

controlled by a microcontroller. This itself can be controlled by the user or from an IEEE bus. Either of these control sources, when activated, generates an interrupt. The main control function is achieved by the microcontroller monitoring the output voltage and adjusting its PWM output stream accordingly. The value of this output voltage is displayed, as well as the operating conditions selected by the user. Under conditions of overload (detected by excessive output current) a further interrupt is generated.

Clearly a number of unsynchronised activities call upon the processor resources. Some are continuous and must be maintained, for example the main control function. Others are continuous but could possibly be suspended, for example updating the display. Others are intermittent, with some flexibility in speed of response, for example new control data from IEEE interface or keypad. Others occur only very rarely, but demand instant response, for example the Over Current Monitor. These activities are tabulated in the figure and categorised according to their frequency of occurrence, and perceived relative importance.

The characteristic features of this system, and many others like it, are as follows:

- the microcontroller must perform a number of activities more or less in parallel
- some activities are synchronous with each other
- the activities have different relative importance
- the activities have differing time demands
- some of the activities must be performed within a very strict time frame

We can go on to observe:

- that some activities are event-driven, like pressing the keypad, or the IEEE bus communication
- that other activities are time-based; we would, for example, expect the main control function to be performed periodically, perhaps every few milliseconds

We will from here adopt the common convention of calling the activities *tasks*, and the relative importance we shall call the *priority*.

The type of programming applied so far in this book is often called sequential. Instructions follow in a sequence which is predetermined by the program, even though there may be some dependence on external inputs. It is in general possible to predict with a high level of confidence the outcome of the program execution. Yet now we are looking at a situation where we appear to need more or less *parallel* activity. A demand is being laid on the processor to perform a number of tasks *at the same time*, rather than one after each other.

It is still possible to attempt to program a system such as Fig. 8.13 using a sequential program structure. Four options are shown in Fig. 8.14. In the first of these each task completes what it needs to do before handing over to the next. The characteristics of this program structure are:

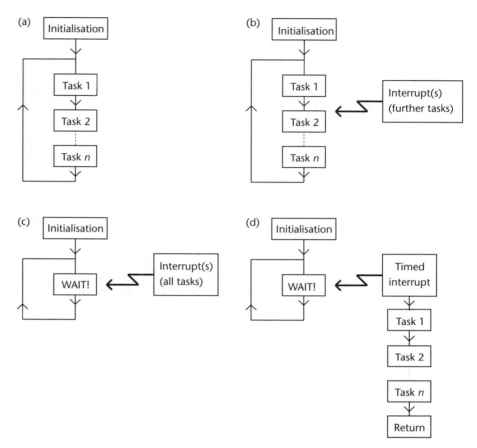

Figure 8.14 Simple multi-tasking program structures.

- while any one task is running, the others are entirely inactive
- all tasks are synchronised to each other

There are several disadvantages with this structure:

- one task is dependent on the others to complete before it gets a chance to be active
- if any one task takes excessive time or gets locked in a loop, the others may be delayed or become entirely disabled
- the loop can be dominated by tasks which need a long time to execute – these may effectively determine loop duration and hence repetition rate
- simple implementation of the loop assumes that all tasks are synchronous, and need equal frequency of processor attention
- if a change is made to the program section of one task, it may have knock-on and unforeseen consequences for the others

An improvement is offered by the option of Fig. 8.14(b). Here we can leave the lower priority tasks in the loop, and give the more urgent and/or asynchronous tasks higher priority by making them interrupt-driven. Some further prioritisation between the interrupt-driven tasks is of course also possible, using techniques introduced earlier in this chapter. Furthermore, some interrupts may be continuously enabled, with others just locally enabled (say for one task only), for example an ADC 'Conversion Complete' interrupt. This is effectively the program structure of many simple to medium complexity embedded systems, and in such applications it can be perfectly successful. An extreme form is option (c), where all tasks are interrupt driven. There are unlikely to be many applications where this is the preferred approach.

A possible disadvantage of (a) or (b) is that the loop repetition rate is dependent on the duration of each task. It is therefore difficult to predict, and may be variable. To avoid this, the structure of Fig. 8.14(d) can be used. Now the whole loop is driven by a timed interrupt, typically from a C/T Interrupt on Overflow. This is the structure of the case study of Appendix C. The loop repetition rate is now predictable, and can be usefully used as the basis for timed activity. The average repetition rate of each task is of course equal to the loop repetition rate, even though there might still be some time variation from one iteration to the next, due to individual tasks varying their timing demands. A further apparent advantage is that, should one task lock up, the interrupt has the chance to free it at the next loop iteration.[3] This structure does depend on there being adequate idle time at the end of the loop to absorb any timing fluctuations in task execution time.

For simple multi-tasking applications, and given a careful timing analysis based on program and ISR execution times, these program structures can work with complete success, and can be developed to quite a level of sophistication. Any significant change in the program may, however, be difficult to implement, as program timings would have to be recalculated. A further increase in complexity would make these timing calculations increasingly arduous and error-prone, and the program would be difficult to develop and manage.

Simple sequential programming applied to multitasking finally falls down in situations when:

- there are many tasks, AND
- tasks have widely varying needs for processor time, OR
- tasks have distinctly different priorities, OR
- tasks are not synchronous with each other.

---

3 This extent of this 'advantage' is worth thinking through. If tasks are all within an interrupt loop, then the next timed interrupt won't be seen unless nested interrupts are permitted. If they are permitted, the interrupt return address is within the locked task, which solves little.

We need therefore to find a systematic way of dealing with the more complex multi-tasking situation, particularly when working to strict time deadlines. This leads us to consider the concepts of the Real-Time Operating System.

8.5 Real-Time Operating Systems

An alternative solution to problems of the sort represented by Fig. 8.13 is to use an operating system to time-share processor resources. The required processor activity is separated clearly into a number of tasks, each of which is developed as an autonomous program strand. The operating system shares CPU time between the tasks (not necessarily in fixed order), and controls which task is to be executed next, as illustrated in simplistic form in Fig. 8.15. Although the operation of this figure may appear superficially similar to Fig. 8.14(a), there are fundamental differences. In Fig. 8.14(a), each task completes its activity before allowing the CPU to pass on to the next. In Fig. 8.15, the operating system may suspend a task which has *not* completed, in order that another higher priority task may take over. But more of that in a moment.

If all tasks get an adequate share of computer time when they need it, then the illusion of parallel processing, i.e. all tasks being dealt with simultaneously, can be achieved. This is known as concurrent programming (actually it is quasi-concurrent, as we only create the *appearance* that it is genuinely concurrent). When the requirements of real-time are also applied to the operating system, it becomes a Real-Time Operating System (RTOS, generally pronounced 'Arr-Toss'!), a programming tool applied to meet the needs of concurrent programming in the real time environment.

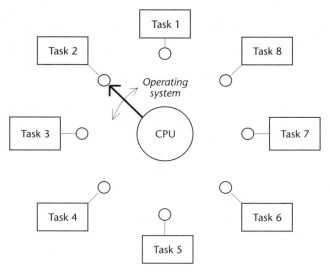

Figure 8.15 Concurrent programming?

An RTOS must perform three essential functions:

- decide which task should run, and for how long
- control the use of resources shared between the tasks, for example memory and hardware peripherals
- provide communication and synchronisation between tasks

These functions are undertaken by the RTOS *kernel*. The kernel is customised by attaching to it the tasks that are specific to any particular application, and by defining for it the application hardware resources. Part of the kernel is the *scheduler*, which determines which task is to run.

The *advantage* of using an RTOS is that it provides a systematic design environment, which should lead to a reliable software product and shorter development time. An RTOS-based program is, however, likely to occupy more memory, and the process of switching between tasks takes CPU time, which is lost to the program function. These added overheads represent the *disadvantage* of using an RTOS.

RTOSs are widely used in the realm of fast and large microprocessors. In their fully fledged form they represent a very sophisticated area of programming, well beyond the scope of this book. RTOSs *are*, however, also found in the small-scale embedded system, sometimes in 'stripped down' form, and it is worth exploring their concepts, even if the version finally applied is a comparatively simple one. Here we give an introduction and flavour; for further details the reader should consult sources such as Refs. 7.3, 8.2 or 8.3.

8.5.1 Scheduling and the scheduler

Figure 8.15 is simplistic because it implies that all tasks require equal amounts of processor time, and that all are of equal priority. In practice, considerable ingenuity is applied to the problem of determining the priority of tasks and ensuring they get CPU time when they need it.

The scheduler determines which task is to be executed next. There are a great variety of schedulers around, and the scheduling strategy applied is one of the features which determines the performance of the RTOS.

The simplest scheduling strategy has already been considered. It is represented by Fig. 8.14(a), and is called *cyclic scheduling*. Each task is allowed to run to completion, and it then hands over to the next. This is an example of *non-pre-emptive* scheduling.

An advance on cyclic scheduling is the *Round Robin* scheduler. Now the operating system is driven by a regular clock interrupt (the clock 'tick'). Tasks are selected in a fixed sequence for execution. On each clock tick, the current task is discontinued and the next is allowed to start execution. Now the scheduler must preserve the context (all flags, registers and so on) of the task being left, so that on its return the task can pick up activity seamlessly. This context switching inevitably adds to the work which the processor is required to do and to the system memory requirements. All tasks are treated as being of equal importance and wait in turn for their slot of CPU time. As tasks are *not* allowed to run to completion, this is an example of a *pre-emptive* scheduler.

An advance on Round Robin scheduling is the *Prioritised Pre-emptive* scheduler. Imagine now that tasks in the Round Robin Scheduler are given priorities. The scheduler is still run by a clock tick, and on every tick it checks whether a task of higher priority has been made ready and should be given CPU time. Under this regime, high-priority tasks are allowed to complete before time is given to those of lower priority. Conversely, an executing low-priority task is replaced by one of higher priority.

8.5.2   Tasks

Tasks are written as if they are to run continuously (even though they do not), normally as a sequential program in an endless loop. Though they run autonomously from each other, they may depend on services provided by each other, and may need to be synchronised in time. They cannot call on a section of another's code, but may be able to access common code, for example C libraries.

In a pre-emptive system tasks move between a number of different states:

- *Active* (or *Running*): the task is the one which is executing
- *Ready*: the task is ready, and will run as soon as it is allocated CPU time
- *Suspended* (or *Blocked*): the task is not immediately ready; perhaps a required resource is not available
- *Dormant*: the task does not need CPU time

It is not always easy to determine which system activities are worthy of being established as tasks. The number of tasks created should not be too great, as for every task switch there is a programming overhead to be paid. In general, a set of activities which are closely related in time should be grouped together into a single task. Similarly, if they are closely related in function and interchange a large amount of data, they should be grouped into a single task. A single task should not, however, attempt to serve more than one distinct time deadline.

The RTOS provides a means for the programmer to set priorities. The programmer must determine these by evaluating the task deadlines, about which the RTOS has no knowledge. High task priority should be given to tasks which are urgent (i.e. tight time deadlines) or critical (i.e. essential to system survival or success). Lower priority is generally given to heavy computational tasks. In the case of *static priority*, priorities are fixed. In the case of *dynamic priority*, priorities may be changed as the program runs.

Interrupts remain valid units in the RTOS environment, and should be used to respond to urgent asynchronous events. In general they are not used as a task, but may change the status of a task. As elsewhere, they should be kept short, passing data back to a task which will respond accordingly.

8.5.3 Data and resource protection

Several tasks may need to access the same item of shared resource. This could be hardware (including memory or peripherals), or it could be a common software module. The way tasks share such resources must be handled with great care. A method for dealing with this is by the *semaphore*. Each shared resource has a semaphore, which is used to indicate whether the resource is in use. In a *binary semaphore*, the first task needing to use the resource will find the semaphore in a GO state, and will change it to WAIT before proceeding to use the resource. Any other task in the meantime needing to use the resource will have to enter the WAIT state. When the first task has completed accessing the resource, it will change the semaphore back to GO. This leads to the concept of *mutual exclusion*; when one task is accessing the resource, all others are excluded.

The *counting semaphore* extends the concept to a set of identical resources, for example a group of printers. Now the semaphore is initially set to the number of units of resource, and as any task uses one of the units, it decrements the semaphore by one, incrementing it again on completion of use. Thus the counting semaphore holds the number of units that are available for use.

As an effect of setting a semaphore to the WAIT state is that another task becomes blocked, they can be used as a means of providing time synchronisation between different tasks.

8.5.4 Implementing the RTOS

Writing an RTOS is a specialised activity, and a number of software companies do only this. The general practice for the programmer needing to use one is to purchase a commercial RTOS kernel, appropriate for his or her hardware platform and anticipated level of complexity. It is then up to programmers to customise the commercial product to their own use by adding to it the tasks. IAR for example offer their 'Tiny-51' for the 8051 family (as part of their 'Windows Workbench'; Ref. 3.2). Some simple kernels are also available as shareware, for example Refs. 8.4 and 8.5.

Commissioning a multi-tasking system can be a very different experience from commissioning a sequential program. In the latter there is a high level of predictability, as everything in the main program always happens in the same sequence. In the multi-tasking situation this is not the case. Slightly different circumstances, for example changes in relative execution speed or in speed of peripheral response, will cause the scheduler to sequence tasks in different orders, with the possibility of different results being reached. Certain errors may therefore occur only on very rare and unpredictable instances, and have the appearance of being random. Testing for such errors is very difficult, and demands careful consideration of anticipated operating conditions. A first step to effective multi-tasking is for the programmer to choose a reputable RTOS, known to be robust and reliable.

Perhaps the most famous case of RTOS malfunction occurred on the Mars Pathfinder mission of July 1997, already mentioned in footnote 5 of Chapter 4. Here a case of priority inversion took place, whereby a lower priority task blocks execution of one of higher priority. The exact combination of computer loading which caused this fault had not been anticipated on Earth. Nevertheless, the programmers responsible were able to identify the cause of the problem and upload a program revision to the spacecraft! The success of this correction was due in large part to the presence of extensive diagnostic facilities in the spacecraft computer. The story is retold in Ref. 8.3, and is well worth reading.

SUMMARY

1. The use of interrupts is more or less unavoidable in systems which have significant time dependence. To use them properly, it is essential to have a clear understanding of the issues of latency and critical regions, and of the impact that interrupt routines may have on each other and on the main program.

2. A wide range of Counter/Timer structures exist, which give very many opportunities in time measurement and generation. These should be explored and used to their fullest extent. In almost every case it is preferable to undertake timing functions in such hardware systems rather than in the software.

3. For simple multi-tasking applications sequential programming techniques can be applied, with judicious use of interrupts.

4. For more demanding multi-tasking in the real-time environment, the use of an RTOS should be considered. These can be applied at varying levels of complexity. An appreciation of their principles will bring benefit even to the comparatively simple multi-tasking situation.

REFERENCES

8.1 Kalinsky, D. (1998) Fundamentals of multi-tasking. In *Proc. of the Embedded Systems Conference*, Europe, 7–9, Sept. San Francisco, CA: Miller Freeman.

8.2 Simon, D.E. (1999) *An Embedded Software Primer*. Reading: Addison-Wesley.

8.3 Briand, L. P. and Roy, D. M. (1999) *Meeting Deadlines in Hard Real-Time Systems*. Los Alamitos, CA: IEEE Computer Society Press.

8.4 *A Real-Time Operating System for PICmicro™ Microcontrollers*. Application Note 585. Microchip Technology Inc. 1997.

8.5 *Simple Real-Time Kernels for M68HC05 Microcontrollers*. Application Note AN1262. Motorola Inc. 1995.

EXERCISES

8.1 A certain microprocessor program has interrupt routines with durations as shown, prioritised in the sequence shown.

Interrupt Routine 1:   550 µs   sends a digit to LCD display

Interrupt Routine 2:   80 µs   organises system shutdown on detection of low power

Interrupt Routine 3:   30 µs   modifies a parallel output port

The microprocessor instruction set has instructions which execute in 1 to 12 machine cycles, and each machine cycle is 1 µs duration. The interrupt call takes 5 µs to execute. The processor can detect and commence response to a pending interrupt during execution of a *return from interrupt* instruction.

(a) What is the interrupt latency:
    (i)   for any interrupt, if only one is enabled?
    (ii) for interrupt 2, if 2 and 3 are enabled?
    (iii) for interrupt 2, if all interrupts are enabled?
(b) On the evidence given, can any improvements be made to this interrupt strategy?

8.2   Context saving on a PIC 16F84 is to be increased to include saving of STATUS and W registers. By how much is the effective latency increased if the clock frequency is 8 MHz?

8.3   A microcontroller has two 16-bit C/Ts. Either can be clocked from an internal clock frequency, selectable within the range 1 MHz to 4 MHz, with the possibility of prescaling by 2, 4, 8, 16, 32, 64 or 128. It is to be used to measure the frequency of a stream of pulses, which ranges from 200 Hz to 25 kHz.
(a) Explain how this requirement can be met.
(b) Explain any changes necessary:
    (i)   if the maximum frequency is increased to 50 kHz
    (ii) if the maximum frequency is increased to 100 kHz
    (iii) if the maximum frequency is 25 kHz, but an alarm must sound if 20 kHz is exceeded

8.4   A microcontroller is being chosen for an application which requires the measurement of the time interval between two events, A and B, each represented by a pulse (on two separate input lines). Event A has a near steady frequency of approximately 12 kHz, while event B follows A within 80 µs. Specify in as much detail as possible a C/T and clock source which can be used to meet this requirement.

8.5   An 80C552 microcontroller is required to generate the repetitive waveforms of Fig. 8.16. Using only the information supplied in this chapter,

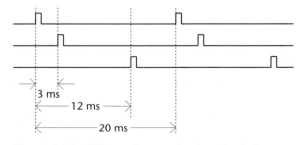

Figure 8.16 Timing diagram for Exercise 8.5.

explain how its Timer 2 can be used for this purpose. Specify a clock frequency, and include a proposed setting for SFR TM2CON.

8.6 The program of Appendix C does not use the 16F84 WDT. If the WDT is to be incorporated, suggest a possible reset rate and show where WDT resets should be included in the program.

8.7 A metronome is designed around a 16F84 microcontroller. It must give out a regular 'tick' (beat), at rates from 40 to 199 beats per minute. The beat rate is shown by two multiplexed seven-segment LED displays, which must be driven in the manner illustrated in Fig. 9.10(b). The displays should be alternated at a rate not less than 50 Hz. The most significant '1' digit is represented by a single LED. The beat rate may be adjusted by two push-buttons, one for 'increase' and one for 'decrease'. All integer beat rates within the specified range must be possible. As the rate changes, the display is updated. The tick is caused by a 50 ± 10 ms logic high pulse from a port bit to a sounder. It must continue at the displayed rate, whatever else the metronome is doing. Draw a block diagram showing a possible allocation of port bits. Describe, with diagrams, how the software could be structured.

8.8 Devise a program structure for the example of Fig. 8.13, giving as much detail as you can. Explain your choice of structure and state any assumptions made.

8.9 Write an evaluation of the Real-Time Operating System for the 16C64 described in the Microchip Inc. Application Note AN585. In doing this you may like to consider the following questions (but you should not limit yourself to them):

- What is the general structure of this program, and to which general model of an RTOS does it conform?
- How are tasks called?
- In how many different states are the tasks placed?
- How is the interrupt structure used?
- What do you think of the general documentation of this program?
- Where and how is context saving done?
- Is there task prioritisation? If so, how is it done?
- Could you reduce the program to a general-purpose kernel? If so, how?
- It is sometimes stated that PIC microcontrollers do not lend themselves well to running an RTOS. To what extent is this statement justified?

University of Derby, Assessed Assignment, February 2001.

8.10 Apply the principle of Exercise 8.9 to the 68HC05 real-time kernels of Ref. 8.5.

Interfacing to external devices

IN THIS CHAPTER

However advanced the hardware and software of a microcontroller-based system may be, the benefit is only obtained when the system communicates with 'the outside world'. This could be the human user or a physical system with which it interacts. In this chapter, therefore, we look at the devices to which microcontrollers are most often interfaced. This includes those devices for human interaction, such as keypads and displays, and those for interaction with a controlled system, including essential sensors and actuators.

The aims of the chapter are:

- to develop further digital interfacing techniques, including interfacing with multiple switches and keypads

- to introduce more complex optical devices and how interfaces may be made with them

- to describe on/off switching of power devices

- to describe continuously variable voltage and current control using PWM and its applications

- to introduce the principles of DC and stepper motors, and their interfacing requirements

9.1 More on the digital interface

In Chapter 2 we saw how the parallel port could be interfaced with switches and LEDs, and explored some of the electrical characteristics of the port pin driver circuit. The level of detail presented allows the design of small, compact and simple systems. It is not, however, adequate for larger or more complex systems. In these the variety of devices to be interfaced will be much greater; single LEDs and switches will give way to keypads and displays, and longer distances of communication may mean that greater attention must be paid to conditioning of the incoming signals.

9.1.1 Interfacing digital input signals

Digital input signals are interfaced to parallel and serial ports and interrupt and timer inputs. The general requirements of all such input signals are:

- they should consist of only 'legal' logic levels; specifically, the input voltage should not lie *between* logic 0 and logic 1, nor *above* the maximum permissible logic 1 level (which will probably be above the power supply voltage), nor *below* the minimum permissible logic 0 (which will probably be below ground voltage)
- they should not be corrupted by voltage 'glitches', even if those glitches themselves constitute a legal voltage level
- they should switch cleanly and fast

Many circumstances, however, generate one or other of these unacceptable conditions. Some of these were introduced in Chapter 6. They are summarised in Table 9.1.

Four simple hardware interfaces, mentioned in Table 9.1, are shown in Fig. 9.1. These may be used alone or in combinations. Schmitt triggers are used alone when a signal has suffered mild degradation, with resulting slow logic transitions or some voltage attenuation. Simple external R–C filters

Table 9.1 Forms of digital signal degradation.

Signal degradation	Possible solution at input
Slow edge (due to slow signal source or low pass filtering effects in transmission medium)	Schmitt trigger
Voltage attenuation (due to attenuation in transmission medium or inappropriate transmitted voltages)	Schmitt trigger
EM interference – 'benign' voltage spikes leading to data error	Analogue filter + Schmitt trigger, digital filter
EM interference – hazardous voltage spikes leading to equipment damage	Current limiting resistor + diode, opto-isolate, other voltage clamp
Switch bounce	Debounce in hardware/software
Earth differential	Opto-isolate

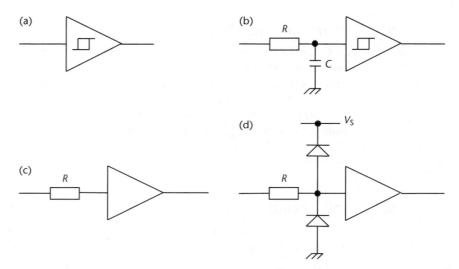

Figure 9.1 Simple interfacing for digital signals, CMOS inputs. (a) Schmitt trigger; (b) Schmitt trigger plus low-pass filter; (c) current-limiting resistor; (d) current-limiting resistor plus protection diodes.

can be used, in conjunction with the Schmitt trigger, to minimise the effects of mild interference; Fig. 9.1(b).

A serious concern if a signal is received from outside an equipment is that it has suffered interference, and may carry voltage spikes beyond the specified absolute maximum. Almost every logic gate input has input protection diodes, designed to clamp the incoming voltage to within the supply rails. The current that these can handle is limited, however, and if exceeded will destroy the diode. Therefore a single series resistor (Fig. 9.1(c)) does provide some protection by limiting current in the internal protection diodes in the case of excess voltage. Values of several hundred ohms are adequate to protect for over-voltages of a few volts. Higher values provide higher protection, but can slow the signal significantly due to the low-pass filtering effect of the resistor plus the gate input capacitance. Improved protection is afforded by providing external protection diodes (Fig. 9.1(d)). Opto-isolation is illustrated in Fig. 9.16 and described in the accompanying section.

Many ICs now include simple digital filtering of input signals, for example within the serial ports described in Chapter 6.

9.1.2 Switch debouncing

When a switch closes, the contacts bounce together for a period of several milliseconds, during which time electrical connection is intermittent. This is illustrated in Fig. 9.2. Depending on the application, this may cause a severe problem.

Three possible hardware solutions to this problem are shown in Fig. 9.3. If a changeover (also called double-throw) switch is used, the circuit of Fig.

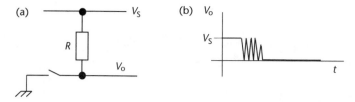

Figure 9.2 Switch bounce. (a) The simple switch interface; (b) V_o on switch closure.

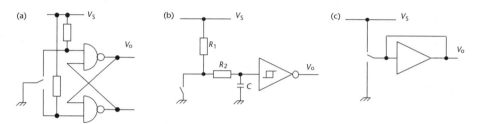

Figure 9.3 Switch debounce circuits. (a) Bistable; (b) Schmitt trigger; (c) buffer.

9.3(a) can be applied. The state of the bistable, made up of the two Nand gates, reverses whenever a logic 0 is applied to one of its inputs. If the switch moves from the lower to the upper position, the bistable changes state on the first switch contact due to the applied logic 0. If the switch then bounces momentarily away from this contact, the upper bistable input flickers between 0 and 1, but the bistable action is to hold its new state. The bouncing is therefore eliminated. The bistable only changes its state again when the switch is returned to its original lower position and a logic 0 is applied to its lower input. At this time a similar debouncing action takes place.

If a single-throw switch is to be used the circuit of Fig. 9.3(b) can be applied, where the Schmitt trigger is a CMOS device. This introduces some delay into the switch action, but has the advantage of using only one logic gate. Here, if the switch is initially closed and then opened, capacitor C will charge up towards the supply voltage with time constant $(R_1 + R_2)C$. If the switch is then closed (and this is where the bounce happens), the capacitor will discharge towards 0 V with time constant CR_2. If $R_1 \ll R_2$, then the time constants are approximately equal. A value for CR_2 of a few tens of milliseconds is adequate for most switches. Suitable values to drive a CMOS input are $R_1 = 10k$, $R_2 = 100k$ and $C = 100n$.

A final alternative is shown in Fig. 9.3(c). Here a buffer drives its own input. The switch forces a change in input and hence output, the buffer holding the new state as the switch bounces. At the moment of switch change the gate output is instantaneously connected to its opposite logic level. This lasts only as long as the gate itself takes to change state, and is normally acceptable in CMOS logic.

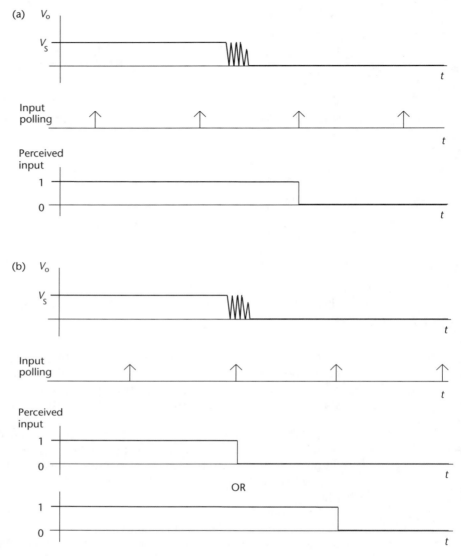

Figure 9.4 Switch debounce achieved through polling (a) Polling misses switch bounce; (b) polling coincides with switch bounce.

Software debouncing is an alternative to hardware. If a switched input is being polled at a rate such that the polling period is much greater than the bounce period, then the input is effectively debounced (Fig. 9.4). In most cases the polling action will miss the switch bounce altogether, as in Fig. 9.4(a). If it does not (Fig. 9.4(b)), then it is uncertain which logic value it will detect at that instant. Whichever it is, the bounce will be eliminated.

Debouncing should be applied to inputs which are dependent on the number of incoming cycles. A good example is counters, which would otherwise count on every switch bounce. In other cases it is not necessary.

9.1.3 Switch arrays and keypads

Interfacing of switches to parallel port inputs was described in Chapter 2. If the number of switches is limited, it is usually adequate just to allocate one to each port bit. These bits can be read by the software as necessary. If they are push switches used for human interaction, then polling at least 10 times per second is normally found to be necessary, otherwise a short switch push will be missed.

If an immediate response to switch depression is required, then some form of interrupt must be invoked. Some processors, like the PIC 16F84, have the useful facility of 'Interrupt on Change', as described in Chapter 2, whereby a logic change on a port bit can be made to cause an interrupt. Failing this, the switch inputs can be 'ANDed' together to provide an Interrupt input (Fig. 9.5).

Keypads are essentially large switch arrays. It is impractical to connect each switch to its own port input, and keypads are normally therefore constructed in a matrix and connected as shown in Fig. 9.6. An ($n \times m$) array of keys can be read by ($n + m$) port bits. The figure shows a 16-key keypad interfaced to an 8-bit port, with four port bits connected to the row lines and four to the column lines. A possible software routine for reading the keypad is shown in the flow diagram of Fig. 9.7.

As keys are likely to be pressed only briefly, however, it is essential to read their value immediately, and an interrupt capability becomes almost essential. Again, the easiest way to implement this is to use a port with an Interrupt on Change capability. The port should be initialised with row bits as output at logic 0 and column bits as input. The processor can then proceed with its other tasks. When a key is pressed an Interrupt on Change is activated, and the outstanding parts of the flow diagram then form the interrupt routine. It is usually advisable to insert a 10 ms delay after a change is detected to allow any switch bouncing to disappear before an attempt to read the keypad is made. On completion of the routine, row and column

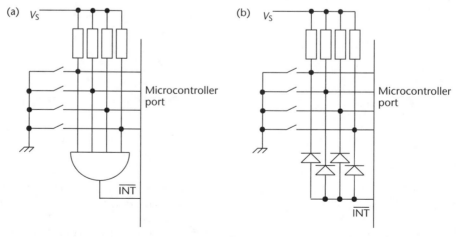

Figure 9.5 Four switches ANDed to produce an interrupt. (a) With logic gate; (b) with diode logic.

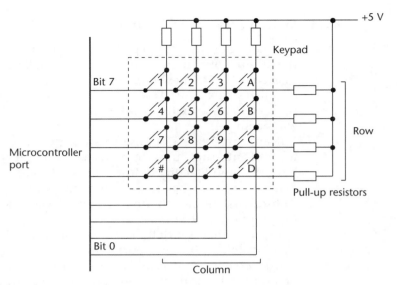

Figure 9.6 A 4 × 4 keypad.

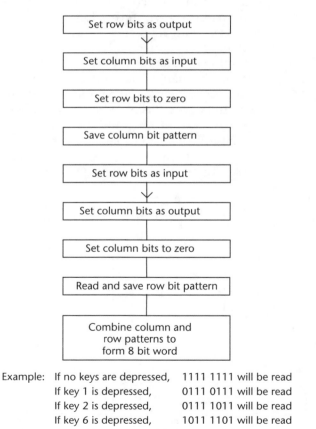

Example:	If no keys are depressed,	1111 1111 will be read
	If key 1 is depressed,	0111 0111 will be read
	If key 2 is depressed,	0111 1011 will be read
	If key 6 is depressed,	1011 1101 will be read

Figure 9.7 A simple routine to read a keypad.

bits should be returned to output and input respectively to await the next keystroke. Interrupt on Change will cause an interrupt both when a key is pressed and when it is released. The release action will have to be detected and then ignored.

9.1.4 Transistor switches

Most electrical loads are too great to be driven directly by the limited current drive capability of a microcontroller port bit, and in this case it is necessary to interface via a power switching device. Both semiconductor and electromechanical devices can be used for this purpose.

Transistor switches represent an easy way of switching DC loads. Bipolar transistors require a small base current to switch a much larger collector current; MOSFETs require a modest gate voltage, with negligible current, to switch a large drain current. Figure 9.8 shows these applied in five simple ways, interfacing a controller port bit to a resistive load R_L. In all circuits

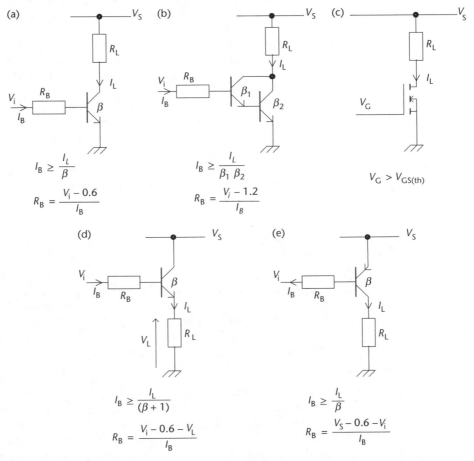

Figure 9.8 Switching resistive loads. (a) npn transistor; (b) Darlington pair; (c) n-channel MOSFET; (d) npn transistor, emitter follower; (e) pnp transistor.

except (e), a logic high from the microcontroller switches current into the load. In (e) it is a logic low which causes current to flow. Note that the load supply voltage V_S does not have to be the same as the microcontroller supply; in many cases it is bigger, for example 12 or 24 V. The formulae given are for the On condition in every case. They are for guidance only, and should be applied with care, noting that transistor characteristics vary widely between devices, and ensuring always that 'worst-case' conditions are catered for. Further design guidance on circuits of this type can be found in Ref. 1.1.

The simplest bipolar transistor switching circuit is shown in Fig. 9.8(a). Applying the formulae shown, a transistor with a forward current gain β of 100, switching a current of 200 mA, will require a base current of 2 mA. This is well within the capabilities of most microcontroller ports, though *not* of the quasi-bidirectional port. Extra current gain can be obtained using the Darlington pair arrangement of Fig. 9.8(b). If each transistor in the pair has a β of 100, the input base current for a load of 200 mA is now only 20 μA.

Unlike the bipolar transistor, the MOS device of Fig. 9.8(c) is purely voltage controlled, and its gate can be directly connected to the port bit. Then the controller output must just comfortably exceed the maximum V_{GS} threshold, i.e. $V_{GS(th)}$, necessary for switch-on. This makes the MOSFET a very attractive option for load switching in the microcontroller environment. Special families of MOSFETs, designed for direct interface to logic levels, are available.

In each of diagrams (a), (b) and (c) one side of the load is connected to the power supply. In some cases it is necessary for the load to be ground-referred, for example when a section of the circuit itself is switched, perhaps for power saving. In this case the circuits of (d) and (e) are options. Circuit (e) is generally the most attractive, as the switched voltage will only be around 0.2 V less than the supply voltage.

WORKED EXAMPLE   **9.1**

A small resistive heater element is to be switched on and off by a PIC 16F84 microcontroller, powered from 5 V. The element has resistance 24 Ω, and is to be driven from a 12 V supply. Applying circuits Fig. 9.8(a) and (c), determine whether each of the transistors detailed in Table 9.2 is suitable for switching this load, and calculate any component values needed.

Solution: With a supply voltage of 12 V and load resistance 24 Ω, we anticipate a nominal load current of 0.5 A. In practice this will be somewhat reduced by any voltage developed across the switching transistor.

For the MOSFET the maximum value of threshold voltage, $V_{GS(th)}$, is 3 V. With 5 V CMOS switching this will be exceeded, and we can expect the Drain–Source resistance $R_{DS(on)}$ to fall to no more than 0.33 Ω. When switched on the current through the heater element will be no less than

Table 9.2 Transistor characteristics.

ZVN4306 MOSFET		2N3704 npn transistor	
V_{DS} max	60 V	β	100 to 300
$R_{DS(on)}$ max	0.33 Ω	V_{CE} in saturation	0.2 V approx.
$V_{GS(th)}$ max	3 V	V_{BE}	0.6 V approx
I_D max	1.1 A	I_C max	800 mA
Max. power dissipation	1.1 W	Max. power dissipation	360 mW

$12/(24 + 0.33)$, i.e. 0.49 A. This is within the maximum permissible Drain current of 1.1 A. The power dissipated in the transistor will be no more than $(0.49^2 \times 0.33)$, i.e. 80 mW, which is well within the maximum permissible power dissipation of 1.1 W.

For the bipolar transistor, applying the circuit of Fig. 9.8(a), we calculate the base current using the worst-case value for β (i.e. 100):

$$I_B = 0.5/100$$

$$I_B = 5 \text{ mA}$$

To calculate R_B, we note from Fig. 2.6(a) that the PIC output voltage will fall to around 4.7 V when sourcing 5 mA. Hence

$$R_B = (4.7 - 0.6)/(5 \times 10^{-3})$$

$$R_B = 820 \text{ Ω}$$

R_B plays a current-limiting role, and there is no need for accuracy in its value. In practice, its calculated value would be reduced somewhat to accommodate possible reduced microcontroller output voltage or transistor β.

The actual load current will be approximately $(12 - 0.2)/24$ A, taking into account the transistor saturation voltage, which gives 0.49 A. From this the transistor power dissipation is given by (0.49×0.2), i.e. 98 mW, which is well within the limit of 360 mW.

Both transistors would undertake this switching task successfully. The MOSFET has the slight advantage, as it draws negligible current from the microcontroller and does not require any equivalent of the base resistor, needed by the bipolar transistor.

9.2 Optical devices 1: displays

The characteristics of the LED were introduced in Chapter 2. While single LEDs are very important as indicators, in many cases we need far more information than one or several units could give. LEDs are therefore widely combined into a range of display formats, including seven-segment, dot matrix and starburst.

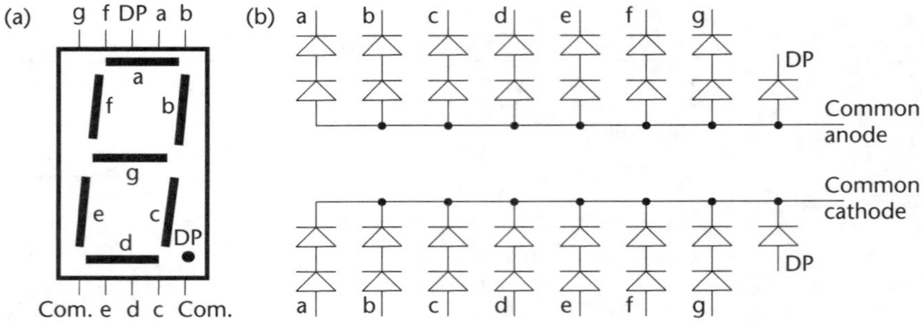

Figure 9.9 The seven-segment LED display. (a) Single-digit format; (b) the common cathode and common anode connections.

9.2.1 The seven-segment LED display

A popular format is the seven-segment display, seen in Fig. 9.9. Depending on the segments illuminated, this unit can display all the digits from 0 to 9 and a limited number of letters. Units like this are available in sizes from 0.3 inches (7.6 mm) high to 5 inches (126.6 mm). The LED chips which form the display are arranged so that either all anodes or all cathodes are connected together, and are hence called *common anode* or *common cathode*. This is shown in Fig. 9.9(b). All except the smallest displays use two or more LEDs in series for each segment, generally with a reduced number for the decimal point. This increases the voltage drive requirement, as each diode contributes its own forward voltage drop.

Seven-segment displays are frequently used in groups to display decimal numbers. The power drive requirement can then become significant. To avoid an excessive number of drive lines, it is common to multiplex the segment drives as shown in Fig. 9.10. For common cathode devices, the cathode of each digit is activated in turn. While each is active the segment drives are switched to the state required. A number of ICs exist specifically to undertake this task, often combined with counter functions (for example the ICM 7212). As each digit in a multiplexed display is on for only a fraction of the time (each digit in this example is activated only one quarter of the time), they must be driven hard when on, otherwise the display will appear faint.

Figure 9.11 shows a PIC 16F84 used as a self-contained two-digit counter driving two seven-segment displays. The Count input is port A bit 2, and counting is done in software. The display units are low-current, high-efficiency common anode type, of typical forward current 2 mA and forward voltage 1.7 V. Each port bit can readily sink the 2 mA required of one segment. However, with all segments 'on' the digit requires 14 mA supplied to its common anode terminal. As this approaches the absolute maximum output rating for port pins on this device and causes significant output voltage drop (Fig. 2.6), it was felt inadvisable for the port bit to source this. Therefore bipolar transistors are used to switch this current, following the pattern of Fig. 9.8(d). Taking into account the base–emitter voltages of the switching transistors, the voltage applied to each common anode when on

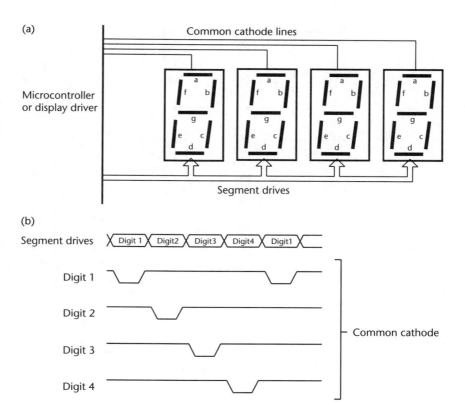

(a)

Common cathode lines

Microcontroller
or display driver

Segment drives

(b)

Segment drives   Digit 1   Digit2   Digit3   Digit4   Digit1

Digit 1

Digit 2

Digit 3   Common cathode

Digit 4

Figure 9.10 Driving multiplexed seven-segment displays. (a) Circuit
connection; (b) drive waveforms.

will be around 4.4 V. Hence the current in each segment, when on, will be
around [(4.4 – 1.7)/1k], i.e. 2.7 mA.

Display switching adapts the pattern of Fig. 9.10. A segment will be on if
its associated common anode control (RA0 or RA1) is high and its cathode
line (one of RB0 to RB6) is low. As the clock oscillator frequency is not critical, a simple $R–C$ oscillator is used.

9.2.2 The liquid crystal display (LCD)

LCDs are of very great importance these days, both in the electronic world
in general and in the embedded system environment. Their main advantages are their extremely low power requirements, light weight and high
flexibility of application. They have been one of the enabling technologies
for battery-powered products such as the digital watch and the laptop
computer, and are available in a huge range of indicators and displays.
LCDs do, however, have some disadvantages. These include limited
viewing angle and contrast in some implementations, sensitivity to
temperature extremes, and very high cost for the more sophisticated graphical display.

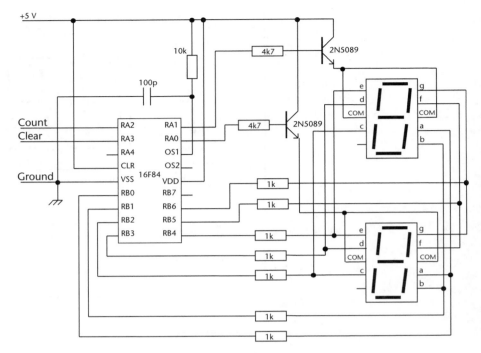

Figure 9.11 A PIC 16F84 used as a display driver.

LCDs do not emit light, but they can reflect incident light or transmit or block backlight. The principle of an LCD is illustrated in Fig. 9.12(a). A small quantity of liquid crystal is contained between two parallel glass plates. The liquid crystal is an organic compound which responds to an applied electric field by changing the alignment of its molecules, and hence the light polarisation which it introduces. A suitable field can be applied if transparent electrodes are located on the glass surface. In conjunction with

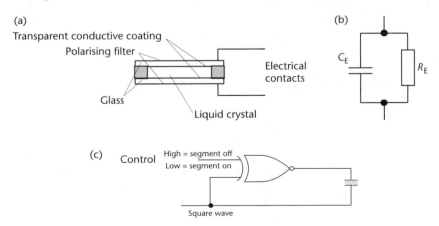

Figure 9.12 The liquid crystal display. (a) Structure; (b) equivalent circuit; (c) simple drive circuit.

the external polarising light filters, light is either blocked or transmitted by the display cell.

The electrodes may be made in any pattern desired. These may include single digits or symbols, or the standard patterns of bar graph, seven-segment, dot matrix, starburst and so on. Alternatively, they may be extended to complex graphical displays with addressable pixels. Some of the methods applied to this sophisticated technology, and the addressing techniques used, are described in Ref. 9.2.

With two conductive surfaces separated by an insulator, the cell behaves essentially as a capacitance in parallel with a high value of resistance, as indicated by its equivalent circuit (Fig. 9.12(b)). It must be driven with an AC source to avoid electrolytic action in the liquid. A simple way of doing this is shown in Fig. 9.12(c). A square wave is applied as shown; when the control input is high, square waves of opposite phase are applied to the cell and it is activated. With the control input low, the same square wave is applied to each side and the cell is not active. In practice this approach is only suitable for the simplest of displays.

Broadly two types of liquid crystal are used: Twisted Nematic (TN) and Supertwisted Nematic (STN). The TN was the first to be implemented, and may be used in transmissive or reflective mode. The reflective mode is used in situations of good ambient light. Unlike most other display technologies, it becomes clearer as the intensity of the ambient light increases. Transmissive mode is used when ambient light is poor, and requires integral backlighting. Once this is introduced the low power advantage of the LCD is inevitably compromised. The STN LCD is in many ways similar to TN, but allows better display of high-density information, with high contrast and wider viewing angles.

When choosing a display, designers may decide to select a standalone LCD unit and design the drive electronics themselves. For this there are a number of dedicated drive ICs available, each committed to a different type or format of display. Alternatively, designers may choose one of the self-contained modules which are now widely available. An example of these is now described.

9.2.3 Hitachi LCD modules and the HD44780 microcontroller

Hitachi and other manufacturers offer a range of easy-to-use dot matrix display modules, which incorporate a dedicated CMOS microcontroller, a driver IC and a display. These are the sort of displays seen on photocopiers, fax machines, telephone systems and so on. Both reflective and backlit types are available. The number of characters that can be displayed ranges at the time of writing from one line of eight characters to four lines of 40 characters.

One LCD controller, the HD44780, is fully detailed in Ref. 9.3. It contains an 80-byte RAM to hold the display data and a ROM for generating the characters. It has a simple instruction set, including instructions for initialisation, cursor control (moving, blanking, blinking) and clearing the display.

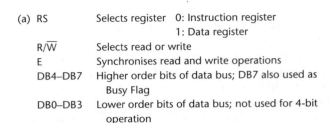

(a) RS   Selects register   0: Instruction register
                                       1: Data register

      $R/\overline{W}$ Selects read or write

      E   Synchronises read and write operations

      DB4–DB7   Higher order bits of data bus; DB7 also used as
                      Busy Flag

      DB0–DB3   Lower order bits of data bus; not used for 4-bit
                      operation

(b)

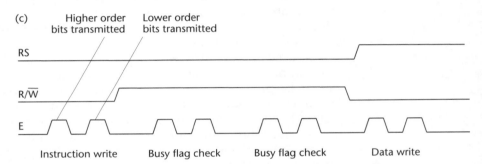

RS	$R/\overline{W}$	E	Action
0	0	⌐↓	Write instruction code
0	1	⊓	Read busy flag and address counter
1	0	⌐↓	Write data
1	1	⊓	Read data

(c)   Higher order   Lower order
       bits transmitted   bits transmitted

RS

$R/\overline{W}$

E

     Instruction write   Busy flag check   Busy flag check   Data write

Figure 9.13 HD44780 interfacing. (a) User interface lines; (b) data and instruction transfers; (c) timing for 4-bit interface.

Communication with the controller is made via an 8-bit data bus, three control lines, and a strobe line (E). These are itemised in Fig. 9.13(a).

Data written to the controller is interpreted either as instructions or as display data (Fig. 9.13(b)), depending on the state of the RS line. An important use of reading data is to check the controller status via the Busy Flag. As some instructions take a finite time to implement (for example a minimum of 40 µs is required to receive one character code), it is essential to be able to read the Busy Flag and wait until the LCD controller is ready to receive further data.

The controller can be set up to operate in 8-bit or 4-bit mode. In the latter mode only the four most significant bits of the bus are used, and two write cycles are required to send a single byte. In both cases the most significant bit doubles as a 'Busy Flag' when a Read is undertaken.

An implementation of a Hitachi display (the LM016L), which incorporates the HD44780 microcontroller, is shown in Fig. 9.14. Here it forms part of an overall user interface, being combined with a keypad and interfaced via two PCF8574 devices (see Fig. 6.11) to the I^2C bus. This combination

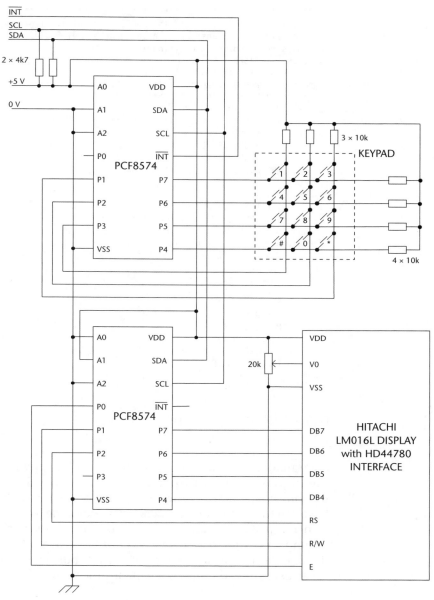

Figure 9.14 A self-contained user interface linked to the I²C bus.

makes a very useful circuit subsystem which can be built into many designs. The Interrupt on Change capability of the '8574 is used to signal a host controller via the interrupt when a key is pressed.

The display is used in 4-bit mode, which allows the whole display interface to be contained within only 8 bits. This is convenient for the hardware, but makes the software somewhat more difficult and time-consuming. A sample subroutine for an 80C552 controller, which applies the timing

```
;************************************************************
;Subroutine to send byte held in STORE2 to display. Display
;control lines are transferred in lower nibble of STORE3,
;which is retained in case string is being sent. STORE1 is
;used as a temporary store.
;Calls SR SENDI2C to transmit I2C data.
;Returns to Program when busy flag cleared.
;************************************************************
OPDISP    MOV I2CNAME,#2H
          MOV A,STORE2 ;establish first nibble
          ANL A,#0F0H
          ORL A,STORE3
          MOV STORE1,A
          ADD A,#1          ;set E high
;and write data to display
          MOV I2COP,A
          LCALL SENDI2C
          MOV I2COP,STORE1  ;same data, with E low
          LCALL SENDI2C
;prepare and send lower nibble
          MOV A,STORE2 ;establish second nibble
          SWAP A
          ANL A,#0F0H
          ORL A,STORE3
          MOV STORE1,A
          ADD A,#1          ;set E high
          MOV I2COP,A
          LCALL SENDI2C
          MOV I2COP,STORE1  ;same data, with E low
          LCALL SENDI2C
;data is sent, wait for busy flag to clear
          LCALL BUSY
          RET
```

Figure 9.15   Data transfer subroutine for Fig. 9.14.

diagram of Fig. 9.13(c) and prepares one character code for transmission to the display, is shown in Fig. 9.15. The actual data transfer is done by a subroutine (SENDI2C) called from the one shown. It can be seen that no fewer than four data transfers must be made via the I^2C bus in order to transmit just one character. The outcome of this is that although this forms a compact design, data transfer is not fast, particularly bearing in mind also the comparatively slow speed of the I^2C bus. Frequent display updates can cause excessive loading of the host processor.

9.3   Optical devices 2: simple opto-sensors

The conduction characteristics of semiconductor material are dependent on a number of external influences, including temperature, mechanical strain and incident light. This has led to the development of a wide range of semiconductor sensors. Here we explore some important semiconductor opto-sensors.

If any piece of semiconductor material is exposed to light, then the incident photons tend to cause hole–electron pairs to be created. If the semiconductor is simply a block of material, then the effect of incident light is to change the apparent resistance of the material. This is the basis of the Light-Dependent Resistor (LDR). It has a rather slow response time (of the order of hundreds of milliseconds), and is of interest for simple light intensity measurements.

Very fast response times are made possible, however, if light falls on a reverse-biased semiconductor junction. Then the diode action is disrupted by the light-induced current, which is approximately proportional to the light intensity.

A slightly slower response, but greater sensitivity, is obtained if light falls on the base of a transistor. Now the photo-induced current is amplified by the transistor action.

9.3.1 The opto-isolator

The opto-isolator is formed from an LED combined with a photo-diode or photo-transistor (Fig. 9.16). When the LED is on, the photo-transistor conducts and the Output line is pulled low. When the LED is off, the transistor does not conduct and the Output line is high. Input and output are powered from two separate power supplies, isolated from each other, labelled here V_{S1} and V_{S2}.

The opto-isolator allows data to be transferred with no electrical connection at all between input and output, and can therefore provide a very high level of protection for either a transmitting or receiving device. This includes situations where there may be a ground differential between transmitter or receiver, where logic signals are being coupled to high-voltage stages in a system (for example in motor drives), or in any situation where there is a risk of an incoming signal carrying hazardous voltages.

Opto-isolators are formed in ICs, often in multiples, with the optical devices hidden within the IC. The Hewlett-Packard HCPL-2630, for example, is a two-channel opto-isolator. Its minimum LED current requirement is 5 mA, and output rise and fall times are 24 ns and 10 ns respectively. The input–output insulation is 2500 V.

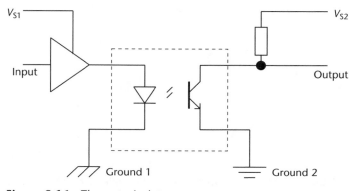

Figure 9.16 The opto-isolator.

9.3.2 Object sensors

These opto-sensors follow the principle of the opto-isolator, i.e. an LED combined with a photo-transistor, but communication between transmitter and receiver depends on the presence or absence of the sensed object. Optical devices are here infrared to minimise interference from visible light, and they are placed behind filters which pass only the light wavelength of interest.

In the reflective opto-sensor (Fig. 9.17(a)), the phototransistor conducts if a reflective surface is placed near to the face of the sensor. The output current depends on the quantity of reflected light, which in turn depends on the LED current, the quality of the reflecting surface and the spacing of the surface from the sensor.

In the transmissive opto-sensor (Fig. 9.17(b)), the transistor will conduct if there is no obstacle in the light path. This sensor often comes in the form of a 'slotted opto-sensor', and can be used to detect paper or tape or for rotary speed measurements (see below).

Either sensor can be driven using the circuit of Fig. 9.17(c). In some applications the Schmitt Trigger buffer may not be necessary. R_1 is calculated to give adequate drive current to the LED. In most cases this is well above the level of current required for a normal display LED. For the reflective sensor, values of LED current in the region of 30 to 40 mA are common. R_2 is calculated to give a well-defined logic 1 in the presence of illumination and a well-defined logic 0 if illumination is absent. The transistor residual *dark current* (which flows even without illumination) should not cause any

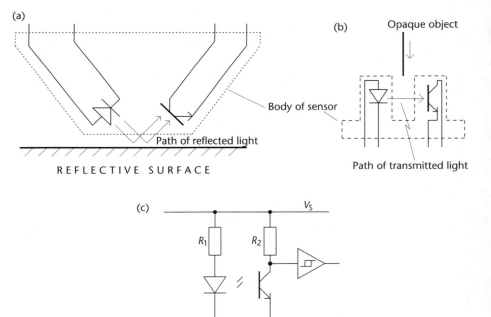

Figure 9.17 Simple opto-sensors. (a) Reflective opto-sensor; (b) transmissive opto-sensor; (c) simple drive circuit for either sensor.

significant voltage drop across it. The current in the presence of illumination can be estimated from manufacturer's data or by experimentation.

9.3.3   Shaft encoders

Optical methods can very conveniently be applied to rotary (and linear) measurements of speed and position. Shaft encoders are devices which give information about the speed or position of a shaft. There are two basic types: the incremental and the absolute.

In an optical incremental encoder a disc is constructed either with regular holes around its rim (Fig. 9.18(a)) or with alternate reflective and dark bands (Fig. 9.18(b)). The first is sensed with a slotted opto-sensor, mounted so that the rim of the disc rotates with the arc of holes passing in the sensor slot. The second is sensed with a reflective sensor. In either case, as the disc rotates a pulse train is developed by the opto-sensor.

This arrangement is very suitable for speed measurement. If the disc has n holes or bands, and the output frequency of the opto-sensor is f, then the angular speed s is given by:

$$s = (f/n) \text{ revolutions per second} \tag{9.1}$$

or

$$s = (60f/n) \text{ revolutions per minute (rpm)} \tag{9.2}$$

If $n = 60$ then the output frequency gives the rpm directly. This is a commonly found arrangement.

A possible disadvantage with this simple arrangement is that it gives no indication of direction. More information can be obtained if there are two bands on the encoder disc, arranged in quadrature, as shown in Fig. 9.19(a). Now the phase relationship of the two rings is dependent on the direction of rotation, and a direction signal can be derived from them (Fig. 9.19(b)), for example by using one to clock the other in an edge-triggered bistable

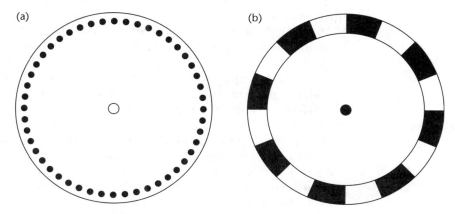

Figure 9.18   Incremental shaft encoders. (a) Holes for slotted opto-sensor; (b) banded for reflective sensor.

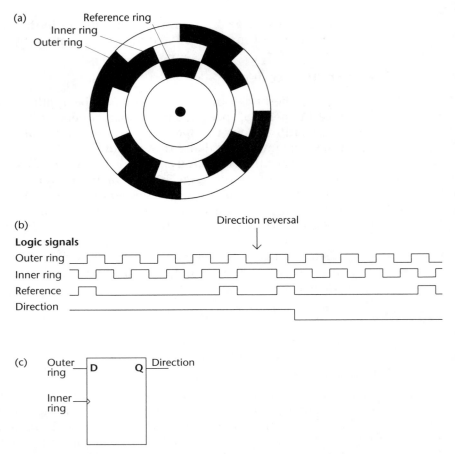

Figure 9.19 Incremental quadrature shaft encoder. (a) Disc pattern; (b) output signals; (c) direction-sensing circuit.

(Fig. 9.19(c)). It is also common to find a reference pulse being generated once per cycle.

Absolute shaft encoders allow angular measurements of shaft position to be made. A number of patterned rings are now made on the encoder disc. Very simple three-band versions are shown in Fig. 9.20. The system resolution is clearly dependent on the number of rings, and most commercial discs have many more than three. For each ring there is an opto-sensor, and all of these are aligned on the same radius. Now whichever the disc position is, the opto sensors will generate a 3-bit word, which will be an indication of the angular position of the shaft. These words are indicated in the diagrams. Unless the opto-sensors are perfectly aligned, there is a real danger of a reading error as pattern boundaries pass them. Therefore the Gray scale pattern, where only one bit changes on any boundary, is important in this context, and is often found.

Shaft encoders suffer minimal wear, and are thus robust and long-lived. They are, however, susceptible to dust and dirt. A variety of low-cost

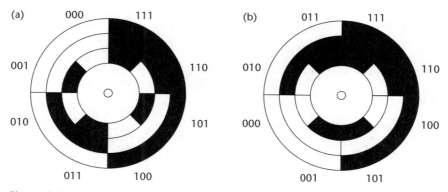

Figure 9.20   Absolute shaft encoders. (a) Binary; (b) Gray scale.

systems can easily be implemented. Up-market encoders with high resolution are, however, costly.

9.4   Inductive loads

9.4.1   On/off switching

Inductive loads form a very important category of electrical device, including as they do all motors, transformers, electromagnetic relays and other actuators. They behave differently from purely resistive loads, which leads to both problem and opportunity!

Voltage and current in a pure inductance are related by the expression $V = L\mathrm{d}I/\mathrm{d}t$. The implication of this simple formula is that the voltage across the inductor is proportional to the *rate of change* of current. In particular, a current in an inductor *cannot* be switched on or off instantaneously. Otherwise the $\mathrm{d}I/\mathrm{d}t$ term, and hence the V term, would be infinite. The explanation for this lies with the energy stored in the magnetic field surrounding the inductance when a current is flowing. This energy is transferred to the field as the current builds up, and returned to the circuit as it decays.

In every practical case, an inductive load, for example a solenoid, DC motor or stepper motor winding, has some associated resistance. Then, when a voltage V is applied, the current rises according to the relationship

$$I = (V/R)[1 - e^{-tR/L}] \qquad (9.3)$$

where I is the current and L and R are the series inductance and resistance. The current rise time will be $2.2L/R$. The applied voltage can be switched using the circuit of Fig. 9.21(a). This diagram illustrates also the current rise and fall.

On switch-off a path must be provided for the current to decay to zero. The *freewheeling diode* D of Fig. 9.21(a) is provided across the inductance so that this decaying current can flow through it. Such a diode must be placed across all switched inductive loads. If it is not, damage will almost certainly be caused to the switching device.

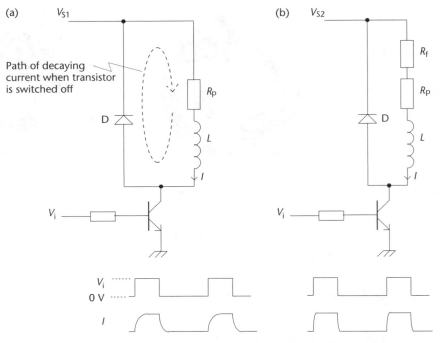

Figure 9.21 Inductive load switching. (a) Simple; (b) with forcing resistor.

In certain situations, a slight delay caused by the current build-up is negligible. In some cases, however, for example when driving stepper motors, it constitutes a major problem. In a stepper motor, as we shall see, current has to be continuously pulsed into the motor windings. The current rise time can be reduced by using the circuit of Fig. 9.21(b). Here a *forcing resistor* R_f is placed in series with the inductive load, and the supply voltage increased by an amount IR_f. The new, reduced current rise time is now $[2.2L/(R_p + R_f)]$. This approach has been widely used, especially for stepping motor drives. It is, however, very wasteful of energy, as is clearly illustrated in Example 9.2.

WORKED EXAMPLE　**9.2**

An inductive load has an inductance of 120 mH and resistance 5Ω. When switched on, it requires a steady state current of 2.4 A.

(a) What is the supply voltage and current rise time if no forcing resistor is used?

(b) What value of supply voltage and forcing resistor must be used if a rise time of 20 ms is to be achieved? What is the power lost in this resistor?

Solution:

(a) If this load is switched using the circuit of Fig. 9.21(a), the current rise time will be $(2.2 \times 120/5)$ ms, i.e. 52.8 ms. The supply voltage must be (5×2.4) V, i.e. 12 V.

(b) The rise time must now be reduced to 20 ms by the application of a forcing resistor, as in Fig. 9.21(b). The new rise time is now given by:

$$t_r = 2.2L/(R_p + R_f)$$

$$20 = 2.2 \times 120/(5 + R_f)$$

$$R_f = 8.2 \ \Omega$$

The supply voltage must be increased by the voltage drop across R_f, i.e. (8.2×2.4) V, leading to an actual voltage of $[12 + (8.2 \times 2.4)]$ V, i.e.

New supply voltage = 31.7V

The power W lost in the forcing resistor is found as follows:

$$W = I^2R$$

$$= 2.4^2 \times 8.2$$

$$= 47 \ W$$

9.4.2 Relays

Relays are switches which are activated by the application of an electrical voltage or current. In their conventional electromechanical form a solenoid is energised to force switch changeover (or closure). In many cases the solenoid is operated from 5 or 12 V, while the switching contacts are capable of switching moderately high voltages and currents, e.g. 240 V, 10 A. Relay size depends mainly on the power-handling capability and on the number of contacts switched. Many miniature PCB-mounting versions are available. Schrack, for example, offers a Double Pole Changeover (DPCO) relay which can switch up to 2 A at 250 V AC with a nominal 5 V, 257 Ω solenoid. Electromechanical relays can easily be controlled using the circuit of Fig. 9.21(a). As mechanical devices they carry the problems of switch bounce and switch contact erosion and wear with time. The switching action and the bounce lead to electromagnetic radiation. There is, moreover, no synchronisation with the switched waveform in AC applications, and some delay in action.

Solid state relays are the semiconductor equivalent of the electromechanical type, and are generally intended for mains voltage switching, although relays which switch DC are also available. They have the important advantages of the semiconductor: no moving parts or wear with time. Moreover, they generally switch as the switched waveform crosses zero. This minimises voltage transients and allows a more predictable switching characteristic. Solid state switches are widely available to switch mains voltages from a control voltage of 5 V or less, and the lower power versions are PCB-mounted.

9.4.3 Reversible switching – the H-bridge

If a bidirectional switching system is required then the bridge circuit of Fig. 9.22(a), sometimes called the H-bridge or Full Bridge, can be used. This is a very versatile circuit, allowing both reversible DC switching and varying voltages and currents to be developed from a single DC supply. A full analysis can be found in Ref. 9.4. The power switching devices (shown here as MOSFETs; other power switching devices can also be used) work in pairs. If top left and bottom right (i.e. Drive B) are switched on simultaneously then current will flow through the load from left to right. If these are switched off, and the load is inductive, current decays by flowing through the free-wheeling diodes associated with the Drive A MOSFETs. Current flow in the reverse direction will take place if Drive A is switched on. If all devices are switched off then no voltage is applied to the load. Level-shifting circuits

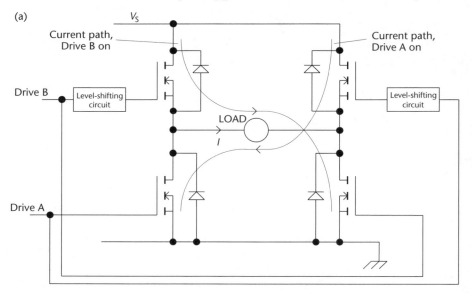

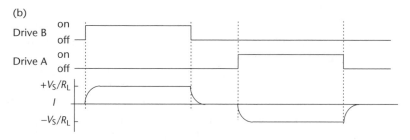

Figure 9.22 The H-bridge. (a) The circuit; (b) switching waveforms.

are needed for the gate drive to the 'high-side' MOSFETs, as the gate voltage applied to these devices must be relative to their source voltage, not to ground. This source voltage is swinging between ground and almost V_S, depending on the state of the associated low-side transistor.

In power systems the power supply V_S will be far greater than the logic supply. Up to a certain power level bridge circuits are available as ICs, with separate supplies for logic and power drive. For example, the SGS L298 has a 5 V supply for logic interface and a main power supply voltage up to 48 V. As the power level and voltage increase further, the power transistors and diodes are frequently found as discrete devices. Modules are also available containing all the power devices. At higher voltages it becomes an increasing challenge to switch the two 'highside' devices on. Various techniques are used to level shift the gate drive to the highside devices, including opto-isolation, a pulse transformer or level-shifting solid-state circuitry. Manufacturers who are active in this area (for example International Rectifier) give good design guidance for this in their Application Notes.

9.4.4 PWM revisited – continuously variable control

Pulse Width Modulation (PWM) was introduced in Chapter 5 as a possible digital to analogue conversion technique, the PWM waveform being adjusted and filtered to a produce an analogue voltage. In power systems PWM comes into its own as a means of controlling load current; in this case it is the inductance of the load which smooths the controlled current.

Figure 9.23 shows a switch S connected to an inductive load. S is opened and closed periodically according to the waveforms shown. When it is

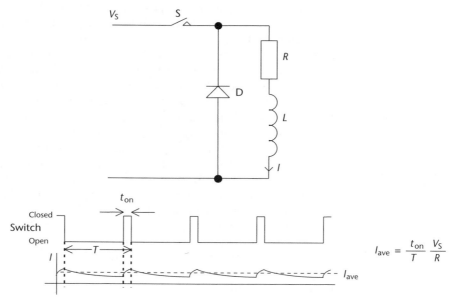

$$I_{ave} = \frac{t_{on}}{T} \frac{V_S}{R}$$

Figure 9.23 Unidirectional current control by PWM.

closed, then voltage V_S appears across the resistor and inductor and the current rises exponentially. When the switch opens, current continues to flow through the freewheeling diode and decays towards zero. If the switch is opening and closing much faster than the current rise and fall times, then the average value of current I is proportional to the (t_{on}/T) ratio of the switching waveform, as shown in the figure. If S is a solid state switch, then a microcontroller can set the load current by controlling a PWM signal to it. With this circuit a unidirectional current can be controlled in an inductive load.

If a PWM signal is applied to the bridge circuit, redrawn in Fig. 9.24(a), then continuously variable control of load current in either direction can be achieved. Now Drive A is on whenever Drive B is off, and vice versa. In practice a little 'dead time' (a few microseconds for example) is usually introduced between activation of each drive to ensure that one drive pair is fully off before the other is switched on. A unity mark/space ratio PWM waveform results in zero average current. By varying the PWM signal a current may be smoothly reduced to zero and then reversed (Fig. 9.24(b)), or an alternating current can be produced (Fig. 9.24(c)).

9.5 The DC motor

For control applications the DC motor has always been popular. It has good torque/speed characteristics, its speed is easily controlled, it is relatively low cost, and it is available in a wide range of sizes. It therefore has an important place in the embedded system environment.

9.5.1 Motor principles

The principle of the motor is easily understood with the diagram of Fig. 9.25(a). A coil (symbolised here as a single turn) is placed in a magnetic field, such that it can rotate. Electrical connection is made to it by a pair of brushes which slide on a commutator. If current is passed through the coil then a second magnetic field is established. With the coil in the position shown the two fields interact to produce a torque, which causes the coil to rotate. Its rest position would be with the plane of the coil perpendicular to the field due to the magnets. If the inertia of the motion carries it just beyond this position, however, then the current direction in the coil is reversed by the action of the commutator. With this new current direction the coil rotates a further 180°, and rotary motion will be sustained.

As it rotates, the coil also generates an emf (in fact, the 'motor' could equally be used as a generator). The voltage generated is known as the back emf, and is in a direction which opposes the applied voltage. It is approximately proportional to speed of rotation.

The equivalent circuit of the motor is shown in Fig. 9.25(b). R and L are the resistance and inductance of the coil winding, and the back emf is symbolised as E. The simple equation derived from this circuit, with current in the steady state, is:

$$V = IR + E \tag{9.4}$$

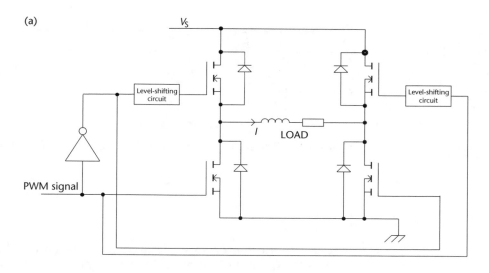

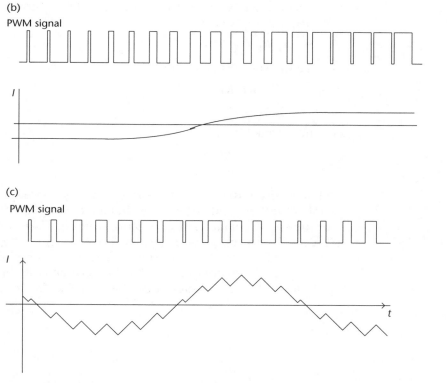

Figure 9.24 Bidirectional current control by PWM. (a) The circuit; (b) controlled current reversal; (c) sine wave generation.

As E is proportional to speed, and torque proportional to current, they can be expressed as:

$$E = nK_E \tag{9.5}$$

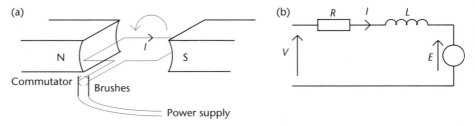

Figure 9.25   The DC motor. (a) Principle of operation; (b) equivalent circuit.

and

$$T = IK_T \tag{9.6}$$

where n is the motor speed, T is the torque and K_E and K_T are motor constants. K_E can be expressed in either volts per rpm, or volts per radian per second. K_T is called the torque constant. Substituting for E and I in Equation (9.4), we get

$$V = nK_E + RT/K_T \tag{9.7}$$

Hence the speed is given by

$$n = \frac{V - RT/K_T}{K_E} \tag{9.8}$$

or the torque by

$$T = K_T \frac{(V - nK_E)}{R} \tag{9.9}$$

Equations (9.8) and (9.9) are useful as they relate speed and torque to the applied voltage and the motor characteristics. Equation (9.9), the motor torque-speed characteristic, is plotted in Fig. 9.26 for the same motor at three different supply voltages. The line gradient is $-K_T K_E/R$, which is independent of supply voltage, speed or torque. The closer R is to zero, the nearer these lines are to the vertical (in which case the motor speed would be independent of torque). It can be seen that for constant torque speed is

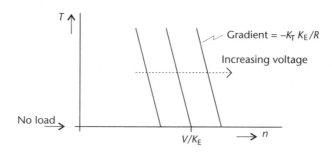

Figure 9.26   DC motor characteristics.

proportional to V. This is a useful result, as it is the basis of our motor speed control – we need to control the applied voltage.

In small DC motors the field is provided by permanent magnets, invariably with more sophisticated geometries than indicated in Fig. 9.25. Considerable research has gone into developing powerful field magnets, of which the 'rare earth' type are the most successful (Ref. 9.5). Small motors are often found with an integral gearbox, giving increased torque and reduced speed.

Large DC motors have an electromagnet instead of permanent magnets for the field. This has the advantage that speed can be controlled by varying the field winding current, which is smaller than the armature current.

The disadvantage of the conventional DC motor lies in the fact that brushes are needed to cause direction change (commutation) of the current in the armature winding. This causes electrical arcing and introduces mechanical wear, hence reducing long-term reliability. *Brushless DC motors* include integral electronics to provide current commutation, and therefore eliminate the need for brushes.

9.5.2 Driving the DC motor

As the equivalent circuit indicates, the DC motor acts electrically as a resistive/inductive load, and can be switched on or off with the circuit of Fig. 9.21(a). If it is to be reversible, then the bridge circuit of Fig. 9.22 can be used. Variable speed drive is obtainable by driving the bridge circuit with a PWM signal. In a servo system the DC motor is normally used with a shaft encoder (see earlier in this chapter) and/or tacho-generator for position and speed measurement.

WORKED EXAMPLE   9.3

A manufacturer gives the following data for a certain small DC motor:

No-load current	15 mA
Nominal voltage	12 V
Terminal resistance	10 Ω
No-load speed at nominal voltage	4000 rpm
Max continuous current	0.5 A

With a 12 V supply, it is found to run at 3000 rpm when delivering a torque of 9 mN m. It is driven from an H-bridge supplied from 12 V, and the losses in the switching devices may be assumed to be negligible.

(a) If it is driven from the full 12 V, what torque can it deliver when drawing maximum rated current?

(b) A no-load speed of 2500 rpm is required. What should the PWM ratio be?

Solution: Let us start by finding the motor constants. The no-load speed is 4000 rpm, and we are already aware, from Fig. 9.26, that this is equal to V/K_E:

$$4000 = V/K_E$$

$$K_E = 12/4000 = 3 \times 10^{-3} \text{ V/rpm}$$

Also

$$V = nK_E + RT/K_T$$

Substituting in values for the loaded motor,

$$12 = 3000 \times 3 \times 10^{-3} + (10 \times 9 \times 10^{-3})/K_T$$

$$3 = (9 \times 10^{-2})/K_T$$

$$K_T = 3 \times 10^{-2} \text{ Nm/A}$$

(a) With a current of 0.5 A, the torque is simply given by Equation (9.6):

$$T = IK_T$$

$$T = 0.5 \times 3 \times 10^{-2} = 15 \text{ mN m}$$

(b) If the no-load speed is to be 2500 rpm, then (noting that T is zero)

$$2500 = V/K_E$$

i.e.

$$V = 2500 \times 3 \times 10^{-3} = 7.5 \text{ V}$$

From the PWM point of view

$$\frac{t_{on}}{T} = \frac{7.5}{12} = 0.625$$

9.6 Stepping motors

9.6.1 Introducing the stepping motor

Stepping motors (also known as stepper motors) are a type of electric motor which are particularly well-suited to interfacing with digital signals. This is because the motor is driven by pulses of current which can be derived directly from a digital pulse stream, *and* because the motor shaft turns through a known angle every time the motor receives such a pulse. Therefore if the motor controller keeps track of the number of steps that have been undertaken, there should be complete knowledge of how much the motor shaft has turned. The motor can therefore theoretically operate under open loop control, without needing positional feedback information. In practice this is not as easy as it sounds, and there are situations

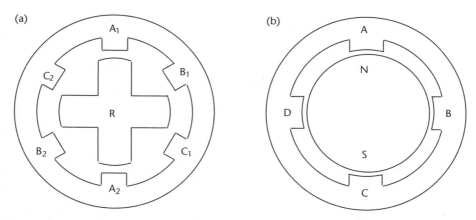

Figure 9.27 Stepping motor types – cross-sectional views. (a) Variable reluctance; (b) permanent magnet.

when the motor does not respond correctly to its electrical input. For example, there are limits on how fast a motor can be accelerated and on its maximum speed, which must be carefully understood and observed.

A stepping motor has a number of coils (or *phases*) and the motor rotates if these are energised in turn. Two motor types are illustrated in Fig. 9.27. In the variable reluctance motor of Fig. 9.27(a), pole pieces A_1 and A_2 are each wound with coils, which are connected in series. B_1 and B_2, and C_1 and C_2, are similarly wound. The rotor R is made of soft iron.

If coils A_1A_2 are initially energised, the rotor will not move, as the line of least reluctance for the magnetic lines of flux is directly from the poles and through the vertical arms of the rotor. If A_1A_2 are de-energised and B_1B_2 are energised, then the lines of flux will pass from the B pole pieces and towards the horizontal arms of the rotor. The effect is to draw these arms towards the B poles, and the rotor will start to rotate anticlockwise. This rotation will continue if coils continue to be energised in the order A, B, C. Each *step* in this instance is 30°. If the energisation direction is A, C, B, then the rotor will turn clockwise. Because the reluctance of each magnetic circuit varies with rotor position this type of motor is called *variable reluctance*. The example shows a three-phase version; other phase numbers are also possible.

In Fig. 9.27(b) the rotor is a permanent magnet, magnetised as shown. Each pole piece is wound with an independent coil. If the phases are energised in turn, the rotor will turn, with a step length of 45°. This motor is of the *permanent magnet* type.

As well as the variable reluctance and permanent magnet motor types introduced here, there is an important combination type, know as *hybrid*.

Clearly both motors shown above develop torque as a new phase is energised. If the rotor is at rest, with one phase energised, and an external torque is applied in an attempt to rotate it, then the motor will exhibit *holding torque*. This is defined as the maximum torque that can be applied to the shaft of a stationary, energised motor without rotation beyond that step

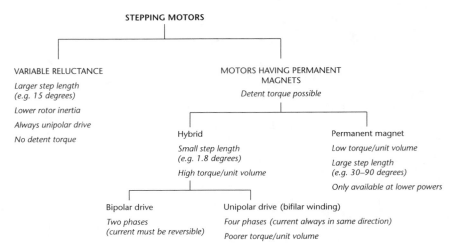

Figure 9.28 The stepping motor family tree.

position taking place (there will inevitably be some small angular displacement from the rest position). Even with no phase energised, motors with permanent magnets display a holding torque effect, known as *detent torque*.

Motors which require coil current to be driven in only one direction are termed *unipolar*; variable reluctance motors are always of this type. Other motors which require that current may have to be driven in either direction are called *bipolar*. Permanent magnet motors are often of this type. It is possible by using *bifilar* windings to render an otherwise bipolar motor unipolar. This approach simplifies the electrical drive requirements of the motor, at the cost of making the motor itself more bulky and expensive.

Each motor type has its own distinguishing characteristics, and these are summarised in Fig. 9.28.

A fundamental limitation of stepper motors arises because the motor phases need to be energised and de-energised rapidly in turn, but the coil inductance opposes that rapid change in current. As the stepping speed increases, the average current in each phase falls and the torque falls, until at some frequency the motor stalls.

A further problem is that as a motor is stepped, it tends to oscillate about each new step position, due to the rotor mass and the 'springiness' of the magnetic field. For certain stepping speeds and load conditions a resonance effect sets in, and the rotor can oscillate (disastrously) out of its intended position.

Motor manufacturers supply speed–torque curves for their motors. Although there is wide variety between motors, all have the general form shown in Fig. 9.29. The motor's torque is approximately constant at slow stepping speeds, and then falls off as the speed increases further. The *pull-in rate* is the maximum stepping rate at which the motor can be started from rest. The *pull-out rate* is the rate to which it can be further accelerated. A region of instability due to resonance effects is also likely to be displayed. These characteristics are of course modified once a load is applied to the

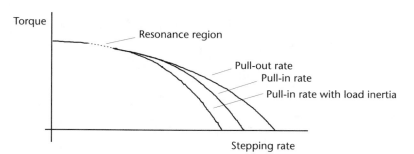

Figure 9.29 Stepping motor torque–speed characteristics.

motor. For example, a frictional load tends to damp resonance. An inertial load reduces the pull-in rate. It is also likely to exacerbate the resonance effect while lowering its frequency.

WORKED EXAMPLE   **9.4**

A stepper motor with a 7.5° step angle is required to turn a load of 30 mN m at a steady rate of 1 revolution every 100 ms. No acceleration time is allowable. What stepping speed is required, and what essential characteristics must the motor have?

Solution: With a step angle of 7.5°, the motor will require 48 steps to complete one revolution. With 10 revolutions per second the stepping rate will be 480 steps per second. It is therefore necessary to select a motor whose pull-in rate allows a torque of 30 mN m at this stepping rate. In practice a safety margin of at least 20% is advisable, leading to a pull-in rate of 36 mN m.

9.6.2 Drive waveforms

The simplest way to envisage the actual drive to a stepping motor is by energising each phase in turn. This is represented, for a four-phase variable reluctance motor, in Fig. 9.30(a). It is known as *wave drive*. There is, however, a torque advantage (at the expense of greater drive current) if phases are energised in pairs, and this is known as *two-phase on*; Fig. 9.30(b). For each of these two the step length is the same, though the rotor rest positions of the two-phase on lie midway between those of the wave drive. It should also be possible to combine the two patterns, moving from one to the other and halving the step length. This is known as *half-stepping*, and is illustrated in Fig. 9.30(c). Half-stepping allows more precise positional control and reduces the risk of resonance. It does, however, provide uneven torque, as steps when a single phase is on have somewhat less torque than

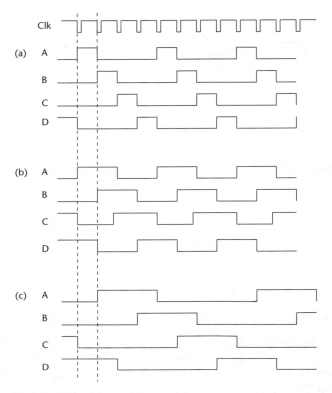

Figure 9.30 Motor drive waveforms.

steps with two phases. A further advance on this is known as micro-stepping, which allows even more precise positional control. This mode will not, however, be considered further here.

9.6.3 Motor speed profiles

In order to run a motor reliably it is necessary to start it at or below its pull-in rate. This can be viewed as effectively an instantaneous jump in speed. The motor *can* then be operated at this speed until the required position is reached, at which point it can be stopped. This approach is represented by line (1) in Fig. 9.31; it is the simplest way of applying a stepping motor.

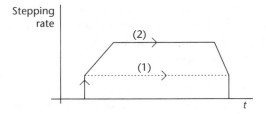

Figure 9.31 Example stepping motor speed profiles.

To minimise the time taken to effect a move the motor is normally started within its pull-in rate, but then accelerated towards its pull-out rate. As the target position is reached the motor is then decelerated back to its pull-in rate, from which it can be stopped at the right moment. This 'velocity profile' is shown in Fig. 9.31, line (2). Generally the deceleration rate can be faster than the acceleration rate, as most loads aid the slowing process.

9.6.4 Switching the motor phases

When driving the motor, each coil will require its own drive circuit. The simplest approach, for the unipolar case, is to use the switching circuit of Fig. 9.21(a). As described in that section, as the stepping rate increases, there comes a point when the phase current cannot reach its steady state value. The circuit of Fig. 9.21(b) has therefore been widely used, resulting in an increased maximum speed, at the cost of reduced efficiency. Where bipolar drive is required the bridge circuit of Fig. 9.22 can be used.

The optimum approach to driving the stepping motor involves chopping the phase current. Here the coil is driven from a voltage much higher than the rated voltage. The current therefore rises rapidly. It is monitored, and when rated current has been reached it is switched off. The current then decays through one or more freewheeling diodes. When it has fallen to just below the required level it is switched back on again. This principle is illustrated, applied to a bridge circuit, in Fig. 9.32. A sensing resistor, of very low value (usually well below 1 Ω), is now included in the earth path. Whichever direction the current flows, it passes through the sensing resistor. The voltage across this resistor is compared with the preset voltage V_{ref} by a

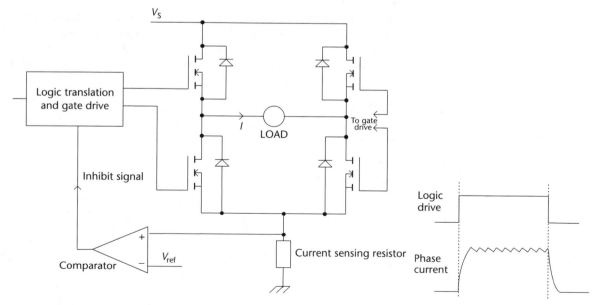

Figure 9.32 Current chopping.

comparator. When the comparator output goes high, drive to the winding is temporarily inhibited.

9.6.5 The motor drive system

There are several distinct parts to a stepper motor drive system, as illustrated in Fig. 9.33, which assumes open loop control. At the controller end lie the basic control functions of determining motor speed, direction and how far it is to rotate. A clock generator is needed, each cycle representing one motor step. For speed ramping the clock generator speed should be controllable. The accumulated steps must be counted, and may be compared with a target value; when that value is reached the motor is stopped. From each clock cycle it is necessary to develop the switching waveform appropriate to the motor. Finally, these waveforms are interfaced with the motor windings through the power output stage.

It is sometimes difficult to determine how much of the stepper motor control should be retained in the microcontroller. While the switching waveforms of Fig. 9.30 may be generated in a microcontroller, it is often preferable to do so in hardware, for which many ICs are available. Major options for microcontroller control include:

1. The microcontroller acts as clock source and translator; external devices are needed only for the power drive stage. The basic clock can be generated from a periodic source such as a Counter/Timer operating in overflow on interrupt mode. This speed can be ramped as necessary.

2. The microcontroller acts as clock and direction signal source; an external IC is used for the translator.

3. The microcontroller controls a voltage-controlled oscillator (VCO), which is the clock source. The VCO can be switched on and off, and simple ramps are developed with R–C networks. A Counter/Timer on the controller, counting output pulses from the VCO, is used to measure distance travelled.

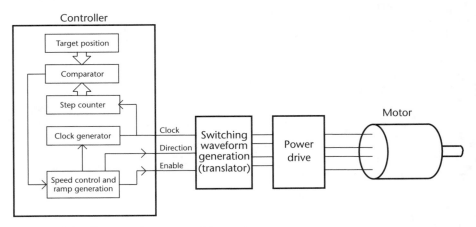

Figure 9.33 The stepping motor drive system.

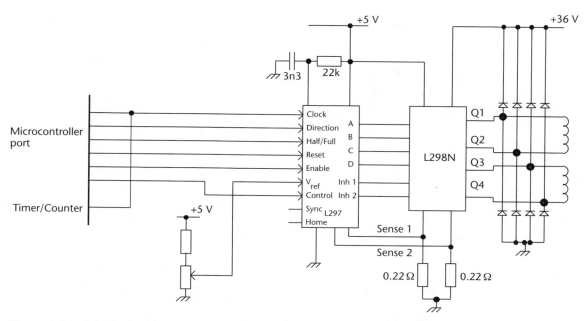

Figure 9.34 Standard ICs applied to the interface between mcrocontroller and motor.

Option 1 is very intensive of controller time and limits other controller activities. It does, however, limit the IC count. Option 2 still commits the controller to developing a time-based signal. Option 3 places the lightest burden on the controller. Its disadvantages are the increased IC count and the loss of close control over the ramp function. Examples of these options are given below.

Figure 9.34 shows an example of Option 2, and is based on a chip set supplied by SGS. The L297 runs from a standard supply and generates the switching waveforms (i.e. four phase and two inhibit controls) from certain control inputs. The L298N is a dual full-bridge driver, running from up to 36 V, in a 36-lead 'Multiwatt' package, which can drive a total per-phase DC current of 4 A. Current through the windings is sensed in R_{S1} and R_{S2} and compared with an input voltage V_{ref}. By controlling V_{ref} the user can control the maximum current flowing in the phase. In most cases the microcontroller interconnection to the L297 will be restricted to clock, direction and possibly inhibit; others may be tied high or low.

An application of Option 3 is illustrated in the circuit of Fig. 9.35(a), which uses the VCO part of a 74HC4046 IC. Two pins from a microcontroller port can be used to control the inputs 'Clock Inhibit' and 'Speed'. The output frequency of the VCO, appearing on pin 4, is controlled by the voltage asserted on pin 9, the VCO voltage input. The frequency range is predetermined by the values of R_1 and C_1, in this case shown as 150k and 100 nF. When the 'Speed' input is at logic low, the VCO input is at around 1.7 V, and at logic high it is 3.3 V. Switching between the two follows an exponential curve of rise time determined by the resistor values and C_2. The clock may be disabled by setting the Clock Inhibit line high.

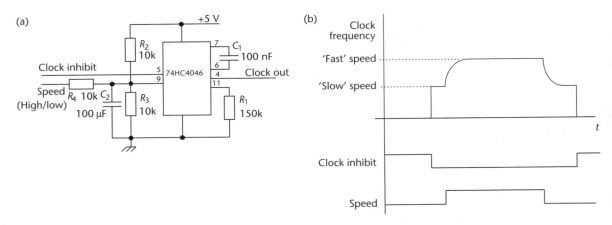

Figure 9.35 Applying a VCO to generate the clock signal. (a) The circuit; (b) the resulting speed profile.

With the component values shown, the slow clock speed is around 300 Hz, the fast one around 600 Hz, and the rise/fall time between them around 0.7 s. Figure 9.35(b) shows how the clock frequency varies with changes in the logic inputs. With the slow speed adjusted to motor pull-in rate and the fast speed to pull-out rate, this motor control is adequate for many applications.

For further design guidance on the stepping motor, references such as Ref. 9.6 should be consulted. Many manufacturers also provide useful technical data. A good example of this is Ref. 9.7.

SUMMARY

1. Incoming signals may be degraded by interference or distortion, or they may be potentially hazardous to the receiving circuitry. Consideration must therefore be given to whether they can be directly connected to a digital input, or whether some form of input protection and/or signal conditioning is needed.

2. Sophisticated LED and LCD displays are available, each having their own interfacing requirements.

3. Optical devices are widely used for both sensing and signal interconnection.

4. Low-power electrical loads can be switched by direct connection to a microcontroller port bit. Standard electronic techniques can be applied to switch larger loads.

5. Continuously variable voltage and current control can be exercised with PWM outputs. A common application is with the DC motor.

6. The stepping motor is an important actuator in the embedded system environment. Its basic characteristics must be understood before applying it.

REFERENCES

9.1 *Implementing Wake-Up on Keystroke*. Application Note AN552. Microchip Technology Inc. 1997.

9.2 Kellar, P. (1997) *Electronic Display Measurement*. Chichester: Wiley.

9.3 *HD44780U Data Sheet*. Hitachi Semiconductors. `http://semiconductor.hitachi.com/products/`.

9.4 Williams, B. (1992) *Power Electronics. Devices, Drivers, Applications and Passive Components*, 2nd edn. Basingstoke: Macmillan.

9.5 Kenjo, T. (1991) *Electric Motors and Their Controls*. Oxford: Oxford Science Publications.

9.6 Kenjo, T. (1994) *Stepping Motors and their Microprocessor Controls*, 2nd edn. Oxford: Oxford Science Publications.

9.7 *Product Selection and Engineering Guide*. Airpax Mechatronics. `http://www.thomsonind.com/airpax/airpax.htm`.

EXERCISES

('+' indicates a simple review question)

+**9.1** Explain the phenomenon of 'switch bounce'. Why is it frequently necessary to 'debounce' signals from switches?

+**9.2** Describe, with the aid of circuit and flow diagrams, three possible ways to debounce a switch, including one software method. Give and explain typical time and component values.

9.3 What precautions would you take when interfacing the following digital input signals to a microcontroller pin? State any assumptions made.

(a) Limit switch on a lead screw; action of switch is to disable further travel.

(b) Optical transmissive object sensor on conveyor belt.

(c) Keypad used for photocopier data entry.

(d) Optical rpm counter on electrical machine, 20 m remote from microcontroller.

9.4 A four-digit common cathode display is made using the seven-segment display unit pictured in Fig. 9.9(b). The forward voltage of each diode when in conduction is 1.9 V, and the current requirement of each segment and the decimal point is 20 mA. It is possible for all segments in the display, but only one decimal point, to be illuminated simultaneously. Calculate the maximum possible power dissipation.

9.5 Two opto-sensor LEDs are to be switched on or off simultaneously from one microcontroller port bit, which is able to source or sink 5 mA. Their current requirement is 35 mA, for which their forward voltage is approximately 1.9 V. A ZVN4306 MOSFET and a 2N3704 transistor are available, with data as shown in Table 9.2.

(a) Indicate with the aid of circuit diagrams (showing any component values) how this switching requirement could most efficiently be achieved:

(i) If the microcontroller port output switches between 0 V and 5 V

(ii) If the microcontroller port output switches between 0 V and 2.4 V

(b) What changes to your circuits would need to be made, if any, if the microcontroller port was quasi-bidirectional and able to sink 10 mA but source only 100 μA.

9.6 The LED of a reflective opto-sensor has a forward voltage of 1.9 V when it is in conduction. Its recommended operating current is 35 mA. The sensor has a maximum dark current of 100 nA, and gives a minimum current of 200 μA when a reflective surface is placed 2 mm from its surface. It is connected into the circuit of Fig. 9.17(c). The sensor output is connected to a PIC16F84 port input of characteristics shown in Table 2.1. Both opto-sensor and microcontroller are powered from 5 V. Calculate the resistor values needed:

(a) neglecting the effect of the port input leakage current (as specified in Table 2.1)

(b) including the effect of the port input leakage current

9.7 A phase winding of the Airpax 35L048B stepper motor is quoted as having a winding resistance of 64 Ω and inductance of 38 mH. Its operating voltage is 12 V.

(a) If the voltage applied to the winding is changed instantaneously from 0 V to 12 V, what is the rise time of the current and what is the steady state current value?

(b) The motor is to be operated so that the phase is pulsed with a square wave voltage of frequency 300 Hz. The loss of torque is found to be unacceptable, so it is decided that the current rise time must be reduced to 0.5 ms by the addition of forcing resistors. What value of resistor should be used, and what will the new supply voltage be?

9.8 Four common-cathode seven-segment LED displays are to be multiplexed, as shown in a general way in Fig. 9.10. Each segment is made up of two LEDs in series, each with a forward voltage in conduction of 1.9 V. The display driver outputs 5 V when active, whatever the current sourced; each segment line is connected to an output of the display driver via a resistor. If illuminated segments must pass an *average* current of 8 mA, what should be the value of these resistors? State any assumptions made.

9.9 In the subroutine of Fig. 9.15, explain how the E line is strobed high and low.

9.10 Sketch a flow diagram for a stepping motor controller which starts the motor at a stepping rate of 200 steps per second and accelerates it to 400 steps per second. Use a Counter/Timer Interrupt on Overflow for timing purposes. Is the acceleration produced in this way linear?

Supplying and using power in a power-conscious world

IN THIS CHAPTER

Napoleon Bonaparte, the famous French general, astutely remarked that an army marches on its stomach. Had he been an embedded system designer of the twenty-first century, he might have remarked that the embedded system marches on its power supply.

The embedded system has a need for power, and for a conditioning system which matches the source of power to those needs. In some cases the choice of power source is obvious; high power requirements will demand single- or three-phase mains electricity, a portable handheld device will have to be battery-powered. In others the choice may not be clear-cut; can a previously mains-powered equipment be converted to battery power, for example? In other cases again a balance of power sources will be considered: a mains powered product may need battery backup, or a battery-powered device may plug into the mains to recharge.

A good and reliable power supply is certainly the beginning of a good and reliable product. The power supply design must proceed in parallel, and interact with, other hardware design. It is not to be regarded as a final chore, something to be added when all the interesting work has been done. An inefficient power supply will render a product uncompetitive, while an unreliable one will render it unviable.

With the incredible growth of handheld and portable electronic appliances and gadgets (consider only the pocket calculator, the laptop computer and the mobile phone) there has been a surge of demand for low-power circuits. It is in this area where developments are being made in power supply design. An important part of this chapter will therefore be power-conscious design. Regular power supply design, with mains input, is detailed in many texts, for example Ref. 1.1.

This chapter aims to present power supply issues that are important to the embedded system designer, focusing on low-power applications. The specific aims are:

- to review mains and battery power sources
- to describe relevant voltage regulation and conversion techniques
- to describe power and battery management techniques
- to investigate the characteristics of CMOS, the low-power technology
- to identify the characteristics of the low-power microcontroller
- to summarise low-power design techniques

10.1 Sources of power

10.1.1 The problem of power

All electronic systems, and hence all embedded systems, require a supply of electrical power to enable them to operate. The choice lies broadly between mains power, batteries and renewable sources, with the latter including solar, wind and hydroelectric.

The overall power supply is pictured in Fig. 10.1. The Target System is the system to be powered. There is a power source and conditioning to match that source to the target. There may also be an alternative power source, should the major source fail or be removed. In almost every case this alternative source is battery-based.

Figure 10.1 suggests that, in supplying power to a system, the problem for the system designer is threefold:

1. to quantify the supply requirements of the target system
2. to identify a source or sources of power
3. to match the power source(s) to the target system

We will start by considering the sources of power. Examples of these, in different combinations and for different applications, are given in Table 10.1.

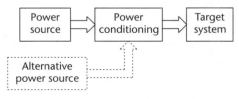

Figure 10.1 The supply of power.

Table 10.1 Example power sources and their combinations.

A mains-powered item; for example a numerically controlled machine

A mains-powered item with battery backup of minimal essential functions; for example a personal computer

A mains-powered item with backup for all functions; for example an office computer with an Uninterruptible Power Supply (UPS)

A battery-powered item

A battery-powered item with mains recharging

An item powered by renewable energy; for example a solar power inverter

An item powered by renewable energy with battery backup; for example an Automatic Weather Station powered by solar panels with an internal rechargeable battery

In most cases, the choice of power source lies between mains power and batteries. The relative advantages of these two are summarised in Table 10.2.

In brief, mains power gives us unlimited power (at least for our purposes!) at low cost, while batteries give strictly limited power at high cost.[1] Yet using batteries frees us from enslavement to the mains power outlet; to gain this freedom it is necessary to explore the characteristics and needs of these special and costly devices.

Table 10.2 A comparison between mains and battery power.

Mains powered	Battery powered
'Unlimited' power	Strictly limited power
Limited portability in operation	Portable in operation
Potential hazard	No electrical hazard
Mains itself is source of electromagnetic interference (EMI)	No EMI from battery
High reliability of supply in most locations	Uncertainties of battery life
Little demand for power-conscious design	Power-conscious design essential
No battery replacement problem	Concern with battery replacement or recharge
Low energy cost	Higher energy cost

1 It has been estimated (Ref. 10.1) that energy from the cheapest commercial battery costs over 1000 times more than energy supplied from the mains! Part of this estimate indicates that a button cell (not the cheapest), provides energy at a cost of £10 000 per kWh!

10.1.2 Batteries

Batteries are classified either as primary cells (non-rechargeable) or as secondary cells (rechargeable). Primary cells are typically used for low power or occasional applications, where replacement is infrequent. Where power drain is higher, secondary cells are more commonly used.

Batteries are available in a huge variety of packages. These include the popular D, C and AA cylindrical packages for consumer applications, and the 9 V PP3, long associated with radio power supply. Familiar also are the button cells used in wristwatches and other small devices. Many other more specialised packages have also emerged as the need to supply power in tighter spaces has become paramount. These include packages for direct PCB (printed circuit board) mounting.

When selecting a battery type, the main considerations are:

- *Battery capacity*: This indicates the capacity of a particular cell or battery, and is generally measured in Amp-Hours (Ah) or milliamp hours (mAh). The inference is that a battery of 500 mAh can sustain a 500 mA current for 1 hour or a 1 mA current for 500 hours. In reality the situation is not so simple: batteries do tend to recover between periods of use, and display somewhat different capacities depending on load current. Capacities are therefore usually quoted for a given discharge current. Multiplying the Amp-Hour capacity by the battery terminal voltage gives an approximate value for the actual energy stored in Watt-Hours.

- *Terminal voltage and discharge curve*: This indicates how the terminal voltage of a particular battery falls during use. Some batteries have very flat discharge curves, maintaining a near constant terminal voltage until they are almost exhausted. Others have a curve which slopes steadily down. There is often a small portion of the curve at the beginning where the terminal voltage falls rapidly. Almost all discharge curves show a steep slope as the battery approaches exhaustion.

- *Energy density*: This indicates how much energy per unit volume or unit mass is available from a particular battery chemistry. The highest energy densities are available from primary cells.

- *Shelf life/charge retention*: This is an indication of the expected life of a battery before use. It is significant not only for the storage of batteries before use, but also for batteries whose application has very low or very occasional current demands, for example a battery backup system.

- *Price*: No battery is particularly cheap, and the consumer recognises this. It is important therefore to consider the relative battery costs and to design products which require the fewest batteries possible and the longest life possible.

Serious contenders for powering electronic equipment include alkaline and lithium batteries (mostly found as primary cells) and sealed lead–acid, nickel–cadmium and nickel–metal-hydride batteries (secondary). These types are reviewed below, with summary data being given in Table 10.3.

Table 10.3 Indicative characteristics of selected battery types.

Battery type	V_{oc}[1]	Energy density		AA cell capacity	Shelf life[2]
		Wh/kg	Wh/litre		
Alkaline	1.58	125	315	1.6 Ah	5–7 years
Lithium thionyl chloride	3.9	375	850	2.1 Ah	5–20 years
Nickel cadmium	1.35	40	120	0.6 Ah	5 year life; 15–20% discharge per month
Nickel–metal-hydride	1.35	65	170	1.2 Ah	5 year life; 15–20% discharge per month
Sealed lead–acid	2.1	30	100		10 year life; 15–20% discharge per 6 months

[1] V_{oc} is the unloaded terminal voltage of a fresh cell
[2] Storage at 20 °C

- *Alkaline*: This is the most commonly used primary cell, with reasonable cost and a good shelf life. While the alkaline cell is well able to deliver large currents, because of its long shelf life it is also suitable for long-term, low-current applications. Rechargeable alkaline cells are also available.

- *Lithium*: Lithium batteries are known for their exceptionally long shelf-life and very high energy density. Dramatic evidence of this was given during the final hours of the 1995 NASA Galileo probe (Ref. 10.1). After travelling for six years and being subject to enormous acceleration forces, the on-board lithium/sulphur dioxide batteries were able to provide a capacity of 18 Ah. Lithium batteries have a flat discharge characteristic and wide operating temperature range. Different chemistries are used to suit different discharge rates. Rechargeable lithium cells are now becoming increasingly available.

- *Lead–acid*: The mainstay of the car battery, this type *can* also be considered for certain portable or static systems. Sealed lead–acid batteries contain an acid gel, so there is no need for water top-up (as in the car battery) and there is no risk of leakage. Like the car battery, they retain the ability to supply large currents and are thus used in the UPS and systems where there may be heavy current demands. Lead–acids should not be stored in a discharged state; if they are to be stored, they should be recharged periodically.

- *Nickel–cadmium (Ni-Cad)*: This remains the most popular rechargeable battery for the handheld device. They are electrically robust, with a discharge characteristic which is moderately flat, being 1.24 V ± 100 mV for most of the discharge cycle. They can be stored for sustained periods (for example over five years) in any state of charge, but have a significant self-discharge rate. Ni-Cads display a 'memory effect' whereby the cell loses capacity if it is repeatedly recharged before it has been completely discharged. This can be avoided by ensuring that a cell is fully discharged before recharging. In memory backup applications the Ni-Cad is normally continuously trickle-charged at a low rate; this normally gives a life of at least five years. Ni-Cads require a constant current charge, and can be overcharged without significant harm. They can be recharged hundreds of times, with a slow reduction in cell capacity. After 500 to 1500 deep discharge cycles the capacity is reduced to around 50%. There is a similar loss after five years of continuous low-level trickle charging.

- *Nickel–metal-hydride*: This technology is similar in performance to the Ni-Cad, with almost identical terminal voltage but higher energy density, leading to a cell capacity up to 40% greater.

10.2 Supply voltage control

10.2.1 Power conditioning

Voltages from the power source, whether it is battery, mains or solar panel, rarely come at values which the circuit designer wants. Even if they did, there is the problem that the power source voltage might not itself be stable. Mains voltage suffers from fluctuations, battery terminal voltages fall, and solar panel output depends on the level of insolation. Therefore a range of techniques have been developed to match the supplied electrical energy to the requirements of the system.

In the context of battery supply, immense ingenuity has been applied to scaling down power electronic techniques into a domain where every microamp counts and where every possible scrap of energy is squeezed from the battery.

Collectively, these techniques offer the capability to:

- regulate voltage
- step a DC voltage up or down
- invert voltage (i.e. from positive to negative or vice versa)
- control the charging of batteries
- switch power on/off, or switch between different power sources
- monitor supply voltages and take action (shut down, for example) if thresholds are passed

The first three of these we go on to consider in this section.

10.2.2 The voltage regulator

In simple terms voltage regulators can be considered to be circuits having three terminals: input, output and earth; they are often supplied as three-terminal ICs, connected as shown in Fig. 10.2. Battery (or source) voltage and current are symbolised as V_B and I_B, voltage and current supplied to the

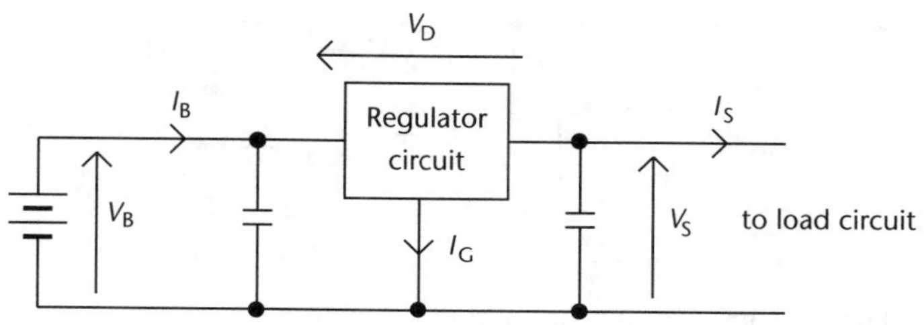

Figure 10.2 The voltage regulator.

load circuit as V_S and I_S, and the voltage differential between input and output voltages as V_D. I_G is called the Ground or Quiescent Current, and is the current drawn by the regulator circuit for its own operation.

The purpose of the voltage regulator is to maintain its output V_S at a fixed value as long as its input voltage V_B lies anywhere within a predetermined range, set by the actual regulator circuit. The lower limit of this input voltage range is always somewhat higher than the output voltage. When the input voltage falls below this limit, the circuit 'drops out' of regulation. The minimum value of V_D for V_S to remain at its regulated value is called the device *dropout voltage*. The upper input voltage limit is determined by the maximum voltage or power dissipation rating of the regulator circuit.

A major feature of interest to the low power designer is the efficiency. Referring to Fig. 10.2, the regulator efficiency is:

$$\text{Efficiency} = \frac{\text{Output power}}{\text{Input power}} = \frac{I_S V_S}{I_B V_B} \times 100\% \qquad (10.1)$$

Clearly, the closer the output power is to the input power, the higher is the efficiency. Let us therefore explore this difference between the two powers, i.e. the 'lost power':

$$\text{lost power} = \text{input power} - \text{output power}$$
$$= V_B I_B - V_S I_S$$
$$= (V_D + V_S)(I_G + I_S) - V_S I_S$$
$$= V_D I_G + V_S I_S + V_D I_S + V_S I_G - V_S I_S$$

Finally,

$$\text{lost power} = V_B I_G + V_D I_S \qquad (10.2)$$

Evidentally, for a given I_S (which is determined by the load circuit), if we are to minimise lost power and hence maximise efficiency we must make every effort to minimise the terms V_B, I_G and V_D.

10.2.3 The linear voltage regulator

Traditionally, regulators have been of linear design, with basic form as shown in Fig. 10.3. Output voltage is controlled by the *pass transistor*,

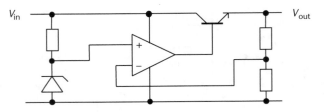

Figure 10.3   The linear voltage regulator.

which lies between input and output. A potential divider is placed across the circuit output, and a voltage derived from it is compared in an error amplifier with a voltage reference. The amplified difference signal is applied to the pass transistor, which sets the output voltage to minimise the difference between the reference and fed back voltages. The dropout voltage for this circuit depends on the output voltage range of the error amplifier, on the pass transistor type and configuration, and on load current.

An example of the linear three-terminal regulator is the popular 78L05, offered by many manufacturers, which has the characteristics shown in Table 10.4.

Consider the 78L05, operating from a 9 V battery and supplying 10 mA at 5 V. The input current is the sum of output and quiescent currents, hence 13 mA. Applying Equation (10.1), the efficiency is:

$$\frac{5 \times 10 \times 10^{-3}}{9 \times 13 \times 10^{-3}} \times 100\% = 43\%$$

Even if the input voltage falls to its minimum possible, i.e. 7 V, the efficiency only rises to 55%.

These are poor figures, and suggest that conventional linear regulators are unlikely to be adequate for battery power. The 78L05 was not, however, designed for the battery-powered market, and it *is* possible to modify the linear regulator design to make it more efficient. In particular, the quiescent current and the dropout voltage can be significantly reduced. Many linear regulator devices have in fact been developed specifically for low power. These devices have very low quiescent current (of the order of 10 μA to 100 μA), and very low dropout voltage (well below 1 V). With low dropout devices, even when regulation is lost, useful battery life can sometimes be extracted by letting V_{out} drift down with V_{in} until some lower limit is reached (e.g. 4.5 V for a 5 V system).

Table 10.4   Summary characteristics of 78L05 voltage regulator.

Input voltage range	7 V to 30 V
Output voltage	5 V ± 0.2 V
Quiescent current	3 mA
Max. output current	100 mA

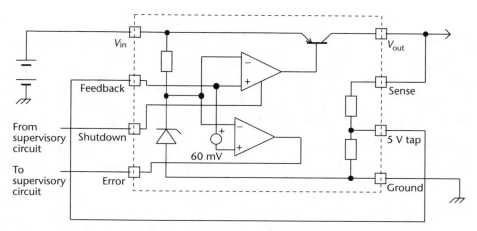

Figure 10.4 The National Semiconductors LP2951, simplified circuit diagram.

Figure 10.4, for example, shows the simplified circuit of a typical low-power regulator, the National Semiconductors LP2951, available as an 8-pin DIL package. While the pattern of this regulator is similar to Fig. 10.3, there are some important differences. The pass transistor is pnp instead of npn, with output taken from the collector instead of the emitter. This configuration reduces dropout voltage at the expense of increasing output resistance. There is an internal potential divider, which can be connected if a 5 V output is required. Alternatively, an external divider may be used. There is also an internal comparator, which monitors the feedback voltage and flags if there is more than 60 mV between it and the reference voltage. The error flag is activated when this threshold is passed, giving a warning that the output is 5% in error (60 mV is 5% of the 1.2 V reference voltage).

WORKED EXAMPLE **10.1**

An LP2951 regulator is needed to supply 1 mA at 5 V ± 10% from a battery source. A low battery condition should be flagged. How should the regulator be configured and what operating efficiencies can be expected? Relevant data is given in Table 10.5.

Table 10.5 LP2951 summary characteristics at 1 mA load current.

Ground pin current	100 μA
Dropout voltage	110 mV
Error flag	Activated when output falls 5% below programmed value

Solution: The acceptable output voltage range is 4.5 V to 5.5 V. We could set the output to be 4.5 V, by using an external potential divider, with some consequent increase in current consumption. In this case the error flag would be activated when the output has fallen to 5% less than this (i.e. 4.275 V), which is unacceptably low. We could set the output just above 4.5 V, so that the error flag comes on as 4.5 V is reached. This would still require use of an external potential divider. For minimum current consumption let us therefore use the internal potential divider. The error flag will then be activated at 4.75 V, which is safely within the permissible range. At this moment, the anticipated battery voltage will be (4.75 V + dropout voltage), i.e. 4.86 V. The circuit will be connected as in Fig. 10.4.

We can go on to calculate approximate efficiencies for various battery voltages, as shown in Table 10.6.

Table 10.6 Regulator efficiencies for various input voltages.

Battery voltage	9 V	6 V	5.11 V	4.86 V
Efficiency	$\dfrac{5\times 1}{9\times 1.1}=50.5\%$	$\dfrac{5\times 1}{6\times 1.1}=75.8\%$	$\dfrac{5\times 1}{5.11\times 1.1}=89.0\%$	$\dfrac{4.75\times 1}{4.86\times 1.1}=88.9\%$

The efficiency is clearly poor for large input to output voltage differentials, as expected. However, it reaches attractively high values for input voltages below around 6 V.

Example 10.1 shows that high efficiency is possible with a linear regulator designed for low-power applications when the input–output voltage differential is low. To this we can add the advantages that the regulator is simple and cheap, with low noise on the regulated output and negligible radiated EMI. Its disadvantage is a lack of versatility. Only voltage reduction is possible, and there is low efficiency for high input voltage.

The quiescent current for a linear voltage regulator is moderately independent of input voltage. For a constant load current, therefore, the linear regulator draws an approximately constant current from the power source, independent of its actual input voltage. It is useful to bear this in mind when estimating battery capacity requirements.

10.2.4 Switching converters

A linear regulator effectively 'wastes' surplus power capability by dissipating it as heat in the pass transistor. There are, however, other approaches to the problem of voltage conversion and control, which come under the general heading of switching converters. These converters store little gulps of battery energy in either a capacitor or inductor and transfer just about all of it to the load circuit, mostly applying feedback techniques to ensure that it is delivered at the voltage required. High efficiency and versatility are the main advantages of such circuits. In different configurations they can not only regulate a voltage, but also step voltages up and down and invert. A fundamental characteristic of their action is that they continuously switch the input voltage and current. This leads to a significant disadvantage: a higher level of radiated EMI when compared with linear regulators and some unavoidable output ripple.

Switching converters are *extremely* important for battery-powered embedded systems, and accordingly we now take time to review some of the important configurations.

10.2.4.1 Step-down (buck) converter

The circuit of Fig. 10.5(a) shows how this converter works. When the switch is closed, current flow builds up continuously in the inductor according to the relationship

$$V_L = (V_{in} - V_{out}) = L \, dI/dt \tag{10.3}$$

If the switch opens after a period of time then, with V_{in} disconnected, there is no further cause for current increase. The sign of the dI/dt term in Equation (10.3) reverses, and V_L therefore changes polarity. The voltage at point A, instantaneously undefined due to the opening of the switch, falls rapidly until, at a voltage a little below ground, the diode becomes forward biased. The inductor current can now continue to flow, decreasing in magnitude, through the diode. It does so until it has fallen to zero or the switch is reclosed. Whether the switch is opened or closed, the inductor current is transferred to the output capacitor and also to any load present. The amount of charge deposited on this capacitor by the inductor current determines the value of output voltage, V_{out}.

This switching pattern, represented in Fig. 10.5(b) in idealised form, is repeated continuously. The longer the switch is kept closed during a switching cycle (t_{on}), the higher will be the peak, and hence the average, current. The current waveform shown here, in which current falls to zero for a period of time in each cycle, is call *discontinuous conduction*. In many

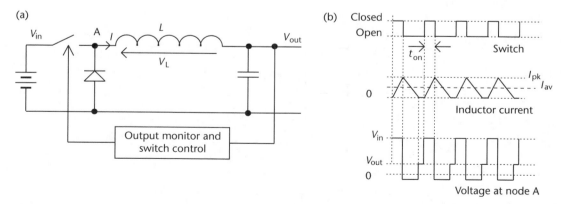

Figure 10.5 The step-down (buck) converter. (a) Circuit; (b) current/voltage waveforms.

cases *continuous conduction* is preferable, as for a given average current the peak current is lower.

To maintain the output voltage stable, it is monitored and compared with a reference. The switching waveform is controlled such that the output voltage is maintained at a predetermined value.

Switching waveforms are usually PWM or Pulse Frequency Modulation (PFM). In PFM the pulse width is constant, and the frequency varied. A variant of PFM is known as 'pulse skipping'. Here a constant frequency pulse generator is enabled if the output voltage is lower than its required value, and disabled if it is equal to or higher. The effect of all these techniques is to vary the average duty cycle.

Most losses in this circuit are caused by component resistances, particularly the battery, pass transistor, output capacitor and circuit interconnections. Keeping these as low as possible improves the efficiency. *Unlike* the linear regulator, efficiency suffers only slightly from increased input voltage.

The equations governing the performance of this type of circuit are complex, and are published in a number of references (for example Ref. 9.4). Fortunately, they need not concern the designer who wishes simply to apply ICs of this sort. Information is presented fully, and largely in graphical form, by the manufacturer. The circuit then becomes comparatively simple to apply.

An example of the step-down switching regulator is the Maxim MAX 639, available as an 8-pin DIL IC package, shown in example configuration in Fig. 10.6. This regulator can deliver up to 225 mA and has a quiescent current of only 10 μA. The switch itself is a MOSFET inside the IC, controlled by a pulse generator, and connected between the V+ and LX IC terminals. The diode, a Schottky device for minimised losses, is external. The battery is decoupled with a large capacitor, which reduces the effect of the battery internal resistance when current is being drawn.

The output voltage is monitored via a potential divider. Like the LP2951, an internal potential divider is provided for 5 V regulation; an external one can be used for other output voltages. When the output voltage falls below a certain tolerance, the pulse generator is activated, the MOS switch starts

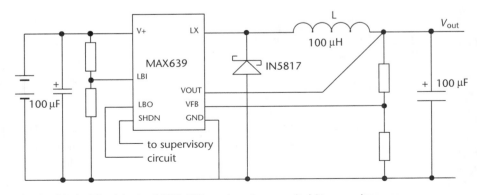

Figure 10.6 The Maxim MAX 639: a step-down switching regulator.

switching and the output rises. When it reaches the required level, the pulse generator is switched off again. This is an example of pulse skipping.

An internal potential divider connected across the battery input is used to compare the battery voltage with the internal 1.28 V voltage reference. When the battery voltage falls below a chosen value, the Low Battery Output (LBO) pin goes low and a supervisory circuit (if included) can take appropriate action. This may take the form of shutting the supply down, which can be done via the SHDN pin.

While full design details are given in the manufacturer's data, the following general points are important for circuits of this sort:

- The inductor must be rated to pass the *peak* current, I_{pk}, without saturation. While, under ideal theoretical conditions, $I_{pk} = 2I_{ave}$, the MAX639 design data suggests as a guideline $I_{pk} = 4I_{ave}$. In this case I_{ave} should be the maximum current supplied to the load circuit. The inductor resistance should be minimised.

- Input and output capacitors are non-critical in value (around 100 μF suggested), but high-quality, low series impedance types are recommended for lowest ripple.

- The diode should be high speed and low forward voltage (i.e. Schottky).

- Great care must be taken both with PCB design, with compact layout and short and broad interconnection. While average currents may be modest, peak ones are not, and any interconnection resistance or inductance arising from poor PCB layout will degrade performance unnecessarily.

For load currents greater than 1 mA the MAX639 can readily achieve an efficiency in excess of 80%. For certain situations it can be above 90%.

10.2.4.2 *Step-up (boost) converter*

If the components of the step-down converter are rearranged, a converter which steps voltage up can be made. In the circuit shown in Fig. 10.7(a), the full supply voltage is applied across the inductor when the switch is closed. The voltage across the inductor opposes the battery voltage. When the switch opens (Fig. 10.7(b)), current flow continues through the diode to the output capacitor. As the current is now falling in magnitude, the inductor voltage reverses and *adds* to the battery voltage. This sum voltage appears at V_{out}, less the diode voltage drop. As with the step-down converter, the switching pattern of the step-up converter can be controlled to regulate the output voltage to a predetermined value.

This configuration has an important advantage over the step-down converter in that the full input voltage is applied to the inductor when the switch is closed. In the step down converter the voltage applied to the inductor is $(V_{in} - V_{out})$, and this approaches zero as the battery voltage falls. With the boost converter, there is the chance of inducing an inductor current even when the battery is very close to exhaustion. One

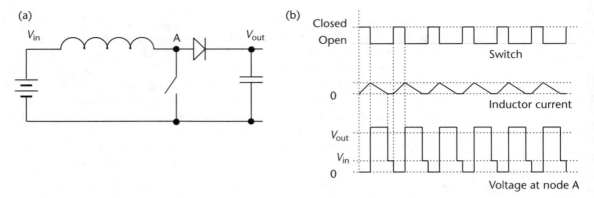

Figure 10.7 The step-up (boost) convertor. (a) Basic circuit; (b) current/voltage waveforms.

disadvantage is that the converter cannot easily be switched off, so short-circuit protection is more difficult to implement.

An example application of this principle is shown in Fig. 10.8, where the voltage from 2 AA cells is being stepped up to 5 V. The MAX856 contains many of the features already seen in the example devices already mentioned. The switching MOSFET is internal to the IC, and connects between terminal LX and ground. Battery voltage can be monitored through LBI, and a low battery condition flagged through LBO. The '856 is not itself a variable regulator: either 3 V or 5 V must be selected as output. Regarding input voltage range, the minimum value for this converter is 0.8 V, while the maximum value is equal to the set output voltage. In this case the switch is left permanently open, and the input is connected continuously to the output via inductor and diode.

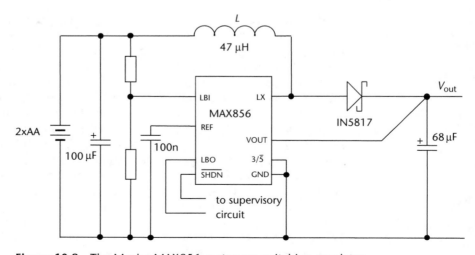

Figure 10.8 The Maxim MAX856: a step-up switching regulator.

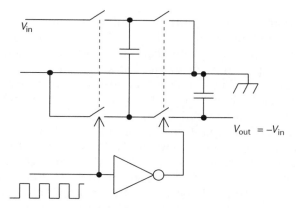

V_{in}

$V_{out} = -V_{in}$

Figure 10.9 The charge pump applied as voltage inverter.

10.2.4.3 *The charge pump*

A charge pump circuit charges a capacitor to the voltage of the input supply and then switches the capacitor terminals to another point in the circuit, where it delivers (some of) its stored energy. Therefore what was a positive voltage can be converted to a negative voltage, or one voltage can be added to another. The circuit shown in Fig. 10.9 inverts the input voltage.

Charge pumps are simple and cheap, with limited radiation of EMI and very high efficiency possible. The power consumed by the charge pump circuitry is dependent on switching frequency, and is very low. In the basic circuit there is no regulating action, so if used alone the input voltage should already be regulated. Due to this lack of regulation it is unsuitable for high current. For very low current applications, however, it can be very attractive. Circuit performance, including load regulation and output ripple, is dependent on switching frequency, capacitor value and the series resistance of the internal MOS switches.

Charge pump inverters are available as standalone ICs (for example the ICL7660, which can convert +5 V to –5 V), for which the user must provide capacitors as external components. They are also integrated into larger scale ICs, together with their switched capacitors, in order to provide extra on-chip voltage rails (for example, EEPROM memory and RS232 driver ICs).

Charge pumps do not have to be used on their own, however. A charge pump could, for example, be followed by a simple linear regulator. If both are operating in their optimum mode, the overall combination can be very effective, giving both high efficiency and low noise. More recent charge pump ICs (Ref. 10.2) take note of the power losses due to switching frequency, and incorporate techniques such as on-demand switching, which gives further power saving.

10.3 Power supply supervision

It is essential for an embedded system to have a reliable power supply. Particularly vulnerable moments are on power-up, when parts of the system may become active at different voltages, and loss of power. Chapter 2 described how power-up could be achieved in an orderly manner. Loss of power carries its own problems. It may be due to intentional power-down, or to a failing battery, or to momentary or partial loss of power (known as 'brown-out'). These latter two conditions are especially dangerous and must be clearly identified. Otherwise parts of the circuit may continue to function while others fail.

To maintain a reliable power supply at all times, or warn the system of unavoidable power loss, we apply a group of techniques called together *power supply supervision*. They include the ability to monitor the 'raw' (i.e. pre-regulated) voltage and the regulated voltage rail or rails. They should be able to distinguish between the noise which is common on digital power lines (and which may produce instantaneous power dips) and a genuine power loss. They are able to take the following actions, as appropriate:

- maintain a system in reset for a specified period after successful power-up
- enable charging of reserve batteries in the presence of the main supply
- give advance warning of power loss, from the state of the raw supply
- cleanly switch off a failing supply
- switch to a reserve supply on loss of main supply, perhaps to limited circuit sections only
- maintain a system in reset when the regulated supply is present but low (i.e. a brown-out)

Because the power supply supervision system takes control of the system reset, it sometimes takes over the Watchdog Timer function, even though this is not directly linked to power supply supervision.

If an impending loss of power is sensed, it may be appropriate to save certain data in Flash memory or battery-backed RAM, and close down certain peripheral devices. Declining battery voltage is a slow process, and if this is the cause of power loss the system can flag the impending loss for a period of time, while still operating normally. If necessary it can then go on to shut itself down. Instantaneous power loss, perhaps through a poor battery connection or loss of mains,[2] gives much less time for a system shutdown.

In the event of instantaneous power loss detected in the raw supply, the system may need to rely on its own reservoir capacitor to sustain it temporarily. To do this, the required shutdown duration, T_{SHDN}, should be estimated. Then, if I_L is the current drawn by the circuit (assumed constant), V_{CC} is the normal supply voltage and V_{th} the threshold voltage (below which the system goes into reset):

---

2 Rapid detection of mains loss is a challenging task. Ref. 10.3 offers a means of doing this.

$$I_L = C \mathrm{d}V/\mathrm{d}t$$

$$= C(V_{CC} - V_{th})/T_{SHDN}$$

$$C = I_L T_{SHDN}/(V_{CC} - V_{th}) \tag{10.4}$$

This represents a minimum value for C, the reservoir capacitance; in practice a larger one should be chosen. The situation is likely to be improved, however, by a capacitor before the regulator, not taken into account in this calculation.

Example circuits in the preceding sections showed how battery supply voltages can be monitored and how the supply can be switched off. These do not achieve the full function of power supply supervision. For this a range of ICs are available, which incorporate the functions listed above in different combinations and at different levels of complexity.

As an example, the MAX690A, a power supervisory IC, is shown in an example application circuit in Fig. 10.10. The main power supply is not itself routed through the IC. It is, however, monitored through the V_{cc} pin. As it rises and passes a reset threshold, a Reset ($\overline{RST}$) pulse is generated. If V_{cc} drops below the reset threshold (i.e. either power-down or a brown-out), then Reset is generated again.

The battery voltage itself can be monitored through the Power Fail Input (PFI). When this goes below 1.25 V, the Power Fail Output (PFO) goes low. In the circuit shown this is used as a non-maskable interrupt (NMI) to the microprocessor. Should the battery voltage fall, the processor can close down the circuit tidily before V_{cc} falls and a Reset disables all further action. V_{out} is an output available for backup supply. Internal switching connects V_{out} to either V_{cc} if it is a valid voltage or V_{bat} otherwise. The source for V_{bat} shown here is a lithium (PCB mounting) battery. A Watchdog Timer capability is also included. If used, the processor should pulse the WDI (Watchdog Input) at least once every 1.6 s; a Reset is generated if it does not.

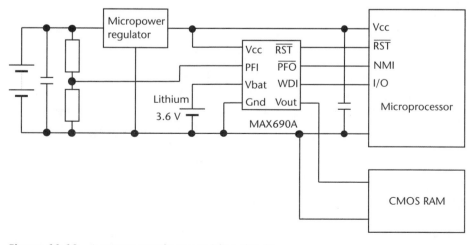

Figure 10.10 A power supply supervision circuit.

10.4 CMOS – the low-power technology

Having considered the supply and regulation of power, we turn now to considering how to achieve a circuit which consumes minimum power.

The only logic family that is suitable for low-power circuits is CMOS (Complementary Metal Oxide Semiconductor). CMOS logic first appeared in the 4000 series, and has over the years become available in a variety of other related families, each with its own advantages. Early CMOS (i.e. the 4000 series) had the potential for fantastically low power consumption, but it was comparatively slow, with poor output drive capability and susceptibility to damage from electrostatic discharge. It was not quite 'taken seriously' by many circuit designers. Later families successfully addressed these problems, and modern CMOS devices (for example the 74HC series) compare well with other logic technologies in terms of speed, yet retain all the low-power capability for which the CMOS family is well known.

To understand how to minimise the power consumption of CMOS, it is necessary first to understand how it consumes that power. An inverting CMOS buffer is shown in Fig. 10.11. When the input is at logic high (i.e. V_{CC}), the lower (NMOS) device is switched on and the output is at logic low. The upper (PMOS) device is fully switched off, and negligible current can flow from V_{CC} to ground. If the input moves to logic low, it will pass a middle region where both devices are partially turned on. When it reaches logic 0 the lower FET will be fully switched off and the upper fully on, and the output will be logic 1. If the input remains some way between ground and V_{CC}, then both MOS devices are partially switched on and significant (and potentially destructive) current can flow from V_{CC} to earth.

The power dissipation of a CMOS gate under normal operation is due to three factors: *quiescent* power dissipation, *capacitive* power dissipation and *transient* power dissipation.

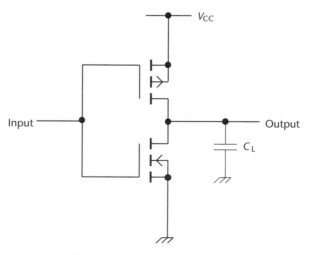

Figure 10.11 A CMOS inverter.

- *Quiescent power dissipation* is due to leakage currents in the circuit when it is not switching. It is very small at room temperature, so that in many cases it can be neglected. In high-temperature situations it can, however, contribute significantly to overall consumption.

- *Capacitive power dissipation* P_C is due to the charging and discharging of load and stray capacitances each time a device switches. This load capacitance, symbolised as a single device C_L in Fig. 10.11, is of course distributed. It exists both as capacitances within the device transistors, as well as the external capacitance of pcb tracks and other inputs to which this output is connected. It is made less by smaller, more recent device geometries. Every time the buffer output swings to logic 1, C_L charges up, with a charge $Q = C_L V_{CC}$. With switching frequency f, This happens f times a second, so the current I_C due to the presence of C_L is:

$$I_C = Qf = C_L V_{CC} f \tag{10.5}$$

Converting this to power P_C by multiplying each side by V_{CC}, we get

$$P_C = V_{CC} I_C = C_L \times V_{CC}^2 \times f \tag{10.6}$$

Note carefully that this power dissipation is proportional to frequency, to the load capacitance and to the square of the supply voltage.

- *Transient power dissipation*, P_T, is due to 'shoot-through' current, which flows through both devices as they are partially turned on during the process of switching. It is given by

$$P_T = C_{pd} \times V_{CC}^2 \times f \tag{10.7}$$

where f is the switching frequency and C_{pd} is a value specified in the manufacturer's data sheet for a particular CMOS IC. P_T is also dependent on the rise and fall times of the input logic signal. If the signal does not switch fast, then the buffer will spend more time with both FETs partially on and P_T will rise. The form of this expression is very similar to that of P_C, and therefore carries the same dependencies on supply voltage and frequency.

WORKED EXAMPLE   **10.2**

A certain 74HC00 IC has the current consumption data of Table 10.7, and is connected as shown in Fig. 10.12. Sketch the IC current consumption against switching frequency, up to 10 MHz, for supply voltages of 3 V and 6 V.

Solution: The total power consumption is *quiescent* plus *capacitive* plus *transient*. At 25 °C:

$$\text{total current} = (2 \times 10^{-9} + 8 \times 3.5 \times 10^{-12} \times f \times V_{CC} +$$
$$4 \times 22 \times 10^{-12} \times f \times V_{CC}) \text{ A}$$

$$= (2 + V_{CC} \times 0.116f) \text{ nA}$$

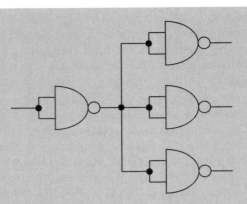

Figure 10.12 NAND gate connection.

Table 10.7 74HC00 current consumption data.

Power supply range	2 V to 6 V
Input capacitance	3.5 pF (per input)
Power dissipation capacitance[1] (t_r, t_f < 6 ns)	22 pF (per gate)
Quiescent current (for 6 V supply)	2 nA typ., 2 µA max. at 25 °C, per IC

[1]tr, tf: logic rise and fall times

This can be tabulated and plotted, as shown in Fig. 10.13. As anticipated, the current consumption is extremely low at low switching frequencies, but becomes significant as the frequency rises.

V_{cc} \ f_c	DC	1 kHz	100 kHz	1 MHz	10 MHz
3 V	2 nA	700 nA	70µ A	0.7mA	7 mA
6 V	2 nA	356 nA	35µ A	0.35mA	3.5 mA

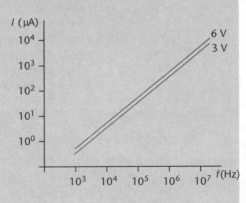

Figure 10.13

Arising from these considerations, the guidelines for minimising power consumption in CMOS circuits, loosely in order of descending importance, are as follows:

● define all inputs clearly as logic 0 or 1

- minimise clock frequencies
- minimise the power supply voltage
- ensure fast logic transitions
- minimise load and interconnection capacitances

Applying these points carefully can save a designer many milliamps!

10.5 The low-power microcontroller

From the preceding discussion, we expect a low-power microcontroller to be CMOS, preferably with a small device geometry, operating with minimum possible clock frequency and power supply voltage.

Some microcontroller features which relate to power consumption are shown in Table 10.8. It is important to bear in mind that the example processors are of quite different complexity and capability, so comparisons on a 'like-for-like' basis are not being made. Included for interest is the Atmel 89C1051. This is one of several 8051 derivatives where the architecture has been redesigned, while remaining compatible with the 8051 instruction set.

The full supply voltage range is indicated in the first row. While some ranges seem dramatically better than others, it must be borne in mind that reduced supply voltage is generally accompanied by reduced output drive capability and reduced operating frequency range.

Table 10.8 Microcontroller characteristics relating to power consumption.

	16F84	68HC11A8	80C552	89C1051
Supply voltage range	2 V to 6 V	3 V to 5.5 V	5 V ± 10%	2.7 V to 6 V
Clock freq. range	DC to 10 MHz	DC to 8.4 MHz	1.2 MHz to 30 MHz	DC to 24 MHz
Clocks per machine cycle	4	4	12	12
Clocks per instruction (CPI)	4, 8	8 to 56[1]	12 to 48	12 to 48
Typical supply current at 5 V	2 mA @ 4 MHz	20 mA @ 2 MHz	35 mA @ 16 MHz	10 mA @ 16 MHz
Low power modes	SLEEP: Oscillator driver turned off, WDT can be enabled; exit by external reset or interrupts, WDT timeout, or EEPROM write complete.	WAIT: processing suspended, clock remains active; exit by enabled interrupt. STOP: internal clock stopped; exit by external interrupt or hardware reset.	(1) IDLE: CPU shut down, all peripherals remain active; exit by enabled interrupt or hardware reset. (2) POWER-DOWN: Oscillator stopped, only contents of on-chip RAM saved; exit by hardware reset only.	Identical to 80C552

[1]Excludes divide instructions

The second parameter is the clock frequency range. As mentioned in Chapter 1, some microprocessors and controllers are fabricated using dynamic logic, which is clocked and has a minimum frequency of operation. The 80C552 is an example of this, having a *minimum* clock frequency of 1.2 MHz. Other controllers are made with a 'fully static' CMOS technology. These can operate down to DC clock frequency (in effect, the clock oscillation can be suspended and restarted without loss of data or continuity).

Clock frequency is not on its own an indication of processor speed. Recall that in Chapter 1 we saw that a certain number of oscillator clock cycles go to make up a machine cycle, and then one or more machine cycles are needed per instruction. An important parameter, cycles per instruction (CPI), can be used to compare the performance of one processor against another. If a processor has a high CPI, it will need to run at a much higher clock frequency than a device with a lower CPI in order to complete the same number of instructions in the same time. For a given instruction time and device technology, the smaller the CPI, the less will be the power consumed for a given rate of instruction throughput. In the table we see that the 80C552 has a high CPI and the 16F84 a low one. Interestingly, Dallas Semiconductor, another company that has adopted the 8051 core, has redesigned it to have a greatly reduced CPI. Further exploration of this topic can be found in Ref. 10.4.

Microcontrollers designed for power-conscious applications generally have one or more low-power modes, usually being called *idle*, *sleep*, *power-down* or similar. In these modes the oscillator is disconnected from some or all of the controller areas, while power is retained. With the clock switched off, transient and capacitive power dissipation immediately disappear. Use of these modes is often the key to achieving extreme low power and battery longevity. The microcontroller, and possibly associated circuitry, can be put into a low-power mode and then re-awoken, for example by the user, a timed function or the receipt of data. Details are given in the table.

There are also other features which introduce unwanted power consumption. The internal weak pull-up of the 80C552, which cannot be disabled, sources 100 μA for a single pin when the input is taken low. This is equivalent to the whole supply current of many present-day low-power ICs.

The current consumptions of the 89C1051 in active and power-down modes are given in Fig. 10.14. This shows the expected dependence on clock frequency. Data similar to this is available for all power-conscious microcontrollers.

10.6 Low-power circuit design

We close this chapter by considering a few design guidelines which help to minimise power consumption in a microcontroller-based circuit. Be aware that applying these guidelines usually involves careful trade-offs. As the listing below indicates, a circuit operating with reduced power may well have reduced bandwidth and increased noise susceptibility. The guidelines are:

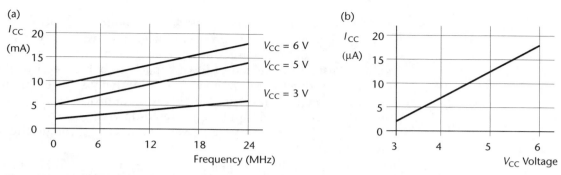

Figure 10.14 Atmel AT89C1051 current consumption characteristics. (a) Typical supply current (I_{CC}), active; (b) typical supply current (I_{CC}), power-down mode.

- *Select technologies which require minimum operating power.* The implication of this is that semiconductors will be CMOS, displays will be LCD, and LEDs (if they must be used) will be high efficiency.

- *Minimise supply voltage(s).* This implies the use of newer, low-voltage devices. These are available both in low-voltage logic families and in micropower op-amps and other analogue devices. Many ICs can run at *very* low supply voltages. Be aware however that reduced supply voltage tends to lead to slower operation, and hence reduced clock frequencies and narrower analogue bandwidths. It also means that noise thresholds are narrowed, so there is increased susceptibility to EMI.

- *Minimise clock signal frequencies.* The reasons for this have been discussed. It will be necessary to have an idea of the time requirements of the system software before selecting the clock frequency. Note, however, that better power consumption figures may be achieved if the system is run fast for a short period and then put to sleep (with an appropriate microcontroller instruction), rather than if it is run continuously at a low speed; see Example 10.3.

- *Scale up resistances.* This leads to a further fall in operating currents. Increasing resistances will, however, increase susceptibility to EMI and reduce output drive capability. If there are analogue parts to the circuit, it will also increase thermal noise and increase time constants, and hence reduce bandwidths (possibly dramatically). It may also increase or bring to light errors due to op-amp bias currents and their temperature-dependent effects.

- *Target devices specifically intended for low-power operation.* Some of these have already appeared in this chapter, and others elsewhere in the book (some of the memory ICs described in Chapter 4 are particularly suitable for low power). Naturally they will be low power, but may also have power-down modes.

- *Minimise circuit complexity.* More components cost more (micro/milli-) amps, and add to line capacitance (and hence capacitive power dissipation).

- *Consider power down of circuit blocks not continuously required*. This heading includes the microprocessor sleep modes already discussed. Other ICs are designed for optional power-down, and micropower regulators usually have a shutdown capability, which easily allows a whole circuit zone to be switched off. Alternatively a power rail may be switched through a transistor, analogue switch or CMOS (HC) logic gate. Ensure that signal inputs to powered-down devices are not themselves active, as they may re-power the device through the input protection diodes.

- *Use low-power crystal oscillator design*. The crystal oscillator circuit itself has the potential for comparatively high power consumption. It is after all likely to be the highest frequency part of the overall circuit, with the input to the oscillator inverter spending a high proportion of its time *not* at a clearly defined logic 0 or 1. Manufacturers' data often gives advice on how to reduce oscillator power.

- *Choose the right battery/regulator combination*. Simplicity remains the key here. First of all assure yourself that a regulator is needed. A simple circuit, operated from a battery with a flat discharge curve, may not need a regulator at all.

WORKED EXAMPLE 10.3 _____

A certain microcontroller is used as part of a data logger system, which makes a measurement every second and then records two bytes of data in RAM. It takes 80 instruction cycles to make this measurement. It requires two instruction cycles to write to memory, and enables the memory for one quarter of one of these. The memory draws 25 mA when activated and 250 μA otherwise. The microcontroller draws $(0.2 + 2 \times 10^{-6}f)$ mA, where f is the instruction cycle frequency. In Sleep mode the microcontroller draws only 50 μA, and it can be woken up by a Timer with no extra current consumption. No other tasks are required of the microcontroller when logging.

What would you recommend for the microcontroller instruction cycle frequency and the use of its Sleep mode? What would be the resulting current consumption?

Solution: 84 instruction cycles are needed per second. The memory access time, one quarter of an instruction cycle occurring twice, will be $2/(4f)$. The total system current consumption I_L will be given by

$$I_L = \text{consumption of controller} \\ + \text{non-active memory consumption} \\ + \text{consumption of memory when accessed}$$

Therefore

$$I_L = [(0.2 + 2f \times 10^{-6}) + 0.25 + 25 \times 2/4f]\ \text{mA}$$

This includes the very slight approximation that the 250 µA non-active memory consumption is a continuous demand. This function clearly has a minimum value. At clock frequencies below this minimum, power consumption is dominated by memory access. Above the minimum it is dominated by microcontroller switching losses. Differentiating gives

$$(\mathrm{d}I_L/\mathrm{d}f) = (2 \times 10^{-6} - 2 \times 25/4f^2)$$

This is zero when

$$2 \times 10^{-6} = 2 \times 25/4f^2$$

$$f = 2.5 \text{ kHz}$$

Then

$$I_L = [(0.2 + 2 \times 2.5 \times 10^3 \times 10^{-6}) + 0.25$$
$$+ 2 \times 25/(4 \times 2.5 \times 10^3)]\} \text{ mA}$$

$$= 0.465 \text{ mA}$$

This value of I_L appears promising. It is certainly very close to the minimum value given by the sum of the memory standby consumption and the microcontroller DC consumption, i.e. 0.45 mA. The clock frequency, however, is very low. It will be difficult to generate and is unlikely to be fast enough for other parts of the software, for example in other operational modes of the logger.

Exploring instead the Sleep mode option, the controller needs to be active for 84 cycles, i.e. a duration of $(84/f)$, and can be asleep for a duration $[1 - (84/f)]$ every second. In this case:

$$I_L = \text{consumption of controller when active}$$
$$+ \text{consumption of controller when asleep}$$
$$+ \text{non-active memory consumption}$$
$$+ \text{consumption of memory when accessed}$$

$$I_L = [(0.2 + 2f \times 10^{-6})(84/f) + (50 \times 10^{-3})(1 - 84/f)$$
$$+ 0.25 + 2 \times 25/4f] \text{ mA}$$

$$= (1/f)(16.8 + 12.5 - 4.2) + 168 \times 10^{-6} + 50 \times 10^{-3} + 0.25$$

$$= [(25.1/f) + 0.3] \text{ mA}$$

The final result of Example 10.3 is very interesting. We cannot get below a consumption of 0.3 mA, as this is the unavoidable minimum steady consumption of memory and microcontroller in Sleep. The consumption beyond this is now inversely proportional to frequency! The implication is that, using Sleep mode, we want to get things done as quickly as possible, even if with a higher instantaneous consumption. The microcontroller can then stay in Sleep mode for as long as possible.

SUMMARY

1. Power for embedded systems is generally supplied from the mains, batteries or renewable sources. Most low-power designs depend on battery supply.

2. A key feature of low-power design lies in battery selection and in the appropriate choice of voltage regulation and battery management techniques.

3. Linear voltage regulators offer high efficiency under certain limited operating conditions, as well as low cost, low noise and low EMI.

4. Switching converters offer great versatility, coupled with very good efficiency. They can be applied to step voltages up or down or to invert. They suffer from radiated EMI and higher output noise.

5. CMOS is the only logic family which is acceptable in power-conscious designs. Even CMOS applications can be optimised to minimise power consumption.

6. Certain microcontrollers are developed with low-power features; these include technological aspects which minimise power consumption under normal use, as well as certain instructions which allow the controller to suspend all or part of its operation.

7. There is a collection of design techniques, founded on an understanding of the mechanisms of power consumption, which leads to the ability to design low-power circuits.

REFERENCES

10.1 Vincent, C. (1999) Lithium batteries. *IEE Review*, **45**(1), 65–8.

10.2 *Charge Pumps Shine in Portable Designs*. Technical Article. Maxim Integrated Products. 2000. http://www.maxim-IC.com/.

10.3 Crowley, T. (1998) Watchdog has extra teeth. *Electronics World*, November, pp. 972–3.

10.4 Piguet, C., Dijkstra, E. and Masgonty, J.-M. (1996) Low-power microcontrollers. *European Microprocessor and Microcontroller Conference*. Miller Freeman.

EXERCISES

('+' indicates a simple review question)

+**10.1** A digital wristwatch requires 5 µW to operate. It is powered from a 3 V lithium cell with a capacity of 20 mAh. If there is no battery self-discharge, for approximately how long will the battery last?

+**10.2** What are the relative advantages of linear and switching voltage regulators/converters, particularly when applied to lower power applications?

+**10.3** CMOS is described as the essential technology for low-power logic applications.

  (a) Explain why this is so.
  (b) Describe briefly the three ways in which power is consumed by a CMOS logic gate.
  (c) Describe how power consumed by a CMOS circuit can be minimised.

+**10.4** You are given the task of designing a battery-powered handheld embedded system. List the design guidelines you would observe to minimise power consumption.

10.5 A microcontroller can operate from 3 V to 6 V. Under a particular set of operating conditions it acts to the power supply as a load of 1 kΩ. It is powered from a 5 V supply and the total circuit decoupling capacitance is 220 μF. If the supply is disconnected, approximately how much time does the microcontroller have to perform a controlled system shutdown if it draws its power during this time only from the decoupling capacitance?

10.6 Each line of an 8-bit data bus is connected to four ICs, each input of which has an estimated capacitance of 8 pF. Each line itself has an estimated capacitance to ground of 6 pF. Each line of the bus is switching between 0 V and 3.3 V at a frequency of 2 MHz. Estimate the power drain due to the activity of the bus alone.

10.7 For a small-scale battery-powered system, running from 5 V, you are given the option of:

  (a) running from two AA cells in series and stepping the voltage up
  (b) running from four AA cells in series and stepping the voltage down
  (c) running from a 9 V battery and stepping the voltage down

  Explain the factors you would take into account when making this decision and recommend a solution strategy.[3]

10.8 A solar panel is characterised at three levels of insolation: bright, light cloud and heavy cloud. For each of these it has a terminal voltage respectively of 9 V, 7 V and 6 V. It is to power a 5 V circuit which draws a constant 50 mA. For each source voltage, estimate the current drawn from the panel and the operating efficiency:

  (a) if an LP2951 linear voltage regulator is used
  (b) if a switching regulator is used, having a constant conversion efficiency of 80%

10.9 A standalone security system 'wakes up' every two minutes to switch on and test its sensors and activate an alarm if any break-in is detected. It

---

3 You may answer this question considering only the techniques described in this chapter. Alternatively, use it as the basis for a survey of other devices and techniques that are available.

records a time-stamped message in CMOS RAM on every wake-up. The wake-up is from a standalone Real-Time Clock, which consumes a steady 120 μA. The system runs from two AA nickel–metal-hydride cells. The sensors require ±5 V supplies at 2 mA per rail, while the rest of the circuit runs from 3 V at 750 μA, continuously powered. Using only devices mentioned in this chapter, draw a block diagram of a suitable power supply system.

10.10 Estimate the active power consumption of an I^2C bus interconnection which runs at 100 kHz. The capacitive loading on both SCL and SDA is estimated to be 150 pF, each line swings between 0 V and 5 V, and the pull-up resistor on each is 4k7. To simplify the calculation assume that each line runs continuously, with a square wave signal. Neglect current consumption within transmitting or receiving devices.

Dealing with numbers

IN THIS CHAPTER

In every embedded system we need to represent numerical quantities, and in almost every embedded system we need to manipulate these quantities mathematically. These statements seem obvious enough, yet each demands careful consideration.

Fundamental decisions regarding numbers have to be made. What number sizes should be used, and what accuracy is required? Will integer arithmetic be adequate, and what will be the resulting errors and approximations?

All but the very simplest of mathematical operations will place demands beyond the microcontroller instruction set, and the programmer will need to apply a subroutine if working in Assembler or depend on a C operator or function.

Libraries of these subroutines and functions are already available in large numbers from manufacturers, third-party suppliers or as shareware. In most situations it is possible to use these without having any detailed knowledge of how they work. There are times, however, when one comes to suspect the operation of a supposedly correct subroutine, or one needs to understand its limitations or adapt it to a different task. Then it becomes essential to have an understanding of its working.

Another important application of number in the embedded system is that of sampled data. In this case an incoming signal is being continuously sampled and processed. In their fullest manifestation, techniques for processing sampled data are grouped into a discipline called Digital Signal Processing (DSP). Such techniques require specialised high-speed processors. It remains appropriate for us to look at simple digital processing of incoming sampled signals to meet the requirements of certain control and instrumentation applications.

Accordingly, the aims of this chapter are:

- to develop further the concepts of fixed point number representation

- to describe fixed point numerical routines for commonly used mathematical functions

- to define the limits of fixed point representation and introduce floating point as an up-market alternative

- to analyse the effects of rounding and truncation

- to explore simple calculations making use of sampled data

11.1 Fixed point number representation

So far in this book we have used the number representations described in Appendix A (unsigned binary, hexadecimal, two's complement and Binary Coded Decimal), with a focus on integer numbers. These representations are adequate as far as they go, and have led to very extensive libraries of mathematical subroutines (for example Refs. 11.1–11.4). To start this chapter we will therefore develop and formalise a little our understanding of the application of binary representation.

11.1.1 Fractional binary numbers

The binary system can be extended, like the decimal system, to represent fractional numbers. The decimal system uses a decimal point. Digits to the left of the decimal point represent multiples of 10^0, 10^1, 10^2 and so on, while those to the right represent multiples of 10^{-1}, 10^{-2}, and further negative powers. For example, the number

$$312.329$$

is made up of

$$(3 \times 10^2 + 1 \times 10^1 + 2 \times 10^0 + 3 \times 10^{-1} + 2 \times 10^{-2} + 9 \times 10^{-3})$$

In a similar way, binary counting can make use of a *binary point*. Bits to the left of the binary point represent positive powers of 2, while bits to the right represent negative powers of 2. For example, the number

$$1001.1101$$

is made up of

$$1 \times 2^3 + 0 \times 2^2 + 0 \times 2^1 + 1 \times 2^0 + 1 \times 2^{-1} + 1 \times 2^{-2} + 0 \times 2^{-3} + 1 \times 2^{-4}$$

In decimal this is

$$8 + 1 + 0.5 + 0.25 + 0.0625, \text{ i.e. } 9.8125$$

It is worth remembering that, even with as many bits in a fractional number as you may wish, a binary number can only represent sums of binary powers. For example, the simple decimal number 0.1 has no complete representation in binary. The equivalent to this is the decimal attempt to represent the simple fraction (1/3), which it does with an endless (literally!) succession of '3's, i.e. 0.3333333333....

Binary points can be included in unsigned binary, as well as in two's complement representation.

11.1.2 Integer and fixed point arithmetic

How then do we make use of this binary point? There is, after all, no apparent chance of representing it in a register or memory location. The simplest option is to ignore it, and treat all numbers as integers; this is sometimes known as *integer arithmetic*. The binary point is then assumed to be just to the right of the lsb.

An alternative is for the programmer to mentally fix a binary point in a number and track it through subsequent calculations. The algorithms themselves are the same whatever the position of the binary point. The binary point is not represented explicitly, so this tracking must be done with care, and there is not complete freedom in its location. Binary points must, for example, be aligned in addition or subtraction operations, whereas their position in multiplication or division is immaterial. For any input data to an algorithm or series of algorithms, the implied binary point at any stage must always be the same, resulting in a known placing of the binary point in the output data. This style of representation and arithmetic is called *fixed point* arithmetic.

To avoid having the need to locate binary points midway within numbers, the convention of *fractional arithmetic* is often adopted. Here the binary point is assumed to be just to the left of the msb.

The application of fixed point arithmetic frequently becomes a problem of scaling. For example, an 8-bit number cannot represent either 0.125, or 1600, so it would be impossible to multiply the two numbers together. If, however, each was scaled, then the multiplication

$$(125 \times 10^{-3}) \times (160 \times 10^{1})$$

could be performed, as both the 125 and the 160 fit well in the 8-bit range. In fixed point arithmetic it is up to the programmer to perform this scaling if necessary before any calculation and then reintroduce it at the end, in this case by multiplying the result of (125 × 160) by 10^{-2}. In practice the scaling would be done in binary powers. An illustration of this forms part of Example 11.3.

11.2 Fixed point numerical routines

We consider now how integer or fixed point routines are applied in the microprocessor world for performing some of the common mathematical functions. We explore each of Addition, Subtraction, Multiplication and

Division. These are the most common maths functions, and others can be built using them. We will also look at conversion between BCD and binary. We assume 8-bit processing power, and that each of these routines takes the form of a subroutine to which operands are transferred in specified locations.

11.2.1 Addition and subtraction

In addition the *addend* is added to the *augend*. The addition of two n-bit numbers results in a number of maximum size $(n + 1)$ bits. In subtraction the *subtrahend* is subtracted from the *minuend*.

For addition and subtraction of numbers greater than 8 bits, it is necessary for an 8-bit processor to use an appropriate sequence of instructions. These will not necessarily form distinct subroutines, as the call and return overhead may not justify the gains. Simple flow diagrams for 16-bit add and subtract, which can be adapted to any controller and extended to 24 or more bits, are shown in Fig. 11.1. Tests are not included for Overflow, but can be simply added.

In Fig. 11.1(a) the 16-bit number in memory locations PHI-PLO is added to the 16-bit number in QHI-QLO. A 17-bit result is possible, which is placed in Carry (msb) and RHI-RLO. Depending on the processor used and the way the flow diagram is applied, either the addend or the augend is likely to be overwritten.

In Fig. 11.1(b), the 16-bit number in memory locations PHI-PLO is subtracted from the 16-bit number in QHI-QLO. If the subtrahend is greater than the minuend a borrow will be generated. The result is placed in Borrow and RHI-RLO. In practice subtractions are often implemented in the hardware by converting the subtrahend to two's complement and then adding it to the minuend.

The PIC 15XXX and 16XXX families do not have add with carry (or subtract with borrow) instructions. These must be grouped from available

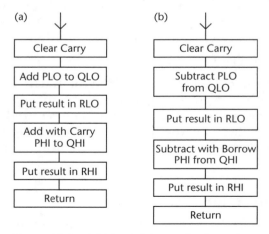

Figure 11.1 16-bit addition and subtraction routines. (a) 16-bit addition; (b) 16-bit subtraction.

```
;Add                            ;Subtract
    movf    PLO,0                   movf    PLO,0
    addwf   QLO,1                   subwf   QLO,1
    btfsc   status,0                btfss   status,0
    incf    QHI,1                   incf    PHI,1
    movf    PHI,0                   movf    PHI,0
    addwf   QHI,1                   subwf   QHI,1
```

Figure 11.2 16F84 16-bit add and subtract implementations.

instructions. Simple 16F84 implementations of the two flow diagrams of Fig. 11.1, which leave the result in QHI-QLO and Carry, are shown in Fig. 11.2.

11.2.1.1 Detecting overflow in unsigned binary and two's complement calculations

The simple calculations of Appendix A showed that two's complement gives consistent results with both addition and subtraction. The same rules of addition and subtraction apply to both unsigned binary and two's complement, but different ranges apply for each, and Carry and Borrow conditions must be interpreted differently. How do we know when ranges are exceeded?

In Unsigned Binary a Carry is generated if an addition leads to the 8-bit range being exceeded, and a Borrow is generated if a subtraction of a larger number from a smaller is attempted. This is simple to detect with the Carry/Borrow flag.

Similar principles apply in two's complement, but range excesses generally lead to a sign bit reversal: for example, adding two positive numbers leads to an apparently negative number. It is worth noting that two's complement overflow can only occur in an addition if the two numbers are of the same sign, or in a subtraction if the two numbers are of different sign.

In an 8-bit addition, where bit 7 is the sign bit, two's complement overflow occurs:

if there is a Carry between bits 6 and 7 but not from bit 7 to the Carry flag; *or*
if there is a Carry from bit 7 to the Carry flag, but not between bits 6 and 7.

In a subtraction, overflow occurs:

if there is a borrow from bit 7 into bit 6, but not from Carry/Borrow into bit 7, *or*
if there is a Borrow into bit 7, but not into bit 6.

WORKED EXAMPLE **11.1**

Give examples of each of the four overflow two's complement conditions described above.

Solution: Two's complement numbers which lie close to the 8-bit range limits can be selected from Appendix A. These additions can be performed:

$$0111\ 1111\ (+127)$$
$$+0111\ 1110\ (+126)$$
$$\overline{1111\ 1101}$$

$$1000\ 0001\ (-127)$$
$$+1000\ 0000\ (-128)$$
$$(1)\overline{0000\ 0001}$$

While both results are correct in unsigned binary, neither is in two's complement. In the first, there has been a carry from bit 6 to bit 7, but not from bit 7 to the Carry flag. In the second there has been a carry into the Carry flag, but not into bit 7.

The following subtractions give a further example of overflow:

$$0111\ 1111\ (+127)$$
$$-1111\ 1110\ (-2)$$
$$(1)\overline{1000\ 0001}$$

$$1111\ 1110\ (-2)$$
$$-0111\ 1111\ (+127)$$
$$\overline{0111\ 1111}$$

In the first of these there has been a borrow from the Borrow flag into bit 7, but bit 6 has not needed a borrow. In the second there has been a borrow from bit 7 into bit 6, but not from the Borrow flag into bit 7. As can be seen by inspection, both are invalid two's complement results.

It is easy to extend these principles to 16-bit or longer words. An Overflow flag is included in the Status/Condition Code/Program Status Word register of some microprocessors (for example 68HC08 and 80C51), which gives direct indication that the two's complement range has been exceeded.

11.2.2 Multiplication

In a multiplication, the *multiplicand* is multiplied by the *multiplier*. An n-bit × m-bit multiplication will result in a maximum of $(n + m)$ bits of output. For example, a 12-bit × 8-bit multiplication will need 20 bits to hold the result.

The simplest form of multiply is simply repeated addition. Multiplication can be achieved if the multiplicand is added to itself by the number of times equal to the multiplier.

The principle of a practical multiply routine is to inspect each bit of the multiplier in turn. If the nth bit is a 1, then 2^n times the multiplicand is added to the partial result. When every bit has been tested and acted on, the final result is complete. This is shown in the 3-bit example of Fig. 11.3(a). The process is also illustrated in flow diagram form in Fig. 11.3(b) for a 16-bit multiplier.

There are two important repeated actions in the main loop of this flow diagram: detecting the value of the nth bit, and determining and adding in the (multiplicand × 2^n) term. Some ingenuity is applied to implementing these in a typical microprocessor implementation. In an 8 × 8 multiply, for example, on each loop the multiplier is shifted right into Carry, and the Carry bit then tested. This is always bit n, which starts as the lsb. If the bit is a 1, then the multiplicand is added into the *higher* byte of the 2-byte partial

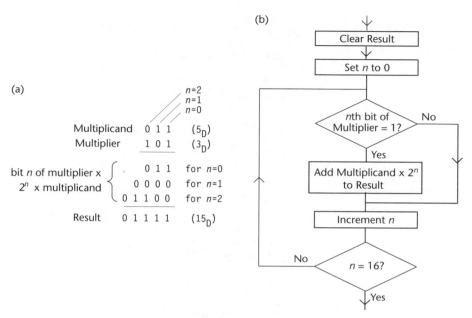

Figure 11.3 The principle of multiplication. (a) Example; (b) flow diagram.

result. On each loop the partial result is then rotated right, so that at the end of the routine each addition of the multiplicand has been rotated to its correct place in the result. This saves the trouble of actually computing the (multiplicand $\times 2^n$) term. This is illustrated in Fig. 11.4. The principle can be readily extended to 16 or 24-bit multiply routines.

Routines of this sort are common in microprocessor applications. They are not directly usable with two's complement representation. If negative numbers are to be multiplied, it is best to determine the number signs and determine the sign of the result, which will be negative if the signs are different, and positive otherwise. Any negative number should then be converted to its unsigned binary form and the multiplication performed. If the result was determined as negative, it can then be converted to two's complement.

11.2.3 Multi-byte multiplication using 8-bit hardware multiplier

Many processors now have an 8×8 hardware multiplier. This can be used to shorten a multi-byte multiply operation. Suppose for example two 3-byte (i.e. 24-bit) numbers are to be multiplied. A 6-byte (48-bit) result will be expected, and the operation can be represented as:

$$(A \times 2^{16} + B \times 2^8 + C \times 2^0) \times (D \times 2^{16} + E \times 2^8 + F \times 2^0)$$

where A, B, C, and D, E, and F are the bytes of the two numbers. This multiplies out to:

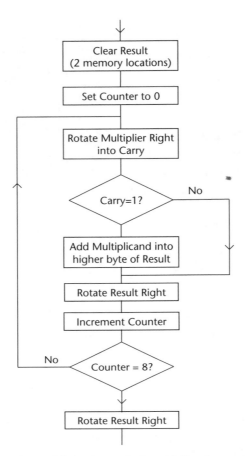

Figure 11.4 A practical multiplication routine.

$$AD \times 2^{32} + BD \times 2^{24} + CD \times 2^{16} + AE \times 2^{24} + BE \times 2^{16}$$
$$+ CE \times 2^8 + AF \times 2^{16} + BF \times 2^8 + CF \times 2^0$$
$$= AD \times 2^{32} + (BD + AE) \times 2^{24} + (CD + BE + AF) \times 2^{16}$$
$$+ (EC + BF) \times 2^8 + CF$$

Each 2-byte product (for example AD) is a 2-byte number. The summation within each pair of brackets should be formed, and the overall result generated by added the weighted sums of all of these.

11.2.4 Division

In a division routine, the *dividend* is divided by the *divisor*. As division is the inverse of multiplication, one might expect the division process to be of comparable complexity and duration, but unfortunately this is not so. While an n-bit × m-bit integer multiply routine leads to an integer result of maximum length $(n + m)$, an $[(n + m) \div m]$ routine does not guarantee an n-

bit result, nor will the result necessarily be an integer. The concept of a remainder must therefore be applied.

Just as multiplication is effectively a repeated shift and add process, so division is a repeated process of subtract and shift. A simple but impractical division routine repeatedly subtracts the divisor from the dividend until the remainder in the dividend is smaller than the divisor. The number of times that the subtraction was made is the result. For example, in decimal, 3 can be subtracted 123 times from 369 to give the result $369 \div 3 = 123$.

Practical divide routines start from this principle, but to save time, the term (divisor $\times R^n$) is repeatedly subtracted, where R is the radix and n is a diminishing integer. After each subtraction it is necessary to test the result. If it is negative then too large a number has been subtracted and it must be added back in. Every time a successful subtraction is made, R^n is added to the intermediate result. The digit n is decremented by 1 when no further subtraction can be made for that value.

Using the example decimal figures applied above (3×10^2) can first be subtracted from 369, giving an intermediate result of 100 and a remainder of 69. A further subtraction of 300 would give a negative result, so would not be a valid action. (3×10^1) can then be subtracted from 69 giving 39 (with an intermediate result of 110), and again, giving a further intermediate result of 120, and a remainder of 9. Finally (3×10^0) can be subtracted from 9 three times, giving a final result of 123, with no remainder.

In a binary routine, the first task is to determine the starting value for n. This is done by shifting the divisor left, while counting the number of shifts made, until it has the same number of bits as the dividend. This is illustrated in the Flow Diagram of Fig. 11.5. Word sizes for dividend and divisor are predetermined. In LOOP1 the divisor is shifted left until its msb is a 1. The location Count is incremented every loop, and on completion it holds the value $(n + 1)$. The main division takes place in LOOP 2. The dividend memory locations are now treated as the Remainder, and the [divisor $\times 2^n$] term is repeatedly subtracted from it. Every time the result is non-negative a further bit is rotated into the result. If the result is negative, the subtraction must be reversed. When Count has fallen to zero, the process is complete.

As with the multiply routine above, this Division routine will not work with two's complement numbers. The division should be made using their unsigned magnitudes and the sign computed separately. If the programmer cannot guarantee a non-zero divisor, then a test for this should be included at the start of LOOP1.

11.2.5 Conversion between binary and BCD

There is frequent requirement to convert between binary and BCD. Data may be input in BCD from thumbwheel switches or similar, or may need to be output to a display.

BCD is a less compact means of number representation than binary, so more bits are needed for the BCD version of any number (except those

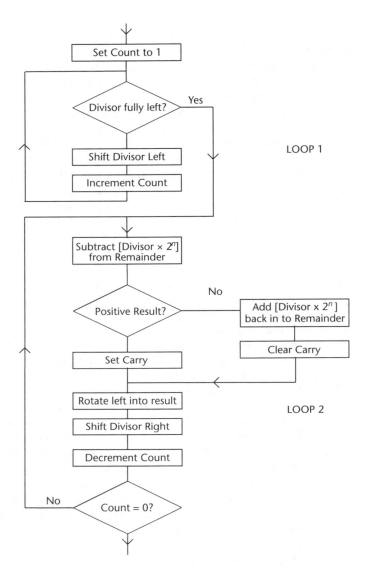

Figure 11.5 A division routine.

which are 9 or less). This is illustrated in Table 11.1, which shows the word size required for packed BCD numbers compared to the word size required for the same number represented in unsigned binary. Note that the most significant BCD digit does not necessarily require the full four bits; knowledge of this can sometimes reduce both storage space and conversion time.

Of some interest when preparing data for display output is how the binary word size is reflected in the resulting decimal number. In this respect 10-bit and 11-bit are useful, as they can be used for 0–999 or 0–1000 output displays and 3.5 digit displays (i.e. with display range 0 to 1999) respectively.

Table 11.1 Comparison of BCD and binary word sizes..

Binary number	Decimal number		
Word size (bits)	Max. no.	Digits	BCD bits
6	64	2	7 (8)
8	255	3	10 (12)
10	1023	4	13 (16)
11	2047	4	14 (16)
12	4095	4	15 (16)
13	8191	4	16 (16)
14	16383	5	17 (20)
16	65535	5	19 (20)

WORKED EXAMPLE   11.2

(a) Convert the BCD number 93 to binary.
(b) Convert the binary number 11 1110 0111 to BCD.

Solution:
(a) Table 11.2 shows the conversion process, following the flow diagram
   of Fig. 11.6(a) (p. 336). Eight shifts and two decimal adjustments are
   made. The resulting binary number, 0101 1101, is the binary equivalent of the starting BCD number.

Table 11.2 A BCD to binary conversion.

BCD number	Result	
1001 0011	0000 0000	
0100 1001	1000 0000	1st shift (right)
0100 0110	1000 0000	Decimal adjust
0010 0011	0100 0000	2nd shift
0001 0001	1010 0000	3rd shift
0000 1000	1101 0000	4th shift
0000 0101	1101 0000	Decimal adjust
0000 0010	1110 1000	5th shift
0000 0001	0111 0100	6th shift
0000 0000	1011 1010	7th shift
0000 0000	0101 1101	8th shift

(b) From this 10-bit binary number we expect a three-digit BCD number, which can be held in 13 bits (Table 11.1). The shifts follow the flow diagram of Fig. 11.6(b), and are shown in Table 11.3. Following the 10th shift, the resulting number can be seen to be 999_{10}.

Table 11.3 A binary to BCD conversion.

Result	Binary number	
0 0000 0000 0000	11 1110 0111	
0 0000 0000 0001	11 1100 1110	1st shift (left)
0 0000 0000 0011	11 1001 1100	2nd shift
0 0000 0000 0111	11 0011 1000	3rd shift
0 0000 0000 1010	11 0011 1000	Decimal adjust
0 0000 0001 0101	10 0111 0000	4th shift
0 0000 0001 1000	10 0111 0000	Decimal adjust
0 0000 0011 0001	00 1110 0000	5th shift
0 0000 0110 0010	01 1100 0000	6th shift
0 0000 1001 0010	01 1100 0000	Decimal adjust
0 0001 0010 0100	11 1000 0000	7th shift
0 0010 0100 1001	11 0000 0000	8th shift
0 0010 0100 1100	11 0000 0000	Decimal adjust
0 0100 1001 1001	10 0000 0000	9th shift
0 0100 1100 1100	10 0000 0000	Decimal adjust
0 1001 1001 1001	00 0000 0000	10th shift

The processes of converting, which are similar, are shown in Fig. 11.6 (p. 336). In each case the original number is shifted repeatedly into the new. After each shift a decimal adjust to the BCD number may be required, as indicated.

Example 11.3 explores the application of fixed point number representation, applying several of the preceding routines to a real-life situation.

WORKED EXAMPLE (11.3)

A system is required to measure and display, in both revolutions per second (rps) and rpm, the speed of a rotating shaft. A sensor arrangement on the shaft emits one pulse per revolution. The speed range is to be from 0.5 to 100 Hz, i.e. 30 to 6000 rpm. It should be displayed to one decimal place in rps and an integer number of rpm. Emergency shutdown should be initiated if the speed exceeds 120 rps. If it is below 0.5 rps then it is acceptable for the display to be blanked.

Display data should be updated as rapidly as possible, and at least once every 3 seconds. The microcontroller to be used has a 16-bit Counter/Timer (C/T) with Interrupt on Overflow capability. This is clocked with a frequency of 921.6 kHz, with the possibility of prescaling by 2, 4, 8 or 16.

The system is to be programmed in Assembler. Determine how the C/T should be used, and what number ranges and arithmetic routines will be needed. Auto-ranging should not be attempted.

Solution: At the slowest shaft speed, the sensor will be emitting one pulse every two seconds. The need for rapid display update, and the slow speed of the shaft, indicate that it will not be possible to measure the speed by pulse counting. Therefore we will measure the period between pulses, and find the frequency, and hence rps, by taking the reciprocal. The rpm can be determined by multiplying this number by 60.

Use of Timer
The 16-bit C/T, being clocked at 921.6 kHz, will have a period measurement capability of $2^{16}/(921\,600)$, i.e. 0.071 s, but we need to be able to measure a period of 2 s. Prescaling the C/T clock by 32 would allow this, but this value is not available. Therefore we will consider using the Interrupt on Overflow capability to update on every interrupt a memory location used as a counter.

If the C/T input clock frequency is f_p, of period T_p, and the C/T count accumulated between pulses is *Count*, then the basic equation used to find the rotational frequency f will be:

$$f = \frac{1}{Count \times T_p}$$

At the low-frequency end the value of *Count* will be very high. We therefore need to investigate what size of C/T will be needed. At the high-frequency end the value of *Count* will be small, and we therefore need to investigate any resolution problems which may occur. To explore possible violations of the specification, let us tabulate some values for extremes of speed, with different prescaler values.

Table 11.4 shows differing count and output display values for some selected trial input frequencies. As the rpm display demands the higher resolution, we try several neighbouring rpm values and see if the Counter will be able to distinguish between them.

With a prescaler value of 1 or 2, displayed values follow trial input values adequately. This is not the case for prescaler values of 4 or greater. A change in *Count* from 2305 to 2304 with a prescaler of 4 leads to a change in rpm display of 5997 to 6000, with the intervening codes never being represented. This does not meet the requirement stated. The situation is worse with a prescaler of 8.

Table 11.4 Computed values of output displays for differing input frequencies.

Prescaler	f_p (kHz)	Actual frequency		Count	Displayed frequency	
		Hz	rpm		Hz	rpm
1	921.6	0.5	30	1,843,200 (21 bit)	000.5	0030
1	921.6	99.98	5999	9217	100.0	5999
1	921.6	100.00	6000	9216	100.0	6000
1	921.6	100.02	6001	9214	100.0	6001
2	460.8	0.5	30	921 600 (20 bit)	000.5	0030
2	460.8	99.98	5999	4609	099.9	5999
2	460.8	100.00	6000	4608	100.0	6000
2	460.8	100.02	6001	4607	100.0	6001
4	230.4	0.5	30	460,800 (19 bit)	000.5	0030
4	230.4	99.96	5997	2305	100.0	5997
4	230.4	99.98	5999	2304	100.0	6000[1]
4	230.4	100.00	6000	2304	100.0	6000
4	230.4	100.02	6001	2304	100.0	6000[1]
8	115.2	0.5		230,400 (18 bit)	000.5	0030
8	115.2	99.91	5995	1153	099.9	5995
8	115.2	99.98	5999	1152	100.0	6000[1]
8	115.2	100.00	6000	1152	100.0	6000
8	115.2	100.02	6001	1152	100.0	6000[1]
8	115.2	100.03	6002	1152	100.0	6000[1]
8	115.2	100.09	6005	1151	100.1	6005

[1]Specification violated

We should therefore accept a prescaler value of 2, implying that we must use the 16-bit Counter in conjunction with 4 bits of overflow in an additional memory location.

Calculations
With a prescaler of 2, we will need to perform the calculation

$$f = \frac{1}{Count \times T_p} = \frac{460800}{Count}$$

with the ability to round to 1 dp. The numerator is a constant which can be embedded in the program. If we use 4 608 000 as numerator (i.e. it is multiplied by 10), then the integer result of the division will have the right number of digits for the rps display, and a decimal point can be inserted between the last two display digits. This effectively reverses the ×10 scaling just introduced. This is an example of Fixed Point scaling.

The numerator is now a 23-bit number, and the denominator will have a maximum of 20 bits. A 24-bit by 24-bit division routine will therefore be appropriate, with the anticipated result when in legal operating range not greater than 10-bit (i.e. representing up to 100.0 rps). If the display output is to be valid up to the alarm limit of 120 rps, then an 11-bit result must be anticipated. The remainder in the division routine can be used for rounding. A binary to BCD routine, 11-bit to 14-bit (see Table 11.1), will be needed to actually produce the display digits.

The rpm result can be deduced by multiplying the rps by 60 or by repeating the above calculation, with the numerator scaled appropriately. Either way this results in a 13-bit binary number (to represent the legal range up to 6000 rpm), and would require a 13-bit to 16-bit binary to BCD conversion routine.

Over and under range
Requirements for over and under range (i.e. speed < 0.5 rps or speed > 120 rps) were given in the problem statement, and these conditions must be detected. The value of *Count* for 120 Hz will be 3840 and for 0.5 Hz it will be 921 600. These values can be detected by comparison on completion of a count period, and appropriate action taken.

11.3 Floating point number representation

Some example ranges of unsigned binary and two's complement representation, are shown in Table A.3 of Appendix A. Fixed point arithmetic is limited because of the limit on numerical range set by the number of bits in the number and the inflexibility of the binary point location. Suppose we need to represent really large or small numbers? Another way of expressing a number, which greatly widens the range and entirely removes the need for scaling, is called *floating point* representation.

In general a number can be represented as $a \times r^e$, where a is the mantissa, r is the radix and e is the exponent. This is sometimes called scientific notation. For example, the number 12.3 can be represented as

$$12.3 \times 10^1 \ or$$
$$.123 \times 10^2 \ or$$
$$12.3 \times 10^0 \ or$$
$$123 \times 10^{-1} \ or$$
$$1230 \times 10^{-2}$$

Floating point notation adapts and applies scientific notation to the computer world. The name is derived from the way the binary point can be allowed to float by adjusting the value of the exponent to make best use of

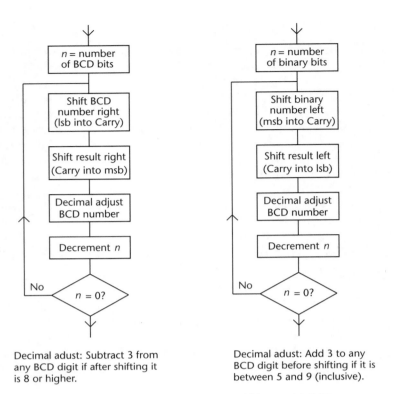

Decimal adust: Subtract 3 from any BCD digit if after shifting it is 8 or higher.

Decimal adust: Add 3 to any BCD digit before shifting if it is between 5 and 9 (inclusive).

Figure 11.6 Conversion between BCD and binary. (a) BCD to binary; (b) binary to BCD.

the bits available in the mantissa. Standard formats exist for representing numbers by their sign, mantissa and exponent, and a host of hardware and software techniques exist to process numbers represented in this way. Their disadvantage lies in their greater complexity, and hence usually slower speed and higher cost. For more complex applications, however, they are essential.

The most widely recognised and used format is the *IEEE Standard for Floating-Point Arithmetic* (known as IEEE 754). In single precision form this makes use of 32-bit representation for a number, with 23 bits for the mantissa, eight bits for the exponent and a sign bit (Fig. 11.7(a)). The binary point is assumed to be just to the left of the msb of the mantissa. A further bit, always 1 for a non-zero number, is added to the mantissa, making it effectively a 24-bit number. Zero is represented by four zero bytes. The number 127_{10} is subtracted from the exponent, leading to an effective range of exponents from –126 to +127. Exponent 255 (leading to 128 when 127 is subtracted) is reserved to represent infinity. The value of the number represented in Fig. 11.7(a) is then

$$(-1)^{\text{sign}} \times 2^{(\text{exponent} - 127)} \times 1.\text{mantissa}$$

This allows number representation in the range:

$$\pm 1.175494 \times 10^{-38} \text{ to } \pm 3.402823 \times 10^{+38}$$

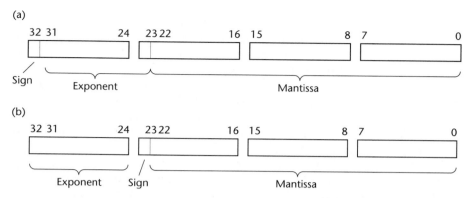

Figure 11.7 32-bit floating point formats. (a) IEEE754; (b) Microchip.

Floating point calculations are found in the small-scale microcontroller world, as part of any C compiler and as libraries of assembler subroutines. IEEE format is not invariably applied. The IAR C compiler (Ref. 7.10) uses it, but the routines of Ref. 11.5 do not. These apply the 32-bit format of Fig. 11.7(b), in which the sign bit has been relocated. This eases processing with the PIC instruction set. The number range is the same as that of Fig. 11.7(a), except that Microchip does not reserve exponent value 128 for infinity, leading to a doubling of the maximum possible number. Ref. 11.5 also offers some routines using a 24-bit format, similar to Fig. 11.7(b), except that the mantissa size is reduced by one byte. This allows faster processing, but with some loss of accuracy.

For small systems of limited memory or functionality floating point arithmetic is usually unnecessary and is best avoided. We shall not explore floating point routines in this chapter; their principles can however be found in Refs. 1.5 or 11.6.

11.4 Truncation and rounding

It is at times necessary to limit the precision of a number to a certain number of digits. In normal life we do this frequently – although the value of π is available to any number of digits, we are usually happy to know it in its four- or five-digit version. In the computer world, we are limited first by the size of memory location available to hold a number. Floating point numbers, for example, have to be constrained to fit into the format of Fig. 11.7. Other restrictions may also apply. Perhaps a display has only a limited number of digits, and the value held must be conditioned to match them, or data must be transferred or stored at a precision less than the value actually held. To reduce the digit count of any number, we can either truncate or round. Both rounding and truncation reduce the precision of the original number and hence introduce error. In a series of calculations, such as those of Example 11.3, errors like this can build up, and it is worth considering how to minimise them.

Truncation is the act of simply removing the unwanted less significant digits. For example, truncating the value of π (3.141592...) to four digits leads to a value of 3.141. Simple truncation always results in the truncated value being equal to or less than the original value – we are effectively subtracting a part of the number, however small. We say that the introduced error is *biased*.

Rounding is the act of removing the less significant digits, but then adjusting the value of the new least significant digit, and possibly other digits as well, so that the rounded number is as close as possible to the original; π rounded to for digits is 3.142. Ideally any rounding method applied should be *unbiased*.

In a binary number, rounding can be introduced by adding 1 to the most significant bit that is to be discarded (the equivalent of adding half of the new LSB), and then truncating. This is close to being unbiased, but is not quite (Ref. 11.6). Furthermore, the rounding action requires a further addition, which slows program operation. It also carries the risk of increasing by 1 the number of digits in the remaining number, and this must be catered for. For example, if the decimal number 999.9 is to be rounded to three digits, then simply removing the .9 and rounding produces another 4 digit number, i.e. 1000.

An alternative rounding method, which adds minimal computation time, simply sets the lsb of the truncated number always to 1. This is sometimes called von Neumann or lsb rounding or jamming. For most numbers it increases the initial error, but gives an unbiased overall error (Ref. 11.6).

If rounding is to be applied at the end of a series of calculations, then the intermediate results must be held to a higher precision than the intended final result. Otherwise it will be impossible to round, and cumulative error may occur. To make this possible in floating point calculations, *guard bits* are applied. These extend the precision of intermediate results, and are used as the basis for any rounding which takes place to form the final result.

The rounding and truncation options described above, for a 4-bit number reduced to two, are illustrated in Table 11.5.

11.5 Operations on sampled data

In many situations a microcontroller is used to sample periodically an incoming analogue signal, as described in Chapter 5, and then undertake some processing of the received data. This might include signal averaging to eliminate noise or interference, integrating or differentiating for a control application, or predicting future samples to allow comparison to be made between incoming data and predicted – a major difference may indicate system malfunction.

It is to be understood throughout this section that an input signal is being sampled continuously at a fixed sampling frequency f_s, with sampling period Δt, and that Nyquist's sampling theorem is being applied.

11.5.1 Averaging

A common requirement when digitising data is to find an average. Commonly this is applied to reduce the effect of interference on a signal in

Table 11.5 Truncation and rounding examples.

4-bit number	Simple truncation		Adding 1/2 LSB		LSB jamming	
	Result	Error	Result	Error	Result	Error
00.00	00	.00	00	.00	01	+1.00
00.01	00	−.01	00	−.01	01	+0.11
00.10	00	−.10	01	+.10	01	+0.10
00.11	00	−.11	01	+.01	01	+0.01
01.00	01	00	01	00	01	0.00
01.01	01	−.01	01	−.01	01	−0.01
01.10	01	−.10	10	+.10	01	−0.10
01.11	01	−.11	10	+.01	01	−0.11
10.00	10	.00	10	.00	11	+1.00
⋮						
11.00	11	.00	11	.00	11	.00
11.01	11	−.01	11	−.01	11	−.01
11.10	11	−.10	100	+.10	11	−.10
11.11	11	−.11	100	+.01	11	−.11

the presence of random or periodic interference. Alternatively, it could be to find the average of a variable over a period of time. A botanist may apply a data logger to study the effects of sunlight on a crop; the average value could give all the information that is needed, the actual variations over the day adding very little of interest. In such an application there is of course further advantage in storing just one daily average rather than an extended sequence of samples.

An average is the sum of all samples under consideration divided by the number of samples. Viewed as a waveform, the average value is the area under the waveform divided by the averaging time.

A simple approach to gaining an approximate average in the presence of random interference is to sum every 2^n samples, where n is an integer. For example, every 16 or 32 samples could be summed. The sum is then divided by 2^n by shifting it right for n bits, or simply truncating the last n bits. This is illustrated in Fig. 11.8 for a value of n of 3. Every eight samples are summed, and a new average is available every 8th sample. The processing overheads are limited, i.e. one addition per sample and the division every 2^n samples. This is suitable for slow-moving data, for example data that is to appear on a numeric display. It effectively produces another set of sampled data whose sampling period is $2^n\Delta t$.

If the intention is to eliminate interference, and the interference is random, then the averaging time should theoretically be infinite. With a finite number of samples averaged, the process does not of course eliminate the effect of the noise altogether, it simply reduces it. It can be shown

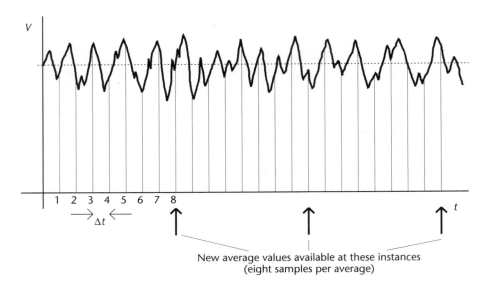

Figure 11.8 Signal averaging in the presence of random noise.

(Ref. 11.7) that random noise is reduced by the averaging process in proportion to the square root of the number of samples taken, i.e. the number of samples must be increased by a factor of four to halve the level of interference.

An approach which gives an average update on *every* incoming sample is the running average. Here a store of the last N samples is maintained. On every incoming sample the last N samples are averaged. Again this is simplest when N is a binary power. In processing terms it requires that the samples are summed every sampling period and the appropriate division undertaken. The samples can be stored in a circular buffer of the form shown in Fig. 3.22(b).

In the presence of periodic interference the advantage gained is dependent on the interference frequency. If samples are averaged over a period T, and T is an integer number of interference cycles, then the interference is ideally averaged to zero. This can be recognised intuitively from Fig. 11.9, where a steady signal (represented by the dotted line) has a

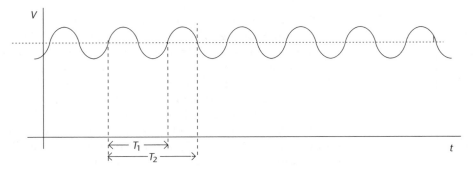

Figure 11.9 Signal averaging in the presence of periodic interference.

sinusoid superimposed on it. If the signal is averaged over period T_1, the effect of the sinusoid disappears altogether. If averaged over an odd number of half cycles, for example period T_2, the advantage is a minimum.

WORKED EXAMPLE   **11.4**

A slow-moving signal is being sampled, in the presence of significant mains hum, 3000 times a second. How many samples should be averaged in order to minimise the effect of the interference?

Solution: The sampling period is (1/3000), i.e. 0.333 ms. If the mains frequency is 50 Hz, of period 20 ms, then 60 samples will be taken in one mains cycle, and these should be averaged. Fifty samples should be averaged if the frequency is 60 Hz. This approach is widely used in data acquisition in the presence of mains interference. The digital voltmeter applies a similar technique – the dual ramp ADC that is uses effectively averages (in hardware) the incoming signal over a period of one mains cycle.

11.5.2 Differentiation

Differentiation allows us to find out the gradient or rate of change of a waveform. This may be for control purposes, for example Proportional + Integral + Differential (PID) Control, or to derive velocity from positional information. Alternatively, it may be for safety purposes; in certain systems excessively fast changes are indicative of actual or impending faults.

Figure 11.10 illustrates a changing waveform being sampled at sampling interval Δt. To find the gradient at any one sampling point the obvious way would be to find the difference between its two neighbouring samples, and divide by the time interval. For example the gradient at sample s_5 can be estimated as:

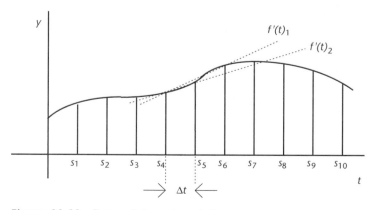

Figure 11.10   Determining rates of change.

$$\frac{dy}{dt} = f'(t) = \frac{s_6 - s_4}{2\Delta t} \tag{11.1}$$

leading to the approximation shown of $f'(t)_1$. A disadvantage of this is that it requires knowledge of a future sample, i.e. s_6, to calculate the gradient at a present sample.

We can therefore attempt a revised method by finding the difference between two neighbouring samples and dividing by the time interval between them. For example, the gradient midway between samples s_4 and s_5 can be calculated as

$$f'(t) = \frac{s_5 - s_4}{\Delta t} \tag{11.2}$$

This calculates the gradient one-half of a sample period previous to the present sample. It is a practical and useful approximation, and gives good results as long as the signal frequency is less than around one sixth of the sampling frequency (Ref. 11.7). It is illustrated by the line $f'(t)_2$ in Fig. 11.10.

Both of these approaches are of course approximations, and both are particularly susceptible to noise, as they use only two samples. We can achieve a better approximation by applying Taylor's series. This relates one point in a curve $f(x)$ to another somewhere else on the curve, at coordinate $(x + \Delta x)$.

$$f(x + \Delta x) = f(x) + f'(x)\Delta x + f''(x)\frac{(\Delta x)^2}{2!} + f'''(x)\frac{(\Delta x)^3}{3!} + \ldots \tag{11.3}$$

This is ideally suited to our purpose. With time as the horizontal axis, we can use this series to relate a point in the curve $y = f(t)$ to one slightly further along the curve, $y = f(t + \Delta t)$, where Δt can have a positive or negative value. To simplify matters, let us reduce this series to just its first two terms, and treat Δt as a negative quantity, i.e. we are looking back in time to a previous sample from time t.

$$f(t - \Delta t) = f(t) - f'(t)\,\Delta t$$

Rearranging this to find $f'(t)$, we get

$$f'(t) = \frac{f(t) - f(t - \Delta t)}{\Delta t} \tag{11.4}$$

This agrees with our earlier expression for $f'(t)$ (Equation 11.2), which is reassuring. Let us, however, include a further term in the series, and refine our approximation.

$$f(t - \Delta t) = f(t) - f'(t)\Delta t + f''(t)\frac{(\Delta t)^2}{2!}$$

Rearranging this to find $f'(t)$, we get

$$f'(t) = \frac{f(t) - f(t - \Delta t)}{\Delta t} + f''(t)\frac{(\Delta t)}{2!} \tag{11.5}$$

To find $f''(t)$, we apply the approximation of (11.4) to get

$$f''(t) = \frac{f'(t) - f'(t - \Delta t)}{\Delta t} \tag{11.6}$$

and note also from (11.4) that the term $f'(t - \Delta t)$ can be approximated by

$$f'(t - \Delta t) = \frac{f(t - \Delta t) - f(t - 2\Delta t)}{\Delta t} \tag{11.7}$$

Hence, substituting for $f'(t)$ and $f'(t - \Delta t)$ back into (11.6),

$$f''(t) = \frac{f(t) - f(t - \Delta t)}{\Delta t^2} - \frac{[f(t - \Delta t) - f(t - 2\Delta t)]}{\Delta t^2}$$

or

$$f''(t) = \frac{f(t) - 2f(t - \Delta t) + f(t - \Delta t)}{\Delta t^2} \tag{11.8}$$

Finally, we substitute (11.8) back into (11.5)

$$f'(t) = \frac{f(t) - f(t - \Delta t)}{\Delta t} + \frac{(\Delta t)}{2!} \frac{[f(t) - 2f(t - \Delta t) + f(t - 2\Delta t)]}{\Delta t^2}$$

which gives

$$f'(t) = \frac{3f(t) - 4f(t - \Delta t) + f(t - 2\Delta t)}{2\Delta t} \tag{11.9}$$

This proves to be a more accurate approximation than Equation (11.4). The sample $f(t)$ can be considered to be the present sample, while $f(t - \Delta t)$ is the previous and $f(t - 2\Delta t)$ the one before that. The value of $f'(t)$ can be achieved without excessive computation.

WORKED EXAMPLE **11.5**

A 2 kHz sine wave signal of peak value 1 V is continuously sampled at intervals of 25 μs, with one sample occurring at $t = 0$. Estimate the gradient at samples around the time $t = 0$, using the approximations of equations (11.4) and (11.9). Find also the precise value given by differentiation.

Solution: This waveform can be represented by $v = \sin(12568t)$V. By standard differentiation its gradient at any point is given by:

$$f'(t) = 12568\cos(12568t)$$

Table 11.6 shows the calculated gradient and the two estimates from applying Equations (11.4) and (11.9) respectively for values of t from −25 to +100 μs. Two earlier samples of $f(t)$ are also given to provide data to make the first estimates. In each case, the error of the estimate, as a difference from the calculated value, is indicated in brackets.

Table 11.6 Comparison of gradient estimates.

t (μs)	$f(t)$	Calculated gradient (Vs⁻¹)	First estimate (Vs⁻¹)	Second estimate (Vs⁻¹)
−75	.80909	–	–	–
−50	.58785	–	–	–
−25	.30906	11953	11152 (−802)	12303 (+350)
0	0	12568	12362 (−206)	12968 (+400)
25	.30906	11953	12362 (+409)	12362 (+409)
50	.58785	10167	11152 (+985)	10546 (+379)
75	.80909	7386	8850 (+1464)	7699 (+313)
100	.95111	3882	5681 (+1799)	4096 (+215)

It can be seen that estimates based on Equation (11.4) give a result of consistently worse value, except around the time $t = 0$. Results based on Equation (11.9) return a modest error which (for these results) peaks at around 3.5%.

11.5.3 Prediction

It is sometimes useful with an incoming waveform to attempt to predict the next value before it is available. For example, the ultrasound ranging system of Fig. 8.9 was used to track a vibrating target. The nature of the target was such that under certain conditions the ultrasound pulse was deflected away from the receiver and not detected. When this happened the Counter/Timer counted to its maximum value and the system output therefore suggested that the target had leapt to the far distance. A simple predictor was therefore set up. Each new value was compared with the predicted one, and if the difference was too great then the sample was rejected and replaced with its predicted value.

The Taylor series, Equation (11.3), can be applied directly to give a simple predictor. Using a two-term approximation, the prediction of the value of the next sample $f(t + \Delta t)$ is given by:

$$f(t + \Delta t) = f(t) + f'(t)\, \Delta t \tag{11.10}$$

The $f'(t)$ term is then substituted using Equation (11.4):

$$f(t + \Delta t) = f(t) + \frac{[f(t) - f(t - \Delta t)]\Delta t}{\Delta t} \tag{11.11}$$
$$= 2f(t) - f(t - \Delta t)$$

A more precise predictor can be developed using Equation (11.9) to substitute for $f'(t)$.

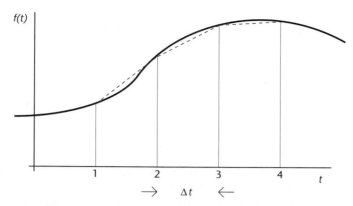

Figure 11.11 Applying the Trapezoid rule to integrate a waveform.

11.5.4 Integration

There are many engineering and instrumentation situations when it is necessary to find an integral.

Geometrically, in integration our aim is to find the area under a curve between certain limits. A first approximation when using sampled data is to apply the trapezoid rule, which replaces the area between each pair of samples with a trapezoid, as shown in Fig. 11.11. This area between samples 2 and 3 for example is given by $(s_2 + s_3)\Delta t/2$. Summing a series of areas, starting at sample s_1 and going on to sample s_n, gives:

$$\text{Total area} = [(s_1 + s_2) + (s_2 + s_3) + (s_3 + s_4) + (s_4 + s_5) + \ldots$$
$$+ (s_{n-1} + s_n)]\Delta t/2$$

$$= (s_2 + s_3 + s_4 + s_5 + \ldots + s_{n-1})\Delta t + (s_1 + s_n)\Delta t/2 \qquad (11.12)$$

Therefore, to integrate a sequence of samples taken at a steady sample rate, a simple approach is to sum the samples continuously. The Δt term can be looked after by a simple scaling value. If the sequence is sufficiently long, it may be possible to disregard the error due to the first and last terms. Alternatively, they can be identified, and halved before adding into the total.

The Trapezoid rule is actually the simplest of a set of formulae (Ref. 11.9) of varying sophistication which are used to estimate integrals. Two important members of this family are *Simpson's Rule* and *Simpson's Three-Eighths Rule*, given respectively below.

$$\int_{t-2\Delta t}^{t} f(t)\,dt = [f(t - 2\Delta t) + 4f(t - \Delta t) + f(t)]\,\Delta t/3 \qquad (11.13)$$

$$\int_{t-3\Delta t}^{t} f(t)\,dt = [f(t - 3\Delta t) + 3f(t - 2\Delta t) + 3f(t - \Delta t) + f(t)]\,3\Delta t/8 \qquad (11.14)$$

Both of these are commonly found as the basis of simple integration algorithms.

As mentioned in the introduction, the full manifestation of operations on sampled data is in the field of Digital Signal Processing (DSP). DSP in its full form is really outside the scope of the small microcontroller or embedded system. Having said that, no boundaries are hard and fast, and Microchip produces at least one Application Note giving guidance on using its microcontrollers for DSP applications (Ref. 11.10).

SUMMARY

1. Fixed point number representation is used for operations of limited number range and complexity. Floating point representation is applied when there is a need to represent wide-ranging high-precision numbers.

2. A huge variety of algorithms exist to implement all standard mathematical functions. These are widely available as subroutine or function libraries. It is sometimes necessary to adapt or check them, or understand the errors they produce. In this case an understanding of their working is essential.

3. The precision of numbers is in practice always restricted, due to limitations of memory locations and processing capability. This may be effected by a process of truncation or rounding. Choosing the optimum technique for the application limits the error that is inevitably part of the process.

4. Small embedded systems often have to handle sampled data. There is a range of standard techniques for dealing with this.

REFERENCES

11.1 *Math Utility Routines*. Application Note 544. Microchip Technology Inc. 1997.
11.2 *Fixed Point Routines*. Application Note 617. Microchip Technology Inc. 1996.
11.3 Motorola Inc. (1983) *M6805 HMOS M146805 CMOS Family Microcomputer/ Microprocessor User's Manual*, 2nd edn. Englewood Cliffs, NJ: Prentice Hall.
11.4 *M68HC08 Integer Math Routines*. Application Note AN1219. Motorola Inc. 1996, revised 1997.
11.5 *IEEE 754 Compliant Floating Point Routines*. Application Note 575. Microchip Technology Inc. 1997.
11.6 Cavanagh, J. (1984) *Digital Computer Arithmetic, Design and Implementation*. London: McGraw-Hill.
11.7 Martin, J. D. (1991) *Signals and Processes, A Foundation Course*. London: Pitman.
11.8 Ayyub, B. M. and McCuen, R. H. (1996) *Numerical Methods for Engineers*. Englewood Cliffs, NJ: Prentice Hall.
11.9 Kreysig, E. (1983) *Advanced Engineering Mathematics*, 5th edn. Chichester: Wiley.
11.10 *Digital Signal Processing with the PIC 16C74*. Application Note 616. Microchip Technology Inc. 1997.

EXERCISES

11.1 Apply the Flow Diagram of Fig. 11.1(a) to add together the two's complement 16-bit numbers given below. Which results (if any) over-flow the 16-bit range, and how is this identified? By brief inspection of the operands, are you able to tell whether the possibility of overflow exists?

(a) 0100 1100 0000 1010 and 0010 0010 0111 0001
(b) 1000 0100 1101 1101 and 1111 0000 0011 1001
(c) 1000 0001 0100 0010 and 0111 1110 1010 0001

11.2 Apply the Flow Diagram of Fig. 11.1(b) to subtract the 16-bit numbers given in Exercise 11.1, where the second number in each pair is the subtrahend. Which results (if any) overflow the 16-bit range, and how is this identified? By brief inspection of the operands, are you able to tell whether the possibility of overflow exists?

11.3 Apply the flow diagram of Fig. 11.4 to multiply together the 8-bit numbers given below. Form the result in 2 bytes labelled ResHi and ResLo. For each cycle of the flow diagram write down in a table the values of the multiplier after rotation, the resulting Carry bit value, ResHi and ResLo both before and after the *Rotate Result Right* action, and the state of the Counter.

(a) Multiplicand = $0011\ 0000_2$ (48_{10}), multiplier = $0000\ 0101_2$ (5_{10})
(b) Multiplicand = $0011\ 0000_2$ (48_{10}), multiplier = $1100\ 1000_2$ (200_{10})

11.4 Apply the flow diagram of Fig. 11.5 to divide together the numbers given below. In LOOP1, determine the value to which Count is incre-mented. In LOOP2, determine and record, for each cycle:

(i) the value of *Divisor* $\times\ 2^n$
(ii) whether the subtraction gave a positive or negative result
(iii) the Remainder value at the *Rotate Left* action
(iv) the Result value after *Rotate Left*
(v) the value of Count after it is decremented

(a) 0111 0101 (117_{10}) dividend divided by 0000 1101 (13_{10}) divisor
(b) 0111 0101 (117_{10}) dividend divided by 0001 0100 (20_{10}) divisor
(c) 0111 0101 (117_{10}) dividend divided by 0 divisor
(d) 0111 0101 (117_{10}) dividend divided by 0111 0110 (118_{10}) divisor

11.5 Devise an algorithm which multiplies together two 2-byte numbers using an 8 × 8 hardware multiply, based on the preliminary information given in Section 11.2.3.

(a) Draw the algorithm as a flow diagram.
(b) For a microprocessor of your choice (having a hardware multiply), develop your algorithm into an Assembler listing and estimate its duration of execution.
(c) For the same microprocessor as (b), find a standard 16 × 16 multiply routine which does not use the hardware multiply facility. Compare its duration of execution to that of your algorithm, as derived in (b).

11.6 Using the flow diagrams of Fig. 11.6, complete the following conversions:

(a) 99 BCD to binary

(b) 256 BCD to binary

(c) 0110 0100 binary to BCD

(d) 7D0 hexadecimal to BCD

11.7 Equation (11.11) gives a simple predictor. Derive a more precise predictor, substituting the estimate for $f'(t)$ of Equation (11.9) into Equation (11.10).

11.8 Explain the reason for MISRA C rule 50, quoted in Table 7.8.

Designing and commissioning the system

IN THIS CHAPTER

We have examined the many and varied aspects of small-scale embedded systems in detail for 11 chapters. We now stand back and try to put the mass of detail into the overall context of the product design and development cycle. This will include surveying the design process for the embedded system, and then considering methods for systematic test and evaluation.

The aims of the chapter are:

- to overview the process of engineering design

- to apply the generalised design process to the design of embedded systems

- to describe development and test procedures

- to describe some of the tools of development and test

12.1 The engineering design process

12.1.1 A definition

The word *design* is widely used, misused and misunderstood. It is applied to art and textiles, to engineering and to a range of activities in between. Even within engineering it is applied in differing senses. Little wonder then that many engineers have surprising difficulty in giving anything approaching a clear definition of *engineering design*. If we are to explore the process, however, it is necessary to have a good idea of what we are looking at. We note that any engineering product is manufactured to a formal definition of the product, traditionally in the form of a set of drawings, but also possibly including software listings (or their equivalent in electronic format), or other means of product definition.[1] We note also that the product starts its life as some statement or perception of a market need. We therefore define engineering design as that group of activities that links these two stages, i.e.

> engineering design is the process whereby a market need is converted into a definition of a product, usually in the form of a set of drawings or electronic files, which can be taken forward to manufacture.

Engineering design is a profoundly complex process, involving not only detailed and expert knowledge of engineering analysis, but also of market forces and legal issues. Above all it has an indefinable creative element – the ability to be able to visualise and define new and possibly revolutionary products, which enjoy success in the marketplace and lead to continuing prosperity for their designers and manufacturers.

12.1.2 The design process

The design process has been widely studied, predominantly in a mechanical design environment. The main principles of these studies can however be readily applied to the design of the electronic and embedded system. A simple, general-purpose and widely used design model (from Ref. 12.1) is shown in Fig. 12.1(a). It shows four design stages, which we will use as the basis for our investigations which follow. The model is developed and expanded, with the inclusion of a first bias towards the embedded system, in Fig. 12.1(b). Here the outcome of each design phase is shown in italics.

Most new products start life because of an identified market need, new possibilities offered by a technological advance, or a novel idea. In every case, if the proposed product is to be viable, these must be translatable into a *customer need* – it must be possible to sell the product once designed and

---

1 The formal definition of the configuration design of a Field Programmable Gate Array, for example, is held as a computer file which defines how the configurable logic blocks of the array are to be set up.

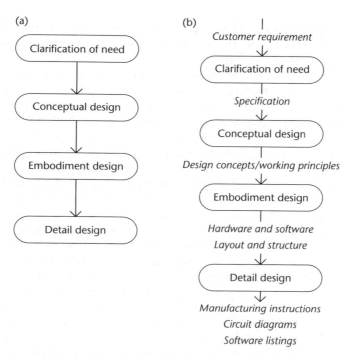

Figure 12.1 Engineering design models. (a) Simple model: Pahl and Beitz; (b) expanded model, showing activity and outcome.

made. The design process thus starts with an expression of this customer need.

Following its identification, the customer need must be understood and defined, a process which leads to the *product specification*. From this the *conceptual design* phase flows, in which the overall concept of the design is determined. At this stage the stress is on generating as many ideas as possible, including both the conventional and the original, and selecting from these the most viable.

When the broad concept or concepts are finalised, the phase of *embodiment design* follows. Here the more generalised ideas of the conceptual design begin to take physical form. In embedded systems terms this implies that the structures of hardware and software become defined. There is an emphasis on subjecting the ideas so far generated to a process of increasingly rigorous engineering evaluation and development. This stage is followed by that of *detail design*. In the case of embedded systems, precise circuit and software details are finalised. At every stage, as the design takes shape, costs should be reviewed.

In brief, it can be said that the specification deals primarily with *needs*, the conceptual design primarily with *ideas*, embodiment design with *technologies*, and detail design with *analysis*.

While the design stages are distinct, and follow each other approximately sequentially, their dividing lines are not rigid. It may, for example, be necessary to undertake a little detail design while in the conceptual design phase in order to check the viability of a particular idea. It may also be necessary to return from embodiment to conceptual design if a concept is thrown into question.

Each design stage requires from the designer a different combination of skills, knowledge and expertise. Some of these can be taught, while others are more dependent on experience. The area most commonly taught is that of detail design. Most novice designers are likely to have some competence in this area; it requires good attention to detail, and it is what engineering and technology departments in universities and colleges spend much of their time teaching. The ability to write a specification does not necessarily depend on a great depth of technical knowledge, and skill in this area can also be readily developed. It is in the middle stages, of conceptual and embodiment design, where experience is likely to play the greatest part.

It is essential to be aware that the early design stages are of *strategic importance*. The danger for novice designers is that they imagine that the whole design process is made up of detail design, and fail to recognise the necessity of the earlier stages. Yet if the specification is wrongly or incompletely drawn up, they will end up solving the wrong problem. If the design concept is not properly thought through, then however good the detail design, the product is unlikely to be competitive. Therefore it is useful to follow this simple design model with some commitment – it encourages a thoughtful and structured approach, and should lead to a much better end result. It should be viewed as a process to channel inventiveness, not as a rigid and unbending framework.

12.1.3 Variant design

The design model of Fig. 12.1 is fully applicable to the design of a new and original product. In practice, most new products are the outcome of *variant design*. That means that they are an upgrade of a previous product made by the company. For example, a car, bicycle or telephone is not reinvented every time a new model is produced. Therefore with variant design, particularly when the variation is small, it may be possible to compress the conceptual design phase. This should, however, be done with caution. This is the phase where strategic innovation is generally introduced, and where old assumptions are questioned; omitting it risks leaving one locked to outdated practices or technologies. Sometimes, after all, the car, bike or telephone *is* reinvented, with radical new concepts or technology.

12.2 The design model applied to the embedded system

The model of Fig. 12.1(a) has been adapted to the electronic design process (Ref. 12.2). Here it is further adapted to embedded system design, as shown initially in Table 12.1. This shows the general activities which may be expected to take place at each stage. Essentially, our design emerges as one of two highly integrated parts: the hardware and the software. While they are drawn out as separate activities in some of the discussion which follows,

Table 12.1 Processes in embedded system design.

Clarification of need	
Understand and define requirements	Identify single phrase function statement Draw up requirements lists – demands and wishes Draw up specification
Conceptual design	

Form underlying concepts	Determine:
	    principal product functions
	    all means of achieving each function
	    optimum solution combination for all functions
	    hence: principal hardware functions
	            principal software functions
	    product appearance and geometry
	    power source
	Estimate cost
	Investigate relevant standards (for example EMC)
	Consider safety, reliability, testability

Embodiment design	
Form block or structure diagrams	Review issues of computation and timeliness Determine solution technology Refine hardware/software division Choose programming language Determine hardware topology Determine data word size and execution speed Define peripherals Determine microcomputer(s) type Determine test and commission strategies Define software structures Define manufacturing requirements Review costs
Detail design	
Convert block diagrams to detail designs	Ensure performance and manufacturability.

their development should be interactive, and should proceed more or less in parallel (a process called *co-design*).

The information in this table is formatted as a design model in Fig. 12.2 (from Ref. 12.3), which shows the interdependence of the design activities. The model focuses mainly on the conceptual and embodiment design phases, as the processes depicted are specific to the embedded system. This model is offered as a framework of activity for the designer of embedded systems.

We will now look at how the design stages can be implemented, both in general terms and in the context of embedded systems design.

12.2.1 Defining the problem: specification development

The initial identification of need might be expressed in a vague and incomplete way. It is therefore necessary to clarify and then define that need. A useful first step is to think up a *single-phrase function statement* – a statement

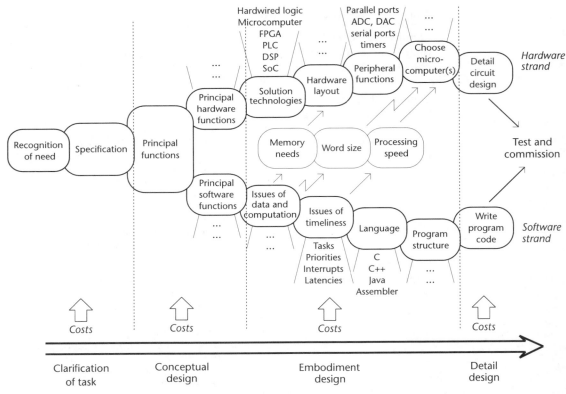

Figure 12.2 A design model for embedded systems. Reproduced by permission of the Institution of Electrical Engineers.

which summarises the main function of the product *without* at this stage saying how that function is to be achieved. Following this a *Requirements List* should be drawn up, showing all the required features and functions of the product. These can be divided into *Demands* (i.e. essential requirements) and *Wishes* (i.e. those that will enhance the product, but are not essential). It is important that, like the function statement, the requirements list is *solution neutral*, i.e. it does not predetermine how the product functions are to be achieved. This comes later.

To illustrate the stages of design, we now embark on an example which explores a product on a much larger scale than any of the other examples in this book. A large-scale project such as this tends to invoke every aspect of the design process, and allows them to be illustrated in an accessible way. A similar, but smaller scale, example is given in Ref. 12.3.

WORKED EXAMPLE **12.1**

Your company has been approached to devise a means of collecting charges in large-scale car parks, for example shopping centres or airports. You may offer advice on the logistics of how this can be done. Initiate the development of the specification.

Solution: As an overall solution-neutral, single-phrase function statement, we will adopt *Collect Parking Charges*. Let us then proceed to a requirements list, as shown in Table 12.2 under the Demands and Wishes headings.

Table 12.2 Requirements list for Example 12.1.

Collect parking charges

Demands	Wishes
Determine charge	Aesthetically pleasing
Indicate charge	Flexible payment method
Collect payment	Accept payment by season ticket
Prohibit car park overflow	Monitor car flow
Physically robust, i.e. Vandal-proof	
Cost competitive	
Accessible to user	
Highly reliable	
Secure	
...	

The requirements list of Table 12.2 is at an early stage of development. As thought and evaluation are applied to the process, generalised statements like 'cost competitive' or 'highly reliable' can become focused and, wherever possible, quantified, in this case with projected maximum project cost and Mean Time Between Failure (MTBF) figures. This process leads ultimately to the specification, a clear and quantitative statement of the intended performance of the system. Primarily it is functionality that we need to specify, but the specification may also take into account how the system is to be tested, as well as issues of reliability. As the specification becomes the basis of all the design work which follows, it is important that it should be all-embracing. It is easy, for example, to leave out the 'obvious'. The headings of Table 12.3 may be used as a checklist.

Like the requirements list, the specification should be solution-neutral, i.e. it should not imply how the functions are to be achieved. Although it becomes the point of reference for subsequent design stages, it should not be viewed as 'set in stone'. Circumstances or the understanding of the design problem may change, or design considerations may force or offer changes. The requirements list and specification may then need to be altered.

Table 12.3 Checklist for embedded system specification preparation.

Functionality	All functions the product is meant to perform,
User interaction	Controls, displays
Physical form	Dimensions, weight, materials, aesthetics
Safety	Legal requirements, special considerations, results of product failure
Operating environment	Target environment, e.g. temperature and humidity ranges, EMC
Signals	Inputs, outputs, response times
Power supply	Source of power, power consumption
Maintenance	Reliability, time between failures
Costs	Development costs, product cost
Schedules	Project time-scales, delivery times

12.2.2 Conceptual design

The conceptual design starts with the product specification, and leads to the overall concept of the design solution. It is first necessary to explore further the function(s) of the proposed product. This can be done with a simple function flow diagram. It can be difficult to draw such a diagram with no idea of the possible solution, so it may be necessary to draw more than one. A possible function flow diagram for Example 12.1 is shown in Fig. 12.3.

Having defined the product functions, it is now necessary to come up with as many ways of performing them as possible. In doing this it is

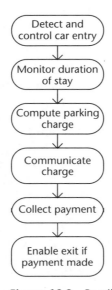

Figure 12.3 Possible functional flow diagram for Example 12.1.

Function	Solution principle				
Detect and control car entry	None	Attendant	Barrier ④		
Monitor duration of stay	None ① ②	Printed card ③	Magnetic strip	Aerial mounted video camera, car tracked by DSP image processing.	Automatic number plate recognition at entry and exit
Compute charge	None (fixed rate)	By driver	With card reader	From automatically generated data	
Communicate charge	Fixed notice-board display	Automated display	Internet-linked in-car navigation system		
Collect charge *Location*	At entry	At exit	Machine in car park	Machine in building	
Method	Coin	Coin/note	Coin/note credit card		
Enable exit if payment made	None	Barrier	Attendant		

Figure 12.4 Possible functional solutions for Example 12.1

important that we do not restrict ourselves either to the conventional or to personal preferences arising from past experience. Ideas can be obtained by intuition, evaluation of existing systems, processes of brainstorming, or analogy with other systems. As we are just 'playing with ideas' at this stage, it is important to be as open as possible. If working as a team, then a humorous, relaxed and non-critical approach should be adopted.

The functions, together with all identified solutions, can then be tabulated. This is shown for Example 12.1, in preliminary form, in Fig. 12.4. The functions from Fig. 12.3 are placed in the first column. All possible design solutions are placed in the corresponding rows; the exact column they occupy is arbitrary. The level of complexity of such diagrams should be controlled. They should be kept reasonably simple, so that it is easy to draw conclusions.

Given a table of this form, it is then possible to identify and evaluate *combinations* of solutions, linking any one set by a line as shown. Clearly some combinations make little sense. In the example it is possible to identify the traditional solution (2) and the 'Pay and Display' solution (1), as well as opportunities to apply novel ideas and new technology.

Acceptable solution combinations can now be identified. Only combinations which satisfy *all* the demands of the requirements list can be accepted. If several combinations meet this requirement, then the final selection may be based on which one achieves most of the wishes as well.

It is useful to revisit the specification when the preferred solution combination has been determined. It may be possible to add or refine detail.

It should now be possible to produce a drawing or sketch of the preliminary conceptual design. A possibility for Example 12.1 is shown in Fig. 12.5. This illustrates solution (3) of Fig. 12.4, and highlights the major system blocks.

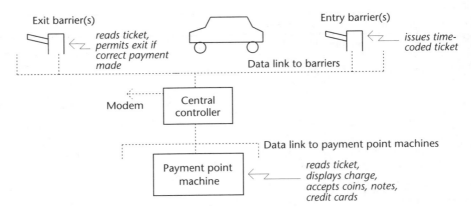

Figure 12.5 Preliminary conceptual design for Example 12.1.

12.2.3 Embodiment design

As the discussion above suggests, the conceptual design phase should lead to a broad understanding of the principal hardware and software functions. As we enter the embodiment design phase, we are going to make those design decisions which will determine the physical and software manifestation of the system. The overall goal will be to produce hardware and software structural designs, probably in the form of detailed block diagrams or software structure diagrams.

For the embedded design, we will need to make the decisions listed under 'Embodiment design' in Table 12.1. The suggested sequencing is indicated in Fig. 12.2, which shows the emergence of software and hardware design strands. Some issues, like designing the software structure or selecting a serial communication protocol, we have considered in detail in an earlier chapter, and therefore just refer to it in passing below. Others issues we consider in greater detail.

12.2.3.1 Choosing the solution technology

Let us step back for a moment from any assumption that the product will be microcontroller-based, and just review the choices available which would allow a control function to be implemented using digital technology. Some options are shown in Table 12.4.

The comparisons shown are subjective and qualitative, and not particularly like-for-like. Unlike a microcontroller, for example, a PLC is a fully fledged engineering sub-system. Nevertheless, they do indicate that the choice is not predetermined. A control function which is fast and computationally intensive may conceivably be better achieved with an FPGA or DSP device. A function that is slower, with high interfacing needs and short development time, may be better implemented in a PLC.

Having made this brief admission that there are other ways of achieving embedded control, we revert to the assumption implicit in this book that the system is to be microcontroller- or microprocessor-based.

Table 12.4 Options for digital controller technology.

	PLC	FPGA	Micro-controller	Micro-processor	DSP
Speed of operation	*	****	**	***	****
Speed of implementation	****	**	***	**	**
Programmability	****	***	***	**	**
Simplicity of application	****	**	***	**	*
Interfacing capability	****	*	***	**	**
Computational capability	*	***	**	****	****

Evaluations: * = poor to **** = excellent
Key: PLC   Programmable Logic Controller
       FPGA   Field Programmable Gate Array
       DSP   Digital Signal Processing

12.2.3.2 Hardware/software trade-offs

It has been demonstrated in several earlier chapters that many functions, including counting, timing, serial data links and PWM generation, can be performed in *either* hardware *or* software. Broad decisions about the allocation of tasks may already have been made in the conceptual design phase. These should, however, be reviewed and refined at this stage. The main relative advantages of placing a function in hardware or software are shown in Table 12.5.

In brief, placing functions in software is favoured by products which have one or more of the following characteristics: small, low cost, low power, produced in quantity and not necessarily of high performance. Functions tend to be placed in hardware in products which are developed fast, high performance (including high speed), less cost-conscious and produced in limited quantity.

In making these decisions the principal advantages of software should be noted, i.e. that once developed it costs nothing to manufacture in quantity, and it is always easier to modify and update than hardware.

12.2.3.3 The software strand

This section aims to place in context some of the aspects of software and program development, covered already in earlier chapters.

The software strand, as Fig. 12.2 proposes, starts with an identification of the software functions and leads on to an evaluation of their requirements. First, the data and computational issues should be considered. What form will the data need to take (i.e. its number representation and word length), what processes will be applied to it, and what quantities of data will need to be processed and stored? The outcome of these decisions will influence the processor word size, the system data memory requirements, and the processing capability and running speed of software and hardware.

Table 12.5 Hardware/software trade-off.

Function in software	Function in hardware
More complex program	Simpler program
Increased program memory size	Reduced program memory size
Potentially increased development time	Potentially reduced development time
Reduced product cost	Increased product cost
Probably slower performance	Faster performance possible
Processor time hogged	Processor time freed for other activities
Lower resolution and accuracy	High resolution/accuracy performance possible
Reduced physical product size	Increased size
More flexibility in microcontroller choice	Microcontroller choice fixed by peripherals needed
May require faster clock	Possible reduction in clock speed
Easier to change/update functionality	Difficult to change hardware design (though easy to reconfigure a hardware peripheral)
Increased overall complexity of software makes further software changes more difficult	Simpler overall software makes further software changes easy
Less power, but note power impact of increased clock speed	Potentially more power-hungry

Consideration of issues of timeliness should then follow, as described in Chapter 8. These will include the responsiveness to external events, the necessity for periodic events, and so on. If the overall timing demands are high, then a Real-Time Operating System should be considered, and tasks and their priorities identified.

Following on from these considerations the designer will be in a good position to choose the programming language, if this is not already predetermined by other circumstances (like company policy). Finally, a program structure can be developed.

12.2.3.4 *Developing a hardware layout*

In this section we consider the general structure of the hardware by first defining its sub-systems. We then decide whether each of these will have its own localised control, and if so, how controllers will communicate with each other. As microcontrollers are now so very inexpensive, it is with ease that we introduce distributed control; indeed, only the very simple systems now have only one microcontroller. We will then want to determine the overall hierarchy, i.e. whether the sub-systems will have a master–slave relationship or peer-to-peer. In determining the data transfer between

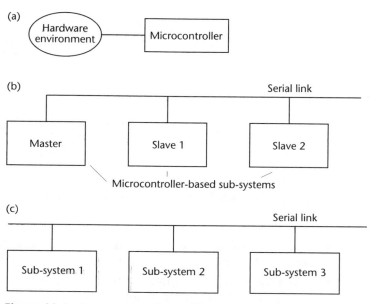

Figure 12.6 Some options for multi-processor topologies. (a) Single controller; (b) distributed control, master–slave; (c) distributed control, peer-to-peer.

them we will want to consider the quantity of data, the speed of data transfer and the reliability requirements.

Fig. 12.6 shows some options for simple small to medium-scale system topologies. The characteristics of each of these may be summarised as follows:

(a) No ready subdivision, and no obvious need for further processing capability.

(b) (i) Simple system, where specific autonomous activities (for example display driver) may be allocated to a separate processor

or

   (ii) more complex, closely linked system, where all sub-systems are under direct control of master processor, for example television and video recorder; data links are simple, for example Microwire, I^2C.

(c) Distributed system, made up of semi-autonomous sub-systems, for example motor vehicle; may have need for high-reliability data link, for example CAN. There is no obvious master.

As the distribution and interconnection of the controlling elements of the hardware are mapped out, it will be necessary to determine other aspects which relate to the overall hardware layout. These include:

● the location and interconnection of all other hardware elements, including peripherals, memory, transducers and human interfaces

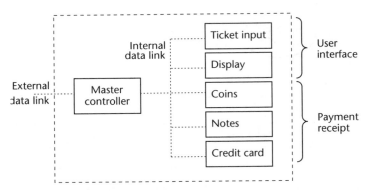

Figure 12.7 Possible hardware block diagram for payment point machine.

- the mode of data transfer between the hardware blocks, whether serial or parallel, and protocol as appropriate
- the power supply configuration, with consideration of power sources and power supply management

Note that although a layout is being considered, we are not finally committing ourselves at this stage to how the elements may be integrated together on one or more ICs, even though we may have a good notion of this. According to this design process, that decision can still wait a little longer.

As an example, Fig. 12.7 shows an intermediate stage in the development of the embodiment design of the payment point machine of Example 12.1. It has been provisionally decided that sub-systems should be created for each of the main user interaction features, and that these should all be linked to a master controller, which itself links to the larger system. This is likely to adopt the model of Fig. 12.6(b). The *overall* system is likely to adopt the model of Fig. 12.6(c).

12.2.3.5 *Defining the peripherals*

Having identified and located them in the overall hardware layout, the peripherals themselves can now be specified. This will include defining the full details on any serial or parallel port to be used, as well as Counter/Timers, ADCs and so on.

12.2.3.6 *Selecting the microcontroller(s)*

In this section we turn our attention (at last) to choosing the microcontroller(s). This is a decision that can prove surprisingly difficult and time-consuming, especially to the inexperienced. In the early days of microprocessors and controllers, it was viewed as a major strategic decision, with very long-term consequences to a company. Nowadays there is perhaps more readiness to switch devices from one product to the next; nevertheless, it remains a strategically important decision.

It is worth noting that Fig. 12.2 makes reference to selecting the *microcomputer*. This terminology is adopted to imply that the computing unit(s) around which the design is based may be microcontrollers designed into a circuit, *or* they may be single-card microcomputers, embedded into the system as complete sub-assemblies. As we are dealing here with small-scale systems, it is assumed that single-card microcomputers are less likely to be applied. For this reason their implementation will not receive further consideration.

The selection cannot be made until an outline specification for each microcontroller (whose number and location in the overall hardware layout will already have been determined) is developed, made up of the items listed below:

- *Word size*. Do the accuracy, range and data-handling demands of the system demand 8, 16, or 32 bit capability? Consider the resolution requirements of the input data and the amount of computation required. While 8-bit devices can handle 16-bit or higher computations, they become slow and the programming cumbersome.

- *Peripherals*. Drawing on the work already done, identify and list all peripherals that you hope to include on-chip. In many cases, as with parallel I/O, this will be entirely straightforward. In others, for example with a complex UART, you may finally decide to use an external device. Allow some extra capacity, at least in parallel I/O, to allow for later upgradability.

- *Memory*. Difficult though it is, attempt to assess the amount of program memory and RAM required. Will there be 'special' data memory requirements, for example EEPROM? Will there be a need for external memory access, therefore implying the need for parallel or serial data communication? Allow some extra memory capacity, as much as 100% of initial estimate, for later upgradability. Are there any constraints on program memory technology (for example insistence on electrical erasability, and in-circuit programmability)?

- *Speed*. Identify the speed requirements, including the fastest response times needed for interrupt. It is impossible to determine the execution speed of software until it is written, so it may be worth sketching out time-critical routines to assess their speed.

- *Power consumption*. If your application is power-conscious, then a low-power device will be required, as discussed in Chapter 10. Consider both steady state power drain and possible sleep or wait states, which can dramatically reduce consumption.

- *Physical size*. Are there dimensional constraints you must consider? Microcontroller sizes range from the older 40-pin DIL monsters to 8-pin surface-mount midgets.

The above issues, purely technical, will allow you to draw up a specification for your microcontroller. You should then be able to scan the offerings of the major microcontroller manufacturers. The points below may guide you towards one or the other.

- *Price and availability*. For every device which appears interesting, check its price and availability, both for unit quantities and anticipated production quantities. Some controllers have remarkable longevity, while others fade with equally remarkable rapidity.

- *Institutional support*. Consider the support which exists in your workplace for one or other microcontroller. This includes both equipment (for example IDE and real-time emulator) and local expertise.

- *Personal experience*. Take into account your own, or the team's, experience, if any. It is not too difficult to change from one microcontroller to another, but a change should not be forced if it is not strictly necessary.

- *Development equipment required*. Draw up a provisional list of the development equipment you would require, appropriate to the scale of the project, for the controller you are considering. Make an estimate of its cost.

- *Vendor support*. Find out what support the microcontroller manufacturer offers. Are there development engineers ready to offer advice and guidance? One of the reasons for the great success of the PIC family, for example, has been the readily accessible support and low-cost tools made available by Microchip Technology.

12.2.3.7 *Design for test*

At this stage of the design process you should make a preliminary decision about the test strategy, and then design accordingly. The following interlinked questions should be considered:

- *What is the initial test and commission strategy?* This implies tests carried out both on the initial prototype and on each new unit as it is made. Exhaustive and searching tests may be required to prove overall performance, as well as the fine details of the specification. Tests are carried out in-house by the manufacturer, with all the company facilities and expertise at hand.

- *What in-service diagnostic features are required?* This implies tests carried out on a unit in the field. Time, equipment and expertise are all likely to be in limited supply. Simple diagnostic tests which lead to pass/fail decisions on sub-systems and which can be operated by semi-skilled staff are preferred. Diagnostic features can be included in the opening phase of the software, for example exercising all principal peripherals and testing for response. Alternatively, a separate test mode can be introduced, called by switch or keypad action. Automatic fault detection and display of fault mode can be considered. Consider also how sub-systems, in both software and hardware, can be isolated for simpler testing.

- *What signals will need to be accessed or displayed?* This may be for either diagnostic or commissioning tests. At the simplest level, a few diagnostic LEDs are of inestimable value, indicating successful power-up and initialisation. Test pins or sockets can be included to ease contact with otherwise inaccessible signals or sub-systems.

- *What on-board features, specifically for diagnostic or monitor purposes, will be needed?* If a monitor is to be included, or Background Debug Mode (for both of these, see later in the chapter), there will be certain hardware requirements.

Finally, plan how program revisions will be introduced. In-circuit programming, as described in Chapter 4, is of course attractive for this. If an EPROM (or microcontroller containing EPROM) is to be pulled in and out, then it may be worth including a Zero Insertion Force socket.

12.3 Detail design

The outcome of embodiment design should be hardware and software designs expressed as block or structure diagrams, with a good level of detail information. From here it should be possible to proceed to the stage of detail design. This applies all the techniques described in the earlier chapters of this book, and needs no further discussion here.

12.4 Commissioning the system

12.4.1 Commissioning strategies

We move forward now to the stage of system test and commission. The design process just described is well under way, and an initial hardware prototype (which may be of the final system or just a trial sub-system) has been constructed. Some software has also been developed and tested in a simulator (as described in Chapter 3). We are now ready to put hardware and software together, and hope for a good result!

While the test phase is of critical importance, it is worth just reminding ourselves of its place in the overall development process. We may hope that the system has been designed for testability. If so, our lives will now be easier. When testing, our only wish is to *verify* our earlier design. That is, we wish to demonstrate that the design, now in its physical form, meets the specification we originally set for it. Therefore we will wish to test every aspect of the performance, as stated in the specification. In so doing, we are likely to come across errors, either in design or manufacture, which block the product meeting the specification. The testing process relies on good design going before it; all we can hope to ensure is that the product is as good as its design; no amount of testing makes a poor product better!

Embedded systems are a particularly difficult category of product to test. The reason for this is suggested in their name. Because they are embedded, they are to a large extent hidden. Unlike a conventional computer, with its keyboard and screen, the embedded system is not usually rich in human interface. The user can't see what's going on, and nor often can the test engineer!

If poorly planned, testing and commissioning can be a time of frustration to the development engineer, and the most open-ended phase of a project. Particularly difficult is this moment when software meets hardware. This is the time when problems of apparently unknown origin occur. Let us therefore start by establishing the essential principles of the test and commission process. These are *not* specific to the embedded system. They apply to any engineering product, and are presented here as three 'golden rules'.

1. *Divide and rule!* We share this rule with all good generals throughout history. In *any* test situation, there is great advantage in being able to isolate system components, and test them individually. When proven, they can then be put together as a system made up of 'known good' sub-systems. The proven sub-systems are of course also available for reuse in later designs.

2. *Guilty until proved innocent!* Here we reverse one of the basic tenets of the law court. In court, the accused is innocent until proved guilty. In testing an engineering product, the product is assumed guilty (faulty) until proved innocent (fully functioning). With this mind set we distance ourselves from the inadequate approach of the lazy student, whose report of an apparently successful test reads 'we switched it on and it seemed to work'. On the contrary, a clear implication is that we must aim to *test everything*, using as our yardstick the quantified statements of the specification. Every aspect of the system performance must be fully examined, for it is the aspects accidentally overlooked in test which cause the greatest problems when it is too late to test. It is accordingly useful to pay attention to the following:

    -   All branches within the software; a laborious yet effective approach is to mark them off on the program listing as they are tested. This is not, however, as simple as it sounds – the comparatively simple program of Fig. 12.8 for example, with its mere 10 conditional branches, can theoretically execute in 2^{10} (just over one thousand) different sequences!
    -   All possible specified input conditions.
    -   All possible specified output conditions.
    -   Conditions close to limits of operation, for example maximum power output, maximum frequency, low and high input levels.
    -   All specified operating environments, including as appropriate extremes of temperature, humidity or vibration.
    -   All possible electrical fault conditions, for example input overload, output short-circuit.
    -   User misuse or abuse, for example multiple simultaneous button pushes.
    -   Power-up, power-down and brown-out performance.
    -   All protection features.

3. *Work strategically, document relentlessly.* There are several aspects to this exhortation. First, the test procedure needs to be planned. We can identify all the sub-systems which can be tested individually, and how these

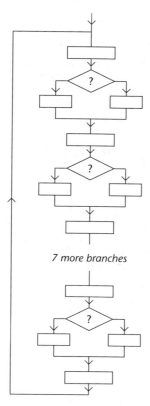

Figure 12.8 A simple program having 10 branches: how should we test?

will be built up into the final testable system. We can also plan how we will achieve the need to test comprehensively. What combination of inputs, for example, whether on simulator or in hardware, allows the program of Fig. 12.8 to be fully examined? If an exhaustive test appears impossible, then either more sophisticated test tools must be acquired or consideration given to simplifying the design.

Second, testing should proceed making reference to excellent documentation. It is pointless attempting to commission a system if the actual design drawings or software listings are incomplete, unavailable or so heavily modified that it is impossible to identify the current version. Drawings, diagrams and software listings should be kept impeccably up to date, with date and revision number clearly indicated. They must be at hand as the test proceeds.

Third and finally, test results should be clearly recorded, giving precise conditions of the test (inputs, test equipment, hardware and software version, test configuration), and the outcome. A result which apparently proves satisfactory performance may be put into doubt by a later test. If we return to the earlier one, but find that the conditions under which the test was made were not recorded, then the earlier test was little more than a waste of time.

START!

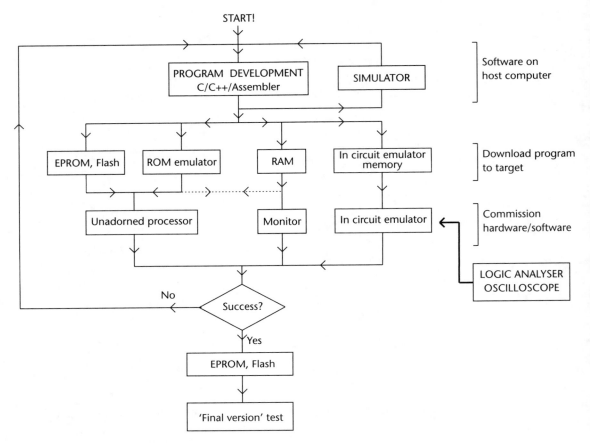

Figure 12.9 Routes for test and commission.

12.4.2 The test procedure

The actual development and test procedure applied will depend on the scale of the project and the tools available. Fig. 12.9 shows in broad terms the development routes available. There is a cycle of program development with testing using a simulator on the host computer. Then the program is downloaded for the first time into the target system, following one of the paths shown. The total system performance is then evaluated, making use of tools which test both hardware and software.

The simplest development path, at the lowest cost, is the route sometimes described as 'burn and crash'. The program is loaded into memory, and system performance is examined with no specialist tools applied. Small programs can be successfully developed in this way, but for complex programs it is not a practical way to operate. Apart from any diagnostic features built into the software or hardware, there is very little control of program operation.

The three major alternatives to this, using a ROM Emulator, Monitor or In-Circuit Emulator, are described in the sections which follow. Whichever of these is chosen, their effectiveness can be increased by using them in conjunction with an oscilloscope and/or logic analyser.

12.5 Further diagnostic tools and their use

The Instruction Set Simulator used for initial software verification was described in Chapter 3. This section describes further diagnostic tools suitable for embedded systems applications. These are sometimes difficult to understand and apply. While they include some standard laboratory equipment, like the oscilloscope, they include much else which is complex, fast-changing and frequently processor-specific.

The overall aim of any diagnostic tool is to give visibility, so that the circumstances surrounding a fault condition can be fully investigated, and a solution sought. The simplest tools, like the conventional oscilloscope or digital voltmeter, can measure and display something as it happens. The next level of sophistication is the tool which can also record the action for later display and analysis. This category includes the storage oscilloscope, the logic analyser and software tools with a trace facility. The next level is the tool which can take control of the action, for example by introducing breakpoints or single-stepping. This facility is sometimes called *run control*, and is displayed by tools such as the In-Circuit Emulator.

While we seek visibility from our test tools, we also want them to be non-invasive, that is, they should not change the system performance that they are being used to measure (unless it is an intended feature, as with run control). Unfortunately, *all* tools are to a greater or lesser extent invasive, and it is important to have an appreciation of where the impact is felt. For example, probes placed on digital lines cause capacitive and resistive loading; the cabling of emulators, however short, causes timing skew in the signals they transmit.

12.5.1 ROM/EPROM emulator

This device is essentially a pod containing a block of RAM, a serial interface, a parallel interface and some control logic. It is designed to replace a standalone memory IC, which would normally be holding program memory. The emulator is thus specific to a particular memory device. The serial interface links to the host computer, which can download the program code to the emulator memory. The parallel interface takes the form of a short ribbon cable, terminated by a DIL connector of the size of the memory which it is to replace. The target system memory is removed, and the emulator connected in its place.

With the ROM emulator, program code can be downloaded and modified easily. As such it is not a diagnostic tool; it simply allows more rapid change of program memory, a significant advantage in itself. Breakpoints can, however, be introduced with some processors in a primitive way by introducing a software interrupt instruction at the line where the break-point is required.

The ROM emulator is moderately low cost, and offers the flexibility of being processor- (though not memory device) independent.

12.5.2 Monitor

In the context of embedded system testing, the terminology *monitor* covers a variety of related test tools, which tend to have a high dependence on the processor architecture. The traditional microprocessor monitor has a dedicated section of program residing in the target system's program memory. A target system serial port is given over to create a serial link with the host computer, and a block of RAM is reserved for use as temporary program memory. This combination of features is designed to allow the host computer to download a test program to the target system RAM. The target system resident monitor program can allow the program to run, but can also introduce limited diagnostic facilities, like breakpoints and single stepping. When halted, say at a breakpoint, it can transfer the contents of register locations back to the host computer for display. The host computer can have the usual development features of assembler and/or compiler, so that the system taken as a whole forms another Integrated Development Environment (IDE).

This approach is common for evaluation board[2] type products. The advantage of the monitor is that it provides a potentially powerful diagnostic tool, allowing a program to be tested at an early stage within the hardware environment. Its disadvantage is that it ties up some of the hardware resources and some of the software. The monitor is also developed for a particular hardware configuration, and will probably need to be redefined in order to adapt it to a changed hardware environment. A target system which accommodates a monitor may therefore form only an intermediate stage in product development.

12.5.3 Background Debug Mode (BDM)

BDM is an evolutionary step on from a simple monitor. In BDM, certain diagnostic features are designed into the microprocessor and are a permanent part of it. Essentially these features are some additional logic which interfaces with the processor core, some microcode within the CPU and a dedicated serial port. Typically the serial port is connected to a small header on the target PCB, and the developer can connect this to a host computer whenever it is required. BDM is a special mode of operation of the CPU, entered via a certain logic combination of the external pins. Through the

---

2 Evaluation boards are offered by many microprocessor and microcontroller manufacturers as a low-cost means of gaining familiarity with a certain device. They are self-contained, encourage the addition of a little hardware prototyping, and allow the performance of the processor to undergo initial evaluation.

BDM the host can take control of the actions of the processor core, allowing single-step, register examination and change, resetting, and restarting.

The advantage of this is that the test facility is embedded for all time directly into the processor. As an integral part of the processor, it is by definition non-invasive. There is no need for the complex and annoying multiple connections of the logic analyser, or some of the niggling uncertainties of the In-Circuit Emulator. Processor cost must of course rise to some extent, as the processor now carries the extra complexity. The test facilities of BDM are comparatively limited, in particular trace is not available.

BDM is implemented in the Motorola 68HC16 microcontroller, and certain other devices. Further information on its implementation may be found in Ref. 12.4.

12.5.4 In-Circuit Emulator (ICE)

The ICE is the most powerful tool for embedded system test and commission. It incorporates the software development power of the simulator, yet does this with the software fully integrated into the target hardware and operating in real time.

The general setup applied with an ICE is shown in Fig. 12.10. The ICE itself consists of an *emulator pod*, which is connected, usually by serial link, to the host computer. Running on the host computer is the software component of the ICE, which effectively amounts to an IDE (for example it may include an editor and assembler). The emulator pod is connected to the target hardware by removing the target processor and plugging in its place a connector linked to the emulator pod. This link is kept extremely short to limit loading effects and signal delays, and is usually a ribbon cable.

Inside the pod is a hardware circuit which should be able to replicate exactly the processor it has replaced. The host computer can download to it the code of the program under test, and as the program runs it is able to monitor register and memory location values. The pod is commonly constructed of a motherboard and personality card. The former is

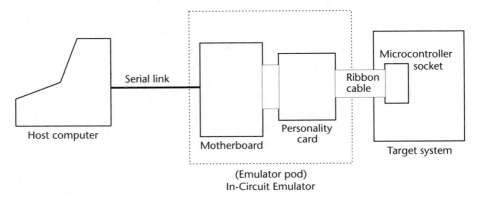

Figure 12.10 In-circuit emulator.

applicable to a family of processors, while the latter may be changed to match exactly the target system processor.

Although the ICE aims to be totally non-invasive in its operation, there are some exceptions to this happy ideal. One problem comes with use of the crystal. Due to the unavoidable length of ribbon cable of the emulator, the sensitive capacitor values around the crystal itself are changed. Therefore the ICE manufacturer may recommend disabling the target system clock and using a clock source supplied within the emulator. This may therefore preclude use of oscillator mode selection, for example in the RC/XT/HS and LP range on PIC microcontrollers. There may also be power supply restrictions. For example an emulator may need to operate at 5 V, whereas the target system is meant to run at 3 V. Power-up, power-down and brown-out conditions are likely to be hard to replicate, as the emulator will certainly influence the characteristics of the power supply. This includes both system power supply and details such as oscillator start-up. It is likely also to be difficult to operate at extremes of clock frequency, particularly at the high-frequency end, but possibly also at low frequency.

The ICE contains all the features familiar in the software simulator: breakpoints, run or single step modes, modifying internal registers and memory locations, and so on. It may also contain certain very useful processor-specific features, for example breakpoint on stack overflow and hardware breakpoints.

An ICE cannot be beaten for the range of capability it brings to the process of testing. However, it tends to be expensive and, being linked to a certain processor, it increases the commitment of a company to one processor. For a new processor, there may not be an ICE available.

12.5.5 The logic analyser

The logic analyser is a general-purpose laboratory tool used in the development of complex digital systems. It allows the user to observe the activity of almost any number of digital lines in a system as long as a connection can be made to them. It is therefore an important tool for the embedded system developer, as it can, given the right circumstances, provide a fairly comprehensive picture of the operation of a system. The need to make physical connection to signals to be monitored is, however, becoming a major limitation. With increasing integration of sub-systems onto an IC, fewer digital lines are being made available for monitor and test. Even if they do appear external to the chip, with devices becoming physically smaller, it is increasingly difficult to access those connections anyway.

Superficially the logic analyser is similar to the storage oscilloscope, in that it captures system signals, stores them in memory and displays them on screen. *Unlike* the oscilloscope, its inputs are subject to a logic threshold, so it is only able to interpret the input as a logic 0 or 1. What is seen is therefore an idealised display of the input waveform, with ringing, rise and fall times, and other imperfections removed. Thresholds may be fixed at TTL or CMOS levels, or in some analysers they are variable. The analyser has multiple inputs (24, 48, 96 or sometimes even more), so that a huge number of lines can be monitored simultaneously. Connection is made to any line by connecting a small 'grabber clip'. These usually come in colour-coded bunches. They can be tiresome to connect, and the connection is not always completely reliable. Special IC adaptors are also available.

The analyser is a complex machine, and is almost certainly an embedded system in itself. Generally it is menu driven, with separate and complex menus to set up display conditions, sampling clock rates and sources, triggers, memory control, and so on.

The analyser operates basically in Timing Analyser or State Analyser mode. In Timing Analyser mode it captures data on its inputs and displays these effectively as a timing diagram. Capture occurs on every cycle of an internal clock, whose frequency can normally be chosen by the user. As it is internal, the clock is of course not synchronised with the incoming data. If the clock is too slow, there is the risk that digital events will be missed. The faster the clock, the higher the timing resolution of input data, but the greater the demands on storage. For safe acquisition the clock should be set at such a speed that the fastest input event should last for two to three acquisition clock cycles.

In State Analysis, the analyser is triggered from an external clock derived from the system under test. Sampling is thus normally synchronous with the sampled data. The clock is chosen so that it will cause the acquisition of events of interest, while possibly completely missing others. For example, Fig. 4.8 shows multiplexed address and data buses. The analyser could be set up to use the line ALE as a clock so that address bus data would be latched while data bus information would be ignored. The data acquired is still sequential, but may not be based on a regular time period. Sophisticated analysers allow qualifiers to be applied to the clock source, gating it perhaps with another input, so that there can be even greater selectivity in the data which is actually captured.

An important feature of the logic analyser is its trigger capability. By using the trigger, both data leading up to a trigger point and data following the point can be examined. Complex trigger conditions can be set up, based on logical combinations of input signals or recognition of a certain input word. For example, if the program memory address bus is being acquired, a certain address can be specified as a trigger. Processor activity before or after occurrence of this trigger can then be explored.

The data stored by the logic analyser can be displayed either as a timing diagram or as a state listing. The latter can appear as binary, octal or hexadecimal. On-screen labels can be added to the display to aid the user in keeping track of the information. While logic analysers are essentially general-purpose, they can be customised to microprocessors in a few useful ways. Some analysers can be fitted with a disassembler, so that incoming data can be interpreted as instructions and displayed as a sequence of mnemonics. Special processor-specific adaptors are also available, which can be plugged into the processor socket, with the processor then plugged back into the adaptor. These allow rapid and reliable connection.

The logic analyser replicates some of the capability of the ICE. Unlike the ICE it is not committed to a single processor, so it is flexible. It is also very fast. It is almost non-intrusive, except for the very small capacitive/resistive loading on the lines monitored and the increase in EM radiation. The general-purpose instrument does not, of course, permit run control. Moreover, if the program crashes, no examination of internal registers can be made.

The logic analyser loses out to the more highly integrated microcontrollers. If address and data buses are only internal to the IC, then the analyser has no chance to access them! In one way this need not worry us, as the designer of the small-scale embedded system is not usually required to undertake fundamental analysis of address and data bus timing relationships. However, it does mean that in such situations we will *not* be able to track program execution using the logic analyser by following (and maybe disassembling) instruction data from the data bus. Despite this, there remain a number of important applications of the logic analyser in this sort of situation. The I/O data will all be accessible, and there are many situations where it is useful to be able to watch a parallel or serial port.

Advanced logic analysers offer many further in-built enhancements, including disk storage, printout facility and an IEEE or other interface. An interesting development is the instrument which offers a combination of logic analyser and oscilloscope capability. This is particularly appropriate for the embedded system designer, as it allows simultaneous display of both analogue and digital information. This can be used either to analyse the electrical characteristics of a selected digital signal (for example rise or fall times, voltage levels, or presence of EMI), or to look at genuinely analogue signals while simultaneously seeing digital activity.

12.6 Some trends in microcontroller technology and applications

Here, in the closing section of the book, we enter the most dangerous part: foretelling the future! Admitting that this is altogether *too* dangerous to be contemplated, let us instead just take note of some developments of today which are likely to be important for progress in embedded systems over the next few years. To keep our feet firmly on the ground, we will use only the present tense for the rest of this section, just describing events which *are* happening, but which are expected to have an impact in the future as well.

From the point of view of technology, devices continue to get smaller, faster and more complex. As part of this, programmable logic devices, for example FPGAs, continue to grow in total gate count. Therefore, in parallel to a continued proliferation of general-purpose ICs, microprocessor and microcontroller designers are turning towards selling their product not only as a piece of silicon, but as Intellectual Property (IP). This the user can download into an FPGA and combine directly with other on-chip designs. The processor core now becomes customisable, so that users can adapt it directly to their requirements. This can lead to dramatically reduced development times, as the prototyping process is much reduced and the hardware/software divide becomes increasingly blurred. With an IP-defined core, the hardware is itself defined by software! As part of this development, System on a

Chip (SoC) technology is becoming popular. Such devices may contain one or more processors, memory and general-purpose data transfer areas.

With memory density forever increasing, opportunities for more complex programs continue to grow, with embedded C++ and embedded Java coming into increasing prominence. Design tools for both hardware design and program generation are also advancing. Automatic code generation allows a program to be developed without the programmer having line-by-line involvement. New hardware design approaches (e.g. Ref. 12.5) are leading to dramatically reduced design time and improved performance.

From the point of view of diagnostic tools, the highly integrated FPGA or SoC devices render many conventional diagnostic tools invalid, as it is no longer possible to probe interconnection in the traditional way. There is therefore a growth in built-in diagnostic facilities, such as the Background Debug Mode described earlier, and new types of hardware tools to receive and process the data transmitted from the built-in diagnostics.

From the point of view of applications, the wave of automation that has swept through both industry and office is now impacting the home, starting with the Internet. At the embedded level, it is possible for every household appliance to be Internet-linked. The first impact of this is mostly in user-to-appliance communication. However, as more and more products become Internet-compatible, they are able to communicate with each other and elsewhere. The TV screen can discreetly display the identification code or number of the person phoning as the telephone rings. Alternatively, it could show the cooker temperature. The washing machine can have in-built diagnostics and call the service company with details of the fault when it is close to breakdown. An Internet-linked lawn sprinkler can check the weather forecast before deciding whether to turn on. The supposed advantage of all of this is increased convenience. There are, however, other benefits. For example, with an increasingly ageing population across Europe and the USA, it is possible to introduce into the home sophisticated and non-invasive monitoring of a person's well-being, with assistance called automatically if needed.

In our personal lives, sophisticated portable or wearable embedded systems are appearing. Smart cards are able to act as a phone charge card or electronic purse, or to carry multiple personal information; one card could suffice for library, driving licence, bank and much more besides. Mobile phones as a minimum provide the telephone function. Over and above this, they are readily Internet-linked and can incorporate computing capability as well as a mass of fixed or reference information, which can include anything from a personal diary to a dictionary.

The final part of this picture is ourselves, the designers. We need huge flexibility to adapt to new tools and technologies, and in this we epitomise the need for continuous professional development. As we continue to play our role in shaping society, it is important that we strive to produce not just things which are of commercial worth, but also devices which are of genuine benefit to those around us.

Happy designing!

SUMMARY

1. There are recognised procedures for undertaking engineering design. These can be adapted to the design of embedded systems, whether large or small.

2. Such design procedures allow a structured approach to the design problem and encourage the designer(s) to gather the right information and make the best possible decisions.

3. As with design, systematic procedures must be applied to the test and commission process. This process can otherwise be one of great frustration and loss of time.

4. A good understanding of diagnostic tools is needed by the system developer. The right combination of tools should be used to suit the diagnostic situation.

5. The future looks bright for embedded designers! There should be ongoing demands for their skills, as long as they are able to adapt to the changing technologies which will be encountered.

REFERENCES

12.1 Pahl, G. and Beitz, W. (1995) *Engineering Design, a Systematic Approach*, 2nd edn. London: Springer-Verlag.

12.2 Culverhouse, P. (1992) A tool for tracking engineering design in action. *Design Studies*, **13**(1).

12.3 Wilmshurst, T. (2001) A design model for embedded systems. *International Symposium on Engineering Education*, January. London: IEE.

12.4 *A Background Debugging Mode Driver Package for Modular Microcontrollers*. Application Note AN1230/D. Motorola Inc.

12.5 Dettmer, R. (2000) Software to silicon. *IEE Review*, **46**(5), 15–19.

EXERCISES

12.1 You work for an innovative company that wants to exploit its expertise in embedded systems. Apply your creative thinking to drawing up a set of product proposals, including the requirements list, specification and conceptual design, for one or more of the product ideas listed.

(a) A greenhouse climate controller.

(b) A bicycle speed and distance measurement system.

(c) A power management system for use in large commercial river barges (for example on the River Rhine). Energy storage is achieved in a battery bank. The supply to this is from a generator (when the motor is running), a solar panel or a small wind turbine. Energy is consumed for the domestic purposes of the operator (who lives on board with his family) and onboard navigational and ancillary equipment.

(d) A domestic central heating control system, in which the temperature profile against time in each room can be preset and controlled individually. Assume that heating is by circulation of hot water to radiators, and that there is a single boiler and water pump.

(e) A certain type of clock found in many English village churches has a pendulum whose mass and geometry control the clock speed. On every hour the clock chimes the number of hours shown on the clock face. Such clocks are not particularly accurate, however, and are subject to seasonal change. A system is required to adjust the clock speed automatically.

Note that it is possible to incorporate a radio receiver to determine the precise time of day. The clock pendulum can also be modified, such that the position of a mass that it carries can be moved up or down by a stepper motor, thereby changing the clock speed. The time displayed by the clock can be detected by counting the hours as they chime.

12.2 A river lock allows boats to move from one river level to another. It consists of two gates, either of which can be opened when there is no water level differential across it. Each gate contains a sluice door which may be opened to allow water to flow through the gate. Simultaneous water flow past two gates (whether due to gate or sluice being open) is not permissible.

A semi-automatic control system is required for the lock gates. The user will have four push-button controls: 'fill lock', 'empty lock', 'open upper gate' and 'open lower gate'. Six sensors give logic level indications of when the lock water is at upper river level or at lower, when either door is open, and when either sluice gate is open. For operational and safety reasons, there is a requirement that if the lock is not used for more than half an hour, the lock water level should be set to its higher level. Being remote, the power supply to the lock is subject to occasional failure.

Assume that this requirement is to be met by a microcontroller-based system. Draw up a requirements list and specification, and undertake completely the conceptual and embodiment design phases.

12.3 You are employed in a small company whose products incorporate PIC microcontrollers. You are the company's only development engineer, and you wish to persuade your manager to agree to the purchase of an In-Circuit Emulator. Your manager, however, knows that you have the MPLAB development system already, and believes (perhaps wrongly) that if you can simulate a program then you don't need the emulator. Write a justification for your proposed purchase, stating clearly the advantages that the emulator would give.

12.4 You have been asked to set up a low-cost microcontroller development system, for example for the development of products based on the Microchip Technology PIC series. List the items of equipment or software you would wish to purchase, stating very briefly what each would be used for.

12.5 List five ideas for embedded systems that would genuinely improve the quality of life for a person or group of people. Consider devices which would give independence to the disabled or vulnerable, enhance education or save energy. Develop one idea into a design, and find or launch a company to market it.

Binary, hexadecimal and BCD

A.1 Numbers and their representation

The number system we are most familiar with, the decimal system makes use of 10 different symbols to represent numbers, i.e. 0, 1, 2, 3, 4, 5, 6, 7, 8 and 9. Each of these symbols represents a number, and we make larger numbers by using groups of symbols. In this case the digit most to the right represents units, the next represents tens, the next hundreds, and so on. For example the number 389 can be thought of as 9 units, plus 8 tens, plus 3 hundreds:

$$
\begin{array}{c}
3\ \ 8\ \ 9 \\
\quad\quad 9 \times 10^0 \\
\quad 8 \times 10^1 \\
3 \times 10^2
\end{array}
$$

We notice that an N-digit number can take 10^N different values. For example a three-digit number can take 10^3, i.e. 1000, different combinations. These are from 000 through to 999.

The *base* or *radix* of the decimal system, just described, is 10. We almost certainly count in the decimal system due to the accident of having 10 fingers and thumbs on our hands. There is nothing intrinsically correct or superior about it. It is quite possible to count in other bases, and the world of digital computing almost forces us to do this.

We can generalise the points just made. A counting system whose base is R, where R is any integer, will require R different symbols to represent its numbers. An M-digit word will be able to take R^M different combinations, which will lie in the range 0 to $(R^M - 1)$.

A.2 Binary basics

The binary counting system has a base or radix of 2. It therefore uses just two symbols, normally 0 and 1. These are called binary digits, or bits.

Numbers are made up of groups of digits. The value that each digit represents depends upon its position in the number. Therefore the 4-bit number 1101 is interpreted as 1 unit, plus 0 twos, plus 1 four, plus 1 eight, i.e. 13.

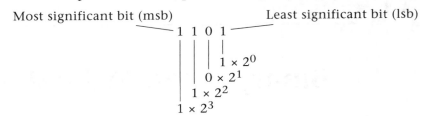

An n-digit binary number can take 2^n different values, with a range from 0 to $(2^n - 1)$. A binary number of any length is sometimes called a word. The bit representing units is called the least significant bit (lsb); the bit representing $2^{(n-1)}$ is called the most significant bit (msb). A common word size is eight bits, which is called a *byte*. The range of numbers that a byte can represent is 0 to $(2^8 - 1)$, i.e. 0 to 255. A 4-bit binary word is sometimes called a *nibble*. Ranges for other commonly used values of n are shown in Table A.2 later in this appendix.

Where numbers become large the prefixes K (representing 2^{10}, i.e. 1024) and M (representing 2^{20}, i.e. 1048576) are sometimes informally used. Thus a 1 Kbyte memory has a capacity of 1024 bytes, or 8192 bits.

A.2.1 Binary addition and subtraction

Binary addition and subtraction are done in the same way as the decimal equivalent, except that there are only two digit symbols to contend with. Bit pairs are added or subtracted. A Carry is generated and added to the next most significant bit pair if a 1 is added to a 1. A Borrow is generated and subtracted from the next most significant bit pair if a 1 is subtracted from a 0. Two examples are shown below. Superscripts in the addition example represent Carries; subscripts in the subtraction example represent Borrows.

$$0^1 \ 0^1 \ 1 \ 0^1 \ 1 \ 1$$
$$\underline{0 \ \ 1 \ \ 1 \ \ 0 \ \ 1 \ \ 0} \quad \textbf{Add}$$
$$1 \ \ 0 \ \ 0 \ \ 1 \ \ 0 \ \ 1$$

$$1 \ \ 1 \ \ 0 \ \ 1 \ \ 1 \ \ 0$$
$$\underline{0_1 \ 1_1 \ 1 \ \ 0 \ \ 1 \ \ 0} \quad \textbf{Subtract}$$
$$0 \ \ 1 \ \ 1 \ \ 1 \ \ 0 \ \ 0$$

A.3 Hexadecimal

Although it is binary numbers that digital computers work with, they are extremely inconvenient for the human. Moreover, they do not convert easily to the decimal system with which we are familiar. A way of making binary numbers palatable to the human mind is by converting them to hexadecimal, which is easily done. Hexadecimal is a counting system with base 16. 16 symbols are therefore used, as shown in Table A.1, each to represent one of the 16 different numbers which a 4-bit binary number can represent.

Table A.1 Hexadecimal counting.

Binary	Hexadecimal	Decimal
0000	0	0
0001	1	1
0010	2	2
0011	3	3
0100	4	4
0101	5	5
0110	6	6
0111	7	7
1000	8	8
1001	9	9
1010	A	10
1011	B	11
1100	C	12
1101	D	13
1110	E	14
1111	F	15

The A–F hexadecimal symbols have been chosen arbitrarily, and upper- and lower-case are acceptable. We can now readily convert long binary numbers into hexadecimal and grasp their significance. With a bit of practice you find yourself even doing mental arithmetic in hexadecimal!

The example below shows how an 8-bit binary number can be represented as a 2-digit hexadecimal number. Bits are grouped into 4s from the right, and the hexadecimal equivalent of each is determined.

$$1001 \quad | \quad 1110$$
$$9 \quad | \quad E$$
$$9 \times 16^1 \quad | \quad E_{16} \times 16^0$$

Because we now have number systems with three different bases, there is considerable potential for misunderstanding. For example, the number 101 may be interpreted as a binary, decimal or hexadecimal number, but obviously has a quite different value in each. The number 93 can be interpreted in hexadecimal or decimal, and again has a quite different value in each. Therefore it is common practice to add an indication of the number base. This is often done by adding the initial or the number base as a suffix, for example 301H, 301_H or 301_{16} for hexadecimal, 301D, 301_D or 301_{10} for decimal, or 101B, 101_B or 101_2 for binary. The dollar sign, as a prefix, is also used to indicate hexadecimal. In this book we will use the radix itself as suffix subscript, for example 110_2, 33_{10}, 579_{16}.

A.4 Binary Coded Decimal (BCD)

Converting a number to or from binary to hexadecimal is easy, as has just been shown. But converting from either of those number systems to or from decimal is rather more tedious, either mentally or in a computer. As we very often need input or output data expressed in decimal (for example in a display or as numbers on a screen) it is sometimes preferable to work in a number system which permits easy conversion to Decimal. Binary Coded Decimal (BCD) plays just this role. It is binary based, but can readily be converted to decimal.

BCD is based on 4-bit binary numbers, which are used to represent the numbers 0 to 9 in the conventional way. It does not, however, permit the equivalents of the hexadecimal numbers A to F to be used. Example BCD numbers are:

BCD	Decimal
0000 0010	2
0001 0011	13
0101 0111	57
1001 1001	99
1001 1000 0111	987
0010 0101 1001 0011	2593

Binary arithmetic in BCD is similar to conventional binary arithmetic, except for the number restriction mentioned. When adding for example, a carry must occur once the number 1001_2 (9_{10}) is exceeded. Many microprocessors and microcontroller have instructions designed to facilitate BCD arithmetic.

A BCD number requires one nibble of binary data for each digit. Two formats are used to achieve this: *packed* format, and *unpacked*. In packed format both the upper and lower nibbles of a byte are used to represent BCD digits. In unpacked format, the lower nibble is used to represent a digit, and the upper nibble is not used at all, or else is used just to represent the sign of the number.

BCD representation allows somewhat easier data interfacing. It is, however, more demanding in storage terms, The number 999_{10}, for example, requires 10 bits to represent it in binary, but 12 in BCD. It also adds some complexity in numerical calculations.

A.5 Representation of negative numbers – offset binary and two's complement

Simple binary numbers allow only the representation of unsigned numbers, which under normal circumstances are considered to be positive. Yet we must have a way of representing negative numbers as well. A simple

Table A.2   Two's complement and offset binary.

Two's complement	Decimal	Offset binary
0111 1111	+127	1111 1111
0111 1110	+126	1111 1110
⋮		
0000 0001	+1	1000 0001
0000 0000	0	1000 0000
1111 1111	−1	0111 1111
1111 1110	−2	0111 1110
⋮		
1000 0010	−126	0000 0010
1000 0001	−127	0000 0001
1000 0000	−128	0000 0000

way of doing this is by offsetting the available range of numbers. We do this by coding the largest anticipated negative number as zero and counting up from there. In the 8-bit range, with symmetrical offset, we can represent −128 as 00000000; 1000000 then represents zero and 11111111 represents +127. This method of coding is called offset binary, and is illustrated in Table A.2. It is used on occasions (for example in analogue to digital converter outputs), but its usefulness is limited as it is not easy to do arithmetic with it.

Let us consider an alternative approach. Suppose we took an 8-bit binary down counter, and clocked it from any value down to, and then below, zero. We would get this sequence of numbers:

Binary	Decimal
0000 0101	5
0000 0100	4
0000 0011	3
0000 0010	2
0000 0001	1
0000 0000	0
1111 1111	−1?
1111 1110	−2?
1111 1101	−3?
1111 1100	−4?
1111 1011	−5?

This gives a possible means of representing negative numbers – effectively we subtract the magnitude of the negative number from zero, within the limits of the 8-bit number or whatever other size is in use.

This representation is called two's complement. Does it give consistent results? We demonstrate that it does by doing the simple sums below. Valid two's complement numbers, representing positive and negative quantities, are added and subtracted, and lead to valid two's complement results. Notice that in reaching these results the Carry that is generated in (b) and (d) and the Borrow in (a) are simply ignored.

(a)		0000 0000	0	(b)	0000 1000	+8
	subtract	0000 0011	−(+3)	add	1111 1110	+(−2)
		1111 1101	−3		0000 0110	+6
(c)		1111 1110	−2	(d)	0000 0101	+5
	subtract	1111 1011	−(−5)	add	1111 1011	+(−5)
		0000 0011	+3		0000 0000	0

The simplest way to arrive at the two's complement representation of a negative number is to complement all the bits of the positive number and add 1. Hence to find −5 we follow the procedure:

original number	complement all	add one
0000 0101 (+5) →	1111 1010 →	1111 1011 (−5 in 2's comp.)

To convert back, simply subtract 1 and complement again.

A.5.1 Range of two's complement

Suppose we wanted to represent −250 in 8-bit two's complement. Following the procedure described above we would do the following:

1111 1010 (+250) → 0000 0101 → 0000 0110 (−250 in two's complement).

There is nothing wrong with this result, except that it also happens to represent +6. Therefore we must agree a convention which eliminates this chance of misunderstanding. We do this very simply by dividing the number range into two and allocating half to negative and half to positive numbers. Conveniently, the most significant bit then becomes a 'sign bit': 1 for negative, 0 for positive. The 8-bit binary range, shown both for two's complement and offset binary, then appears as shown in Table A.2.

Table A.3 Number ranges.

Number of bits	Unsigned binary	Two's complement
8	0 to 255	−128 to +127
12	0 to 4095	−2048 to +2047
16	0 to 65 535	−32 768 to +32 767
24	0 to 16 777 215	−8 388 608 to +8 388 607
32	0 to 4 294 967 295	−2 147 483 648 to +2 147 483 647

Where 8-bit representation is inadequate (and often it is), we can work with a larger number range. In general, the range of an n-bit two's complement number is from $-2^{(n-1)}$ to $+(2^{(n-1)} - 1)$. In every case the most significant bit is the sign bit, and all other bits are used for magnitude. Table A.3 summarises the ranges available for some commonly used values of n.

16F84 instruction set

Byte-oriented file register operations

```
13              8   7   6               0
  ┌──────────────────┬───┬────────────────┐
  │     OPCODE       │ d │   f (FILE #)    │
  └──────────────────┴───┴────────────────┘
```

d = 0 for destination W
d = 1 for destination f
f = 7-bit file register address

Bit-oriented file register operations

```
13            10  9    7   6               0
  ┌──────────────┬───────┬────────────────┐
  │    OPCODE    │b (BIT #)│   f (FILE #)  │
  └──────────────┴───────┴────────────────┘
```

b = 3-bit bit address
f = 7-bit file register address

Literal and control operations

General

```
13              8   7                   0
  ┌──────────────────┬────────────────────┐
  │     OPCODE       │     k (literal)     │
  └──────────────────┴────────────────────┘
```

k = 8-bit immediate value

CALL and GOTO instructions only

```
13        11  10                        0
  ┌────────────┬──────────────────────────┐
  │   OPCODE   │       k (literal)         │
  └────────────┴──────────────────────────┘
```

k = 11-bit immediate value

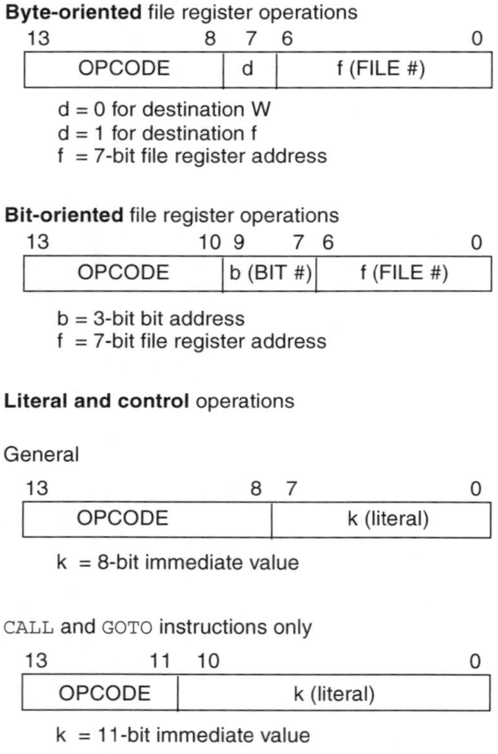

Figure B.1   General format for 16F84 instructions. Reprinted with permission of the copyright owner, Microchip Technology Incorporated © 2001. All rights reserved.

Table B.1 16F84 instruction set summary. Reprinted with permission of the copyright owner, Microchip Technology Incorporated © 2001. All rights reserved.

Mnemonic, Operands		Description	Cycles	14-Bit Opcode MSb ⟶ LSb				Status Affected	Notes
BYTE-ORIENTED FILE REGISTER OPERATIONS									
ADDWF	f, d	Add W and f	1	00	0111	dfff	ffff	C,DC,Z	1,2
ANDWF	f, d	AND W with f	1	00	0101	dfff	ffff	Z	1,2
CLRF	f	Clear f	1	00	0001	1fff	ffff	Z	2
CLRW	-	Clear W	1	00	0001	0xxx	xxxx	Z	
COMF	f, d	Complement f	1	00	1001	dfff	ffff	Z	1,2
DECF	f, d	Decrement f	1	00	0011	dfff	ffff	Z	1,2
DECFSZ	f, d	Decrement f, Skip if 0	1(2)	00	1011	dfff	ffff		1,2,3
INCF	f, d	Increment f	1	00	1010	dfff	ffff	Z	1,2
INCFSZ	f, d	Increment f, Skip if 0	1(2)	00	1111	dfff	ffff		1,2,3
IORWF	f, d	Inclusive OR W with f	1	00	0100	dfff	ffff	Z	1,2
MOVF	f, d	Move f	1	00	1000	dfff	ffff	Z	1,2
MOVWF	f	Move W to f	1	00	0000	1fff	ffff		
NOP	-	No Operation	1	00	0000	0xx0	0000		
RLF	f, d	Rotate Left f through Carry	1	00	1101	dfff	ffff	C	1,2
RRF	f, d	Rotate Right f through Carry	1	00	1100	dfff	ffff	C	1,2
SUBWF	f, d	Subtract W from f	1	00	0010	dfff	ffff	C,DC,Z	1,2
SWAPF	f, d	Swap nibbles in f	1	00	1110	dfff	ffff		1,2
XORWF	f, d	Exclusive OR W with f	1	00	0110	dfff	ffff	Z	1,2
BIT-ORIENTED FILE REGISTER OPERATIONS									
BCF	f, b	Bit Clear f	1	01	00bb	bfff	ffff		1,2
BSF	f, b	Bit Set f	1	01	01bb	bfff	ffff		1,2
BTFSC	f, b	Bit Test f, Skip if Clear	1 (2)	01	10bb	bfff	ffff		3
BTFSS	f, b	Bit Test f, Skip if Set	1 (2)	01	11bb	bfff	ffff		3
LITERAL AND CONTROL OPERATIONS									
ADDLW	k	Add literal and W	1	11	111x	kkkk	kkkk	C,DC,Z	
ANDLW	k	AND literal with W	1	11	1001	kkkk	kkkk	Z	
CALL	k	Call subroutine	2	10	0kkk	kkkk	kkkk		
CLRWDT	-	Clear Watchdog Timer	1	00	0000	0110	0100	TO,PD	
GOTO	k	Go to address	2	10	1kkk	kkkk	kkkk		
IORLW	k	Inclusive OR literal with W	1	11	1000	kkkk	kkkk	Z	
MOVLW	k	Move literal to W	1	11	00xx	kkkk	kkkk		
RETFIE	-	Return from interrupt	2	00	0000	0000	1001		
RETLW	k	Return with literal in W	2	11	01xx	kkkk	kkkk		
RETURN	-	Return from Subroutine	2	00	0000	0000	1000		
SLEEP	-	Go into standby mode	1	00	0000	0110	0011	TO,PD	
SUBLW	k	Subtract W from literal	1	11	110x	kkkk	kkkk	C,DC,Z	
XORLW	k	Exclusive OR literal with W	1	11	1010	kkkk	kkkk	Z	

Note 1: When an I/O register is modified as a function of itself (e.g., MOVF PORTB, 1), the value used will be that value present on the pins themselves. For example, if the data latch is '1' for a pin configured as input and is driven low by an external device, the data will be written back with a '0'.

2: If this instruction is executed on the TMR0 register (and, where applicable, d = 1), the prescaler will be cleared if assigned to the Timer0 Module.

3: If Program Counter (PC) is modified or a conditional test is true, the instruction requires two cycles. The second cycle is executed as a NOP.

A versatile microcontroller-based digital panel meter

Almost every scientific or laboratory instrument contains one or more panel meters, giving a read-out of essential information. There is a wide choice of commercially available digital panel meters available, of 3.5, 4.5 digits or more, with liquid crystal or LED displays. While apparently offering good versatility, their use in practice is sometimes restricted by the ranges, annunciators and other features predetermined by the manufacturer. This article describes a versatile digital panel meter which is intended to give very flexible performance. Being microcontroller-based it is possible to customise its performance in a very precise way to meet the needs of the instrument for which it is being designed.

The circuit is based on the widely used successive approximation ADC (Analogue to Digital Converter). This ADC is described in Chapter 5, and its block diagram appears as Fig. 5.18. This diagram is adapted as Fig. C.1. Such converters are very commonly available as integrated circuits, many with a microprocessor interface. This circuit, however, embeds intelligence within the conversion process itself, by replacing the logic part of the ADC, enclosed in dotted lines in Fig. C.1, with a microcontroller. It then uses a standalone 16-bit DAC (Digital to Analogue Converter) and a comparator to complete the circuit, and outputs data to a liquid crystal display (LCD).

C.1 Circuit description

The circuit diagram implements the block diagram of Fig. C.1 and is shown in Fig. C.2. The input passes first through an INA114 instrumentation amplifier (Ref. C.1). This performs a scaling and buffering function, and converts the input from differential to single-ended. The gain of the INA114 is set by a single resistor connected between pins 1 and 8; the three values quoted give gains of 1, 10 and 100, leading to three possible fixed input ranges: 10 V, 1 V and 100 mV respectively.

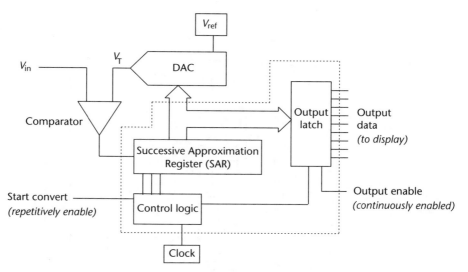

Figure C.1 The successive approximation ADC annotated for panel meter.

An OP42 high-speed op-amp (Ref. C.2) is used to implement the comparator. The maximum DAC output (i.e. V_T in Fig. C.1) is 5 V. Therefore the maximum acceptable input reaching the inverting comparator terminal is 5 V. The 10 V full-scale INA114 output voltage is therefore accurately divided in half by the potential divider to match this 5 V maximum.

The comparator is used with a very small amount of positive feedback to eliminate oscillation as its input voltages converge. So that this does not impair the resolution, the hysteresis introduced must be less than half of one lsb. If a 10 V input range (–5 V to +5 V) is being quantised to a 16-bit number, then one lsb is equivalent to $10/2^{16}$, i.e. 153 μV. With the comparator output swinging approximately to +12 V, the hysteresis is given approximately by (12 V × R_x/470k). This must be less than half one lsb, or 76 μV. Therefore R_x must be less than around 3 Ω. In practice it was found that the output resistance of the DAC provided adequate source resistance, and R_x was shorted out. The comparator output is clamped at the microcontroller input to limit it to the range 0 to 5 V.

The DAC is a 16-bit serial device, the Burr Brown DAC714 (Ref. C.3). It is possible to use this device either in unipolar mode, with output 0 to +10 V, or bipolar mode , with output –5 V to +5 V. Here it is connected in bipolar mode, in which case its digital input must be coded in two's complement. The gain and offset potentiometers adjust the DAC output, and are available to adjust the setting of the system as a whole. The DAC interface requirements are summarised in Fig. C.3. If input A0 is at a logic low, then serial data can be clocked in to the input shift register of the DAC on the rising edge of the clock, msb first. With A0 high and A1 low, data is transferred on a rising clock edge from the shift register to the DAC buffer, and a digital to analogue conversion commences.

The microcontroller, a PIC 16F84, takes over the logic functions of the ADC. It thus receives one input from the comparator and outputs data to

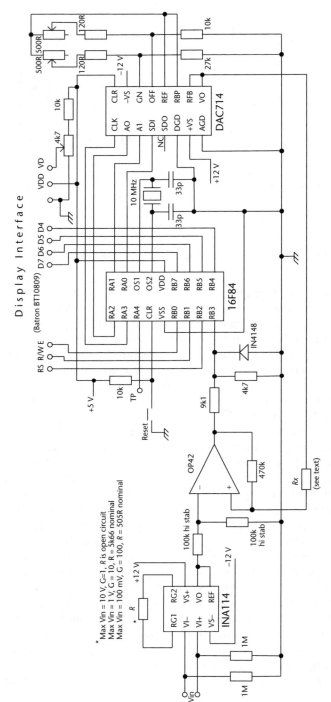

Figure C.2 Versatile panel meter circuit diagram.

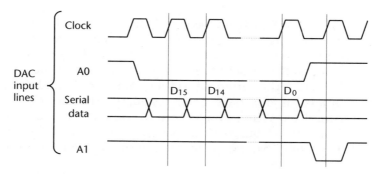

Figure C.3 DAC714 interfacing requirement.

the DAC and display. The Successive Approximation Register (SAR) itself is formed by two memory locations within the PIC.

The output is displayed on an LCD. In this case, an 8-digit Batron BT10809 (Ref C.4) was used. This is a large face display, and is adequate to display the five output digits, along with decimal point, unit and polarity indication. Along with many other displays, this unit uses the Hitachi HD44780 lcd microcontroller (Ref. 9.3), a controller specially designed for LCD interfacing and control. This controller can receive and interpret instructions (for example for relocation of the cursor, clearing the display) as well as character data. It is described in Chapter 9. A 4-bit interface is used in this design, with the most significant bit doubling as a 'Busy Flag' when a Read is undertaken. The advantage of 4-bit operation is that the whole unit may be interfaced with only seven bits (i.e. four data bits and three control bits). The disadvantage is slower speed, and slightly more complex software.

C.2   Program description

A reduced program listing is given at the end of this appendix. It operates the ADC section of the circuit, with unipolar voltage input, but does not include sections for data manipulation or display.

A digital to analogue conversion is initiated at regular intervals using the Interrupt on Overflow capability of the 16F84 Timer. The main conversion process follows the flow diagram of Fig. 5.19 very closely, and takes place within an Interrupt routine. In this program version, the Counter/Timer prescaler is set to divide by 256. With a clock frequency of 10 MHz the prescaler is clocked every 400 ns, and the interrupt rate is therefore once every $400 \times 256 \times 256$ ns, i.e. every 26.2 ms.

The SAR is set up in two memory locations: sarhi and sarlo. The position of the current bit is indicated by the contents of the two locations curbithi and curbitlo. All bits in these locations are logic 0, except for the current bit, which is logic 1. In this unipolar version bit 15 is not used, as this would initiate a negative DAC output value. The initial value of these locations is therefore set to be 40_{16}, 00_{16}, i.e. the first current bit is bit 14. The current bit of the SAR is set high at loop1 by an inclusive OR operation (iorwf) between the SAR registers and the current bit registers. The digital value so formed is transferred to the DAC with subroutine dacsend and the comparator output tested. If low, the current value is cleared by subroutine curbitcl. The current bit is then rotated right and the program loops again unless all bits have been tested. In the latter case the conversion is complete. When conversion is complete the data held in the SAR is converted to Binary Coded Decimal (BCD) and then output to the display. These have been implemented in the full software version.

REFERENCES

C.1   *INA114 Data Sheet*: http://www.burr-brown.com/.
C.2   *OP42 Data Sheet*: http://www.analog.com/.
C.3   DAC714 Data Sheet: http://www.burr-brown.com/.
C.4   Display Data: http://www.batron.com/.

Program listing

```
;**********************************************************
;Program to run Panel Meter ADC.
;Binary to BCD, and display driver routines not included.
;Program operates in unipolar mode, ie negative input
;voltages are not recognised.
;ADC6                                              2.11.98
;**********************************************************
;
;HARDWARE ALLOCATION
;Clock frequency = 10MHz. Clock mode = HS.
;WDT disabled.
;
;RB0 = E (Display)            RB4 = D4 (Display)
;RB1 = R/|W (Display)         RB5 = D5 (Display)
;RB2 = RS (Display)           RB6 = D6 (Display)
;RB3 = Comparator output      RB7 = D7 (Display)
;
;RA0 = Serial Data (DAC)      RA3 = Clk (DAC)
;RA1 = A1 (DAC)               RA4 = Testpoint
;RA2 = A0 (DAC)
;
;Timer Overflow Interrupt enabled

;PROGRAM SUMMARY
;After initialisation, program waits in loop "wait" until next
;timer overflow interrupt. Within ISR a new A to D conversion
;is made, applying standard Successive Approximation technique.
```

```
                    LIST    P=16F84
      ;specify SFRs
      indf      equ     00
      timer     equ     01
      status    equ     03
      fsr       equ     04
      porta     equ     05
      trisa     equ     05
      portb     equ     06
      trisb     equ     06
      intcon    equ     0B


      ;name memory locations
      sarhi     equ     10   ;successive approx. reg.
      sarlo     equ     11
      counter   equ     12   ;used in DAC output routine
      dacoutlo  equ     13
      dacouthi  equ     14
      delcntr1  equ     15   ;used in delay routine
      delcntr2  equ     16
      curbithi  equ     20   ;current bit of conversion, high byte
      curbitlo  equ     21   ;current bit of conversion, low byte
      temp      equ     27   ;for very short term storage
      ;
      ;Reset Vector
            org 00
            goto start
      ;
      ;Interrupt Vector
            org 04
            goto loop0
      ;
            org 0010
      ;
      ;***********************************************
      ;Programme Starts - Configure System
      ;***********************************************
      ;set port bits
      start   bsf     status,5   ;select register bank 1
              movlw   08
              movwf   trisb      ;all portb bits output
                                     ;except 3
              movlw   00
              movwf   trisa      ;all porta bits output
              movlw   07
              movwf   option     ;sets OPTION register
              bcf     status,5   ;select bank 0 for later
                                     ;memory transfers
```

```
              bcf      intcon,2    ;clear pending interrupts
              bsf      intcon,5    ;enable Timer Overflow int.
              bsf      intcon,7    ;set global interrupt enable
wait     goto     wait            ;wait for next Interrupt
;
;*********************************************
;Interrupt Routine - Main Loop
;*********************************************
;
;clear SAR and set current bit to msb (bit 14, as
;DAC uses 2's complement)
loop0    movlw    40
         movwf    curbithi
         clrf     curbitlo
         clrf     sarhi
         clrf     sarlo
         bsf      porta,4  ;set "busy" output
;set current bit to 1
loop1    movf     curbithi,0
         iorwf    sarhi,1
         movf     curbitlo,0
         iorwf    sarlo,1
;output new value to DAC
         movf     sarhi,0
         movwf    dacouthi
         movf     sarlo,0
         movwf    dacoutlo
         call     dacsend
;comparator output high?
         call     delay1   ;let it settle!
         btfsc    portb,3
         call     curbitcl ;comparator high,
                           ;so clear bit
;rotate current bit right
         bcf      status,0 ;clear carry
         rrf      curbithi,1
         rrf      curbitlo,1
;loop again if current bit still valid
         btfss    status,0
         goto     loop1
         bcf      porta,4
;
;conversion is complete, insert data output or display here
;
         bcf      intcon,2 ;clear interrupt flag
         retfie
```

```
;*********************************************
;SUBROUTINES
;*********************************************
;increments successive approx. register
incsar  incfsz  sarlo,1
        goto    incend
        incf    sarhi,1
incend  return
;
;called from main loop, clears current bit
;current bit word is complemented, and ANDed
;with SAR word.
curbitcl  comf  curbithi,0
          andwf sarhi,1
          comf  curbitlo,0
          andwf sarlo,1
          return
;
 ;introduces delay of 500us
delay1  movlw   0fa
        movwf   delcntr1
del1    nop
        nop
        decfsz  delcntr1,1
        goto del1
        return
;
;outputs to DAC714 16-bit word held in dacouthi-dacoutlo
dacsend movlw   8
        movwf   counter
        bcf     porta,2 ;clear A0
dac1    rlf     dacouthi,1
        btfsc   status,0 ;skip if carry clear
        bsf     porta,0 ;set data bit
        btfss   status,0 ;skip if carry set
        bcf     porta,0 ;clear data bit
        bcf     porta,3 ;clock bit in
        bsf     porta,3
        decfsz  counter,1
        goto    dac1
        movlw   8
        movwf   counter
dac2    rlf     dacoutlo,1
        btfsc   status,0 ;set data bit
        bsf     porta,0
        btfss   status,0
        bcf     porta,0
        bcf     porta,3
```

```
                bsf     porta,3
                decfsz  counter,1
                goto    dac2
                bsf     porta,2  ;set dac line A0
                bcf     porta,1  ;clear A1
                bcf     porta,3  ;clock in word
                bsf     porta,3
                bsf     porta,1
                return
        ;
                end
```

Addressing modes, address decoding and the memory map

D.1   Addressing modes

Many instructions in a microcontroller or microprocessor program are associated with an address, or a piece of data, on which the instruction is to be performed. An address may indicate, for example, the location of operand data in RAM, *or* of a constant embedded within program memory, *or* it might be the target address of a branch, *or* the instruction might simply indicate that the Stack is to be used (precise address not specified). Clearly many different styles of addressing are needed. The different types of addressing available in any one microprocessor are called its *addressing modes*. They depend very much on the internal structure of a processor, and hence can differ very much from one processor to another.

The 68HC05 has a very wide range of addressing modes. These form a good introduction to the topic, and are summarised below. Certain variants (for example the various implementations of indexed addressing) are not fully described.

- *Extended addressing:* the two bytes following the instruction hold the full 16-bit address; any memory location within a 16-bit range can be addressed. This is the most general-purpose addressing mode. When the processor detects an instruction in Extended addressing mode, it will then go on to interpret the next two bytes as address information. An instruction in Extended addressing mode appears as the following sequence of three bytes:

  Instruction op code

  Address, higher byte

  Address, lower byte

- *Immediate addressing*: the operand data is held in the byte immediately following the instruction byte. This gives a way of embedding constant data into a program. When the processor detects an instruction in Immediate addressing mode, it will then go on to interpret the next one byte as an operand. An instruction in Immediate addressing mode appears as the following sequence of two bytes:

```
[ | | | | | | | ]   Instruction op code
[ | | | | | | | ]   Operand ('Immediate') data
```

- *Direct addressing*: the single byte following the instruction is interpreted as the lower 8 address bits. The higher bits are assumed to be zero. This is similar to Extended addressing, but allows more compact coding and faster instruction execution time. Only the first 256 bytes of memory ('page zero') can be addressed. An instruction in Direct addressing mode appears as the following sequence of two bytes:

```
[ | | | | | | | ]   Instruction op code
[ | | | | | | | ]   Less significant address byte
```

- *Indexed addressing*: the operand address is held in the *Index Register*, which may be easily incremented or decremented; this is particularly attractive for accessing lists of data in memory, e.g. from 'look-up tables'. Indexed Instructions often also allow (or indeed require) an 'offset' to be added to the Index Register; this is a useful extension of the mode, and allows data to be read from one table, processed, and then easily transferred to another table. The 'HC05 includes Indexed addressing with no offset, and with 8-bit and 16-bit offsets. With no offset, the instruction is one byte only. With 8-bit offset, instructions in Indexed addressing appear:

```
[ | | | | | | | ]   Instruction op code
[ | | | | | | | ]   Offset (to be added to Index Register contents)
```

- *Inherent addressing:* internal registers are 'addressed', for example in instructions which affect accumulators or the Index Register. Instructions in Inherent addressing mode are single byte.

- *Relative addressing:* this is different from the addressing modes above, in that it is not used to identify the location of an operand. Instead, it determines a new value for the Program Counter when a branch instruction takes place. The new Program Counter value is determined by adding the byte which follows the instruction (the 'relative address', which has been determined by the programmer) to the Program Counter, which is then updated. The program then continues from the new location. Because the relative address is coded in two's complement, both forward and backward jumps can be made, but may only be within the fairly restricted range of the 8-bit relative address. An instruction in Relative addressing mode appears as the following sequence of two bytes:

Instruction op code

Relative address

● *Bit Set/Clear*: the address of the byte holding the target bit is identified with Direct addressing, in the byte following the instruction code. The bit itself is identified in three bits embedded within the op code. Any bit within the first 256 bytes of memory can be addressed.

Instruction op code (includes bit identification)

Direct address (of byte holding addressed bit)

● *Bit Test and Branch:* the bit under test is addressed in the same way as Bit Set/Clear. The target address of the branch is contained in a third byte, expressed as a relative address.

Instruction op code (includes bit identification)

Direct address (of byte holding addressed bit)

Relative address

D.2 Address decoding

Address decoding was a very important art in the early days of microprocessors, when every peripheral was an external IC and memory alone could be a whole array of ICs. In the world of the modern microcontroller, it is less important. In some cases, when a system consists of only one standalone controller, it is of no importance at all. Nevertheless there come times when some decoding has to take place or a memory map has to be interpreted applying an understanding of decoding techniques. Therefore the address decoding summary which follows is included. The description is based around the conventional microprocessor, with address and data buses bonded out.

A 16-bit address bus can address 2^{16} (= 65 536) locations. It is most unlikely that we would want to allocate all this memory space to just one memory device. More likely, a variety of different devices would be needed, for example 8 Kbyte of program memory, 2 Kbyte of RAM, 256 bytes of EEPROM or a Digital to Analogue Converter. Each of these addressable devices has a different number of address line inputs, in every case less than the full 16. But each memory location must be able to recognise its own unique address, even though it cannot be connected to all 16 lines of the bus. Therefore most addressable ICs are designed with one or more *chip enable* inputs. These can be enabled by suitable logical combinations of those address lines not directly connected to the memory IC.

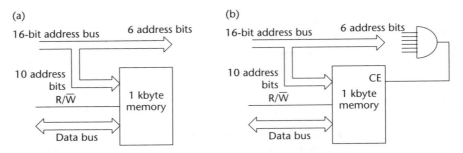

Figure D.1 (a) Incomplete address decoding; (b) full address decoding.

The memory device in Fig. D.1(a), for example, only has the lower 10 bits of the address bus connected to its address inputs, but many combinations of the address bus will access any one memory location. For example, the memory location with the lowest address will be accessed by:

0000 0000 0000 0000

or 0000 0100 0000 0000

or 0000 1000 0000 0000

or 0000 1100 0000 0000 etc.

This problem is overcome by the circuit change in Fig. D.1(b). Now all the higher order address bits are ANDed together, and the AND gate output connected to the Chip Enable (CE) input of the memory. Now the lowest memory location can be addressed only by

1111 1100 0000 0000

and the memory will occupy the address range

1111 1100 0000 0000 (i.e. $FC00_{16}$) to

1111 1111 1111 1111 (i.e. $FFFF_{16}$)

We say the memory has been 'mapped' into the location $FC00_{16}$ to $FFFF_{16}$.

We can divide the memory space systematically into a number of blocks or sub-blocks, each block identified by its own unique enable line. It is common practice to decode the more significant address bits to identify these blocks. A suitable 3-to-8 line decoder IC, the 74HC138, is shown in Fig. D.2. For each of the eight combinations of the three input address lines, one of the output lines is selected.

Suppose we wanted to divide the available 64K of memory space into eight 8 Kbyte blocks. Within each 8K block 13 address lines (2^{13} = 8192) would be required, which would be the lower 13 bits of the bus. The upper 3 bits could be decoded using the 74HC138 decoder. If we further wanted to break down one of the 8K blocks so created we could use another decoder.

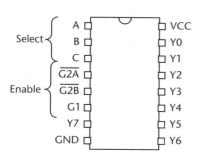

ENABLE		SELECT			OUTPUTS							
G1	G2A + G2B*	C	B	A	Y0	Y2	Y1	Y3	Y4	Y5	Y6	Y7
X	1	X	X	X	1	1	1	1	1	1	1	1
0	X	X	X	X	1	1	1	1	1	1	1	1
1	0	0	0	0	0	1	1	1	1	1	1	1
1	0	0	0	1	1	0	1	1	1	1	1	1
1	0	0	1	0	1	1	0	1	1	1	1	1
1	0	0	1	1	1	1	1	0	1	1	1	1
1	0	1	0	0	1	1	1	1	0	1	1	1
1	0	1	0	1	1	1	1	1	1	0	1	1
1	0	1	1	0	1	1	1	1	1	1	0	1
1	0	1	1	1	1	1	1	1	1	1	1	0

*(G2A + G2B) is the logical OR of inputs G2A and G2B

Figure D.2 The 74HC138 address decoder.

WORKED EXAMPLE D.1

A microprocessor has a 16-bit address bus and 128 bytes of on-chip RAM located in the memory range 0000_{16} to $007F_{16}$. The system requires 8 Kbyte of external program memory, which must be placed at the top of the memory map, and a further 1K of RAM. The RAM has two active low Chip Enable inputs. The system further requires one input/output (I/O) port, having three address inputs and one (active low) Chip Enable input. The RAM can be placed at any suitable memory location. The I/O port should be placed so that it can be accessed with direct addressing. Devise a memory decoding circuit and draw the memory map.

Solution: One possible way of meeting this requirement is shown in Fig. D.3. A 74HC138 divides the memory space into eight blocks, each of size 8K. The highest of these is fully occupied by the program memory. The next lowest (i.e. block $C000_{16}$ to $DFFF_{16}$) is used, incompletely, to enable the RAM. The positioning of the RAM chip in memory space is more precisely defined by the further decoding of the A12, A11 and A10 lines, such that it is enabled only when all of these lines are low. This RAM is thus located from $C000_{16}$ to $C3FF_{16}$. The port is located just above the internal RAM. Because address bits 3, 4, 5 and 6 are not used in its address decoding, its location in memory space is not restricted to just one location. For example, both addresses 0000 0000 1000 0000 and 0000 0000 1000 1000 will access its lowest address location. Incomplete address decoding is common practice, and does not constitute a problem as long as no other devices need be placed in the memory map.

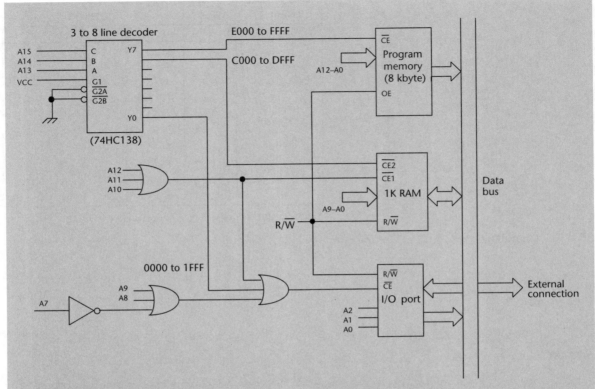

Figure D.3 Example address decoding.

The resulting system memory map then appears as in Fig. D.4.

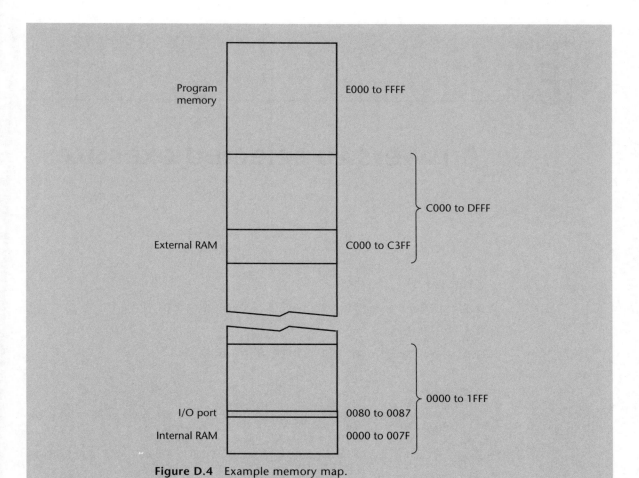

Figure D.4 Example memory map.

Answers to selected exercises

Chapter 1

1.5   Order of speed is A (0.8 μs), B (1.0 μs), C (1.2 μs)

1.6   30 MHz

1.8   (a) 4096; (b) 49 152; (c) –2048 to +2047

Chapter 2

2.4   For 25 mA sink: 0.8 V approx. 20 mW. For 20 mA source: 3.2 V approx. 36 mW

2.5   A current of 2.75 mA approx. flows. Both LEDs will be dimly illuminated

2.6   XX0X 0111

2.7   (a) OPTION set to XX0X 1XXX, bits GIE and T0IE must be set to 1 in INTCON register
(b) OPTION set to XX0X 0011, INTCON as above.
(c) OPTION set to XX0X 0001, INTCON as above. Preload TMR0 with 6 at start of every interrupt

2.8   14.7 ms. Assumes constant current throughout charging period

2.9   Crystal source just acceptable

Chapter 3

3.1
```
goto  0010
bsf   status,5
movlw 08
movwf trisb
movlw 00
movwf trisa
bcf status,5
```

3.6 Total duration is 502 µs. Enter 0f9 into delcntr1 for exact delay

Chapter 4

4.2 (a) 131 072; (b) 32 768 or 32 Kbyte. 15 address lines needed

4.6 Average of just under one (0.86) electron a day

4.7 30.5 pA

4.8 Initial ave. current consumption = 0.702 mA; New ave. current consumption = 4.8 µA

4.10 14 880 bytes needed for 31-day month. 16 Kbytes (16 384 bytes) would accommodate this, EEPROM or Flash

4.12 3.4 µA (HM6264); 1.98 mA (X24165); (Note: These results can be refined using the full data sheets)

Chapter 5

5.1 3 bits of resolution

5.3 V_p = 2.1 V

5.4 6th order filter with turnover frequency at 4 kHz

5.6 (a) 256; (b) 768

5.9 ±9.77 mV; ±39.0 mV

Chapter 6

6.4 (a) Microwire: 80 µs; RS232: 100 µs; I^{2}C: 200 µs (assumes start and stop conditions are exactly 10 µs)
(b) Microwire: 6400 µs; RS232: 8000 µs; I^{2}C: 7310 µs (same assumption as above)

6.9 172 cycles (including all looping, and subroutine call and return), i.e. 68.8 µs; 430 Hz approx.

Chapter 8

8.1 (a) (i) 17 µs, (ii) 35 µs, (iii) 585 µs; (b) Recommend interrupt 2 is made highest priority, and interrupt 1 is reduced in duration, moving some of its activity into main program

8.2 Increase is effectively 3 cycles, i.e. 1.5 µs with this clock frequency

Chapter 9

9.4 2166 mW

9.6 (a) $R_1 = 89\ \Omega$ (preferred values of 82 or 100 Ω can be applied), 21 kΩ < R_2 < 26 MΩ; (b) $R_1 = 89\ \Omega$ (preferred values of 82 or 100 Ω can be applied), 21 kΩ < R_2 < 2.4 MΩ

9.7 (a) 1.3 ms, 187.5 mA; (b) 103.2 Ω (preferred value of 100 Ω can be applied), 31.35 V

9.8 $R = 37.5\ \Omega$ (preferred value of 39 Ω can be applied). Assumes each digit is illuminated for exactly one quarter of the time, and that there are no voltage drops elsewhere in circuit

Chapter 10

10.1 500 days

10.5 112 ms

10.6 6.63 mW

10.8 (a) 50.1 mA: 55.4%, 50.1 mA: 71.3%, 50.1 mA: 83.2%; (b) 34.7 mA: 80%, 44.6 mA: 80%, 52.1 mA: 80%

10.10 6.06 mW

Index